SMART COMPOSITES

Mechanics and Design

Composite Materials: Analysis and Design

Series Editor

Ever J. Barbero

PUBLISHED

Finite Element Analysis of Composite Materials Using ANSYS,® Second Edition, *Ever J. Barbero*

Smart Composites: Mechanics and Design, *Rani Elhajjar, Valeria La Saponara, and Anastasia Muliana*

Finite Element Analysis of Composite Materials with Abaqus, *Ever J. Barbero*

FRP Deck and Steel Girder Bridge Systems: Analysis and Design, *Julio F. Davalos, An Chen, Bin Zou, and Pizhong Qiao*

Introduction to Composite Materials Design, Second Edition, *Ever J. Barbero*

Finite Element Analysis of Composite Materials, *Ever J. Barbero*

SMART COMPOSITES

Mechanics and Design

Edited by
Rani Elhajjar
Valeria La Saponara
Anastasia Muliana

CRC Press
Taylor & Francis Group
Boca Raton London New York

CRC Press is an imprint of the
Taylor & Francis Group, an **informa** business

CRC Press
Taylor & Francis Group
6000 Broken Sound Parkway NW, Suite 300
Boca Raton, FL 33487-2742

First issued in paperback 2017

ISBN-13: 978-1-4398-9591-7 (hbk)
ISBN-13: 978-1-138-07551-1 (pbk)

Library of Congress Cataloging-in-Publication Data

Smart composites : mechanics and design / editors, Rani Elhajjar, Valeria La Saponara, Anastasia Muliana.
pages cm. -- (Composite materials : design and analysis)
Includes bibliographical references and index.
ISBN 978-1-4398-9591-7 (hardback)
1. Smart materials. 2. Composite materials. I. Elhajjar, Rani, editor of compilation.

TA418.9.S62S455 2013
620.1'18--dc23 2013040632

Visit the Taylor & Francis Web site at
http://www.taylorandfrancis.com

and the CRC Press Web site at
http://www.crcpress.com

Contents

Section I Materials

Section II Structures

Section III Sensing

Series Preface

Half a century after their commercial introduction, composite materials are of widespread use in many industries. Applications such as aerospace, windmill blades, highway bridge retrofit, and many more require designs that assure safe and reliable operation for 20 years or more. Using composite materials, virtually any property, such as stiffness, strength, thermal conductivity, and fire resistance, can be tailored to the user's needs by selecting the constituent material, their proportion and geometrical arrangement, and so on. In other words, the engineer is able to design the material concurrently with the structure. Also, modes of failure are much more complex in composites than in classical materials. Such demands for performance, safety, and reliability require that engineers consider a variety of phenomena during the design. Therefore, the aim of the *Composite Materials: Design and Analysis* book series is to bring to the design engineer a collection of works written by experts on every aspect of composite materials that is relevant to their design.

Variety and sophistication of material systems and processing techniques have grown exponentially in response to an ever-increasing number and type of applications. Given the variety of composite materials available as well as their continuous change and improvement, understanding of composite materials is by no means complete. Therefore, this book series serves not only the practicing engineer but also the researcher and student who are looking to advance the state of the art in understanding material and structural response and developing new engineering tools for modeling and predicting such responses.

Thus, the series is focused on bringing to the public existing and developing knowledge about the material–property relationships, processing–property relationships, and structural response of composite materials and structures. The series scope includes analytical, experimental, and numerical methods that have a clear impact on the design of composite structures.

Ever J. Barbero
West Virginia University, Morgantown

Preface

Smart composites as we now know them are increasingly a key factor in scientific and technological achievement of materials. Recent advances in design and optimization of composite structures have played a significant role in the current development of smart materials and structures. Working with smart materials requires going beyond mechanics. Researchers and engineers find themselves needing to have an interdisciplinary knowledge to understand, predict, and model the properties of smart materials having unique structural, processing, and sensing abilities. The new generation of smart materials will consist of not only interacting components and microstructural morphologies but also materials that respond differently under combined external influence. The ability to then combine mechanical, thermal, electromagnetic, and other responses becomes critical not only at the material level but also at the structural scale. The materials are not only expected to bear mechanical loadings but also are designed with inherent capability lending itself for structural health monitoring or nondestructive sensing capabilities. At the same time, these new technologies have to support one another in a symbiotic way.

With this book, we have attempted to present a selection of the latest research in the field of smart materials. In the first section of the book, we discuss composites topics in smart materials related to design of electrically conductive, magnetostrictive nanocomposites and active fiber composites. These discussions include assessment of techniques and challenges in manufacturing smart composites and characterizing their coupled properties; we also present the latest research in analysis of composite structures at various length and time scales undergoing coupled mechanical and nonmechanical stimuli considering elastic, viscoelastic (and/or viscoplastic), fatigue, and damage behaviors. The second section of the book is dedicated to a higher-level analysis of smart structures with topics related to piezoelectrically actuated bistable composites, wing morphing, and multifunctional layered composite beams. Finally, the third section examines topics related to sensing and structural health monitoring, recognizing that multifunctional materials can be designed to both improve and enhance the health-monitoring capabilities and also enable effective nondestructive evaluation. The main point being that sensing abilities can be integrated within the material and provide continuous sensing.

Considering the various new directions, ideas, and methods, in this book we present a unique selection of current topics related to the understanding and design of smart composites from experts across the range of disciplines in smart or multifunctional materials. The editors are thankful for the support of the contributing authors in performing this task. It is clear that in the vastness of this field, it is impossible to contain all the information in one task. However, the editors hope that this book will become a useful source of information for both the academic and practitioner alike. The materials in this book are presented in a tutorial style, with emphasis on examples and practical application. The examples in each chapter and the suggested exercises are intended to make this book suitable for use as a textbook in smart materials or a follow-up course to an introductory composites course in aerospace, civil, materials, and mechanical engineering.

In conclusion, the editors of this book would like to thank the publisher, CRC Press, for pursuing the idea of a textbook in this area. We would also like to acknowledge Professor Ever Barbero who encouraged us in pursuing this project. The editors also thank the following colleagues for their valuable comments on the book chapters: Dr. Francis Avilés Cetina, Centro de Investigación Científica de Yucatán; Dr. Wahyu Lestari, Embry Riddle Aeronautical University; Dr. Salvatore Salamone, University of Buffalo and Dr. Thomas Schumacher, University of Delaware. The editors would also welcome any constructive comments and will take them into account, if possible, in future editions of this book.

Rani Elhajjar
Milwaukee, Wisconsin

Valeria La Saponara
Davis, California

Anastasia Muliana
College Station, Texas

Editors

Dr. Rani Elhajjar is currently a faculty member at the Departments of Civil Engineering and Mechanics and Materials Science and Engineering at the University of Wisconsin-Milwaukee. Before this appointment, Dr. Elhajjar was structural analyst at The Boeing Company, working on the 787 Dreamliner program as part of the Fuselage Methods, Repair, and Major Tests organization. His group was responsible for certification of the composite fuselage in the new airframe design. He was responsible for several test programs, including the Effects of Defects Program on the carbon fiber–based composite fuselage. Dr. Elhajjar received his PhD from the Georgia Institute of Technology and his MS from the University of Texas at Austin where he was focused on research in fracture of pultruded composite materials. He is a registered professional engineer and an active member in the American Society of Mechanical Engineering Experimental Mechanics committee. His current areas of research are in smart composites, structural analysis of composite structures, and full-field strain analysis methods.

Valeria La Saponara received a BS summa cum laude in aeronautical engineering from the University of Naples "Federico II" in Italy, and MS and PhD degrees in aerospace engineering from the Georgia Institute of Technology (Atlanta, GA). She joined the faculty of the University of California, Davis, and received tenure and promotion. Dr. La Saponara received a CAREER Award from the National Science Foundation. She is an active member of the American Society of Mechanical Engineers' Materials Division and Applied Mechanics Division, and conducts research on durability aspects of fiber-reinforced polymer composite structures.

Anastasia Muliana is an associate professor and Gulf Oil/Thomas A. Dietz Career Development Professor II in Mechanical Engineering, Texas A&M University. She obtained her PhD degree in civil engineering from Georgia Institute of Technology in 2004. She is a renowned researcher and educator in the area of micromechanics and viscoelastic response of composites. Her work deals with nonlinear multiscale modeling of coupled heat conduction, moisture diffusion, and deformation in heterogeneous viscoelastic bodies, micromechanics modeling of piezoelectric composites, rate-dependent and thermal stress analyses, multiscale modeling of biodegradable polymers for stent applications, large-scale nonlinear structural analysis, i.e., buckling and delamination, and experimental creep tests. Dr. Muliana has been recognized at the university level, through a TEES Select Young Faculty Award as well as at the national level through the National Science Foundation CAREER Award, US Air Force Young Investigator Program (YIP), and a Presidential Early Career Award for Scientists and Engineers (PECASE).

Contributors

Amir Barakati
Department of Mechanical and Industrial Engineering
The University of Iowa
Iowa City, Iowa

Onur Bilgen
Department of Mechanical and Aerospace Engineering
Old Dominion University
Norfolk, Virginia

Chris Bowen
Department of Mechanical Engineering
University of Bath
Bath, United Kingdom

John A. Capriolo II
Boeing Space and Intelligence Systems
El Segundo, California

Michele D. Dorfman
Lockheed Martin Tactical Aircraft Systems
Fort Worth, Texas

Sukanya Doshi
Department of Civil Engineering
Texas A&M University
College Station, Texas

Rani Elhajjar
Department of Civil Engineering and Mechanics
University of Wisconsin-Milwaukee
Milwaukee, Wisconsin

Nikolas L. Geiselman
Goodrich Aerostructures Group
Chula Vista, California

Eric v. K. Hill
Aura Vector Consulting
New Smyrna Beach, Florida

Daniel J. Inman
Department of Aerospace Engineering
University of Michigan
Ann Arbor, Michigan

H. Alicia Kim
Department of Mechanical Engineering
University of Bath
Bath, United Kingdom

Kevin B. Kochersberger
Department of Mechanical Engineering
Blacksburg, Virginia

Valeria La Saponara
Mechanical and Aerospace Engineering
University of California
Davis, California

Chiu Law
Department of Electrical Engineering and Computer Science
University of Wisconsin-Milwaukee
Milwaukee, Wisconsin

Yirong Lin
Department of Mechanical Engineering
The University of Texas at El Paso
El Paso, Texas

Kenneth J. Loh
Civil and Environmental Engineering
University of California
Davis, California

Bryan R. Loyola
Sandia National Laboratories
Livermore, California

Anastasia Muliana
Department of Mechanical Engineering
Texas A&M University
College Station, Texas

Mohammad Naraghi
Department of Aerospace Engineering
Texas A&M University
College Station, Texas

J. N. Reddy
Departments of Civil Engineering and Mechanical Engineering
Texas A&M University
College Station, Texas

Piervincenzo Rizzo
Department of Civil and Environmental Engineering
University of Pittsburgh
Pittsburgh, Pennsylvania

Henry A. Sodano
Department of Mechanical and Aerospace Engineering
Department of Materials Science and Engineering
University of Florida
Gainesville, Florida

Amir Sohrabi
Department of Mechanical Engineering
Texas A&M University
College Station, Texas

R. Andrew Swartz
Department of Civil and Environmental Engineering
Michigan Technological University
Houghton, Michigan

Olesya I. Zhupanska
Department of Mechanical and Industrial Engineering
The University of Iowa
Iowa City, Iowa

Section I

Materials

1

Field Coupling Analysis in Electrically Conductive Composites

Amir Barakati and Olesya I. Zhupanska

CONTENTS

1.1 Introduction

Recent advances in the manufacturing of multifunctional materials have provided opportunities to develop structures that possess superior mechanical properties along with other concurrent capabilities, such as sensing, self-healing, and electromagnetic and heat functionality. The idea is to fabricate components that can integrate multiple capabilities in order to develop lighter and more efficient structures. Composite materials are ideal candidates for realization of the concept of multifunctionality because of their multiphase nature and inherent tailorability. At the same time, advancements in the design of the multifunctional composite structures require significant strengthening of the scientific base and expansion of our understanding of complex interactions of multiple physical phenomena that lead to the desired multifunctionality. In this context, composite materials present rich possibilities for the development of multifunctional and functionally adaptive structures where multifunctionality can be achieved through interaction of mechanical, electromagnetic, thermal, and other fields.

In the present chapter, the effects of coupling between the electromagnetic and mechanical fields in electrically conductive anisotropic composites are discussed. The work is a continuation of the recent studies of Zhupanska and co-workers (Zhupanska and Sierakowski 2007, 2011; Barakati and Zhupanska 2012a,b) on electro-magneto-mechanical coupling in composites. The analysis is based on simultaneous solving of the system of nonlinear partial differential equations (PDEs), including equations of motion and Maxwell's equations. Physics-based hypotheses for electro-magneto-mechanical coupling in transversely isotropic and laminated composite plates and dimension reduction solution procedures for the nonlinear system of the governing equations are introduced in Section 1.2 to reduce the three-dimensional (3D) system to a two-dimensional (2D) form. A numerical solution procedure for the resulting 2D nonlinear mixed system of hyperbolic and parabolic PDEs is presented in Section 1.3 and consists of a sequential application of time and spatial integrations and quasilinearization. Extensive computational analysis of the response of the carbon fiber–reinforced polymer (CFRP) composite plates subjected to concurrent applications of different electromagnetic and mechanical loads is presented in Section 1.4.

1.2 Formulation of Governing Equations for Anisotropic Solids with Electromagnetic Effects

In this section, 3D governing equations and 2D plate approximation for electro-magneto-mechanical coupling in laminated electrically conductive composites are presented.

In a general electro-magneto-mechanical coupling problem, Maxwell's equations for the electromagnetic field and the equations of motion for the mechanical field need to be solved simultaneously. Maxwell's equations in a solid read as (Panofsky and Philips 1962)

$$\begin{aligned}
&\text{div}\boldsymbol{D} = \rho_e,\\
&\text{curl}\mathbf{E} = -\frac{\partial \boldsymbol{B}}{\partial t},\\
&\text{div}\boldsymbol{B} = 0,\\
&\text{curl}\mathbf{H} = \boldsymbol{j} + \frac{\partial \boldsymbol{D}}{\partial t},
\end{aligned} \tag{1.1}$$

where **E** and **H** are the electric and magnetic fields, respectively; $\boldsymbol{D}$ is the vector of electric displacement, $\boldsymbol{B}$ is the vector of magnetic induction, ρ_e is the charge density, and $\boldsymbol{j}$ is the electric current density vector. Furthermore, the relationships between the electromagnetic parameters are defined by the electromagnetic constitutive equations, which for an electrically anisotropic, magnetically isotropic, and linear solid body in International System (SI) units have the form (Zhupanska and Sierakowski 2005, 2007)

$$\begin{aligned}
&\boldsymbol{D} = \boldsymbol{\varepsilon}\mathbf{E} + \mu(\boldsymbol{\varepsilon} - \varepsilon_0 \cdot \mathbf{1})\left(\frac{\partial \boldsymbol{u}}{\partial t} \times \mathbf{H}\right),\\
&\boldsymbol{B} = \mu\mathbf{H} - \mu\frac{\partial \boldsymbol{u}}{\partial t} \times ((\boldsymbol{\varepsilon} - \varepsilon_0 \cdot \mathbf{1})\mathbf{E}),\\
&\boldsymbol{j} = \boldsymbol{\sigma}\left(\mathbf{E} + \frac{\partial \boldsymbol{u}}{\partial t} \times \boldsymbol{B}\right) + \rho_e \frac{\partial \boldsymbol{u}}{\partial t},
\end{aligned} \tag{1.2}$$

where $\boldsymbol{\varepsilon}$ and $\boldsymbol{\sigma}$ are the electric permittivity and conductivity tensors, ε_0 is the vacuum permittivity, μ is the magnetic permeability (single-value constant and is the same as in vacuum), **1** is the unit tensor of second order, $\boldsymbol{u}$ is the displacement vector, and t is time. The velocity terms in Equation 1.2 represent the effect of the rate of deformations of the solid body on the electromagnetic parameters. On the other hand, in the presence of an electromagnetic field, a coupling body force enters the equations of motion that modify it to the form

$$\nabla \cdot \mathbf{T} + \rho(\boldsymbol{F} + \mathbf{F}^{\mathrm{L}}) = \rho\frac{\partial^2 \boldsymbol{u}}{\partial t^2} \tag{1.3}$$

where ∇ is the gradient operator, **T** is the mechanical stress tensor, $\boldsymbol{F}$ is the vector of the mechanical body force per unit mass, ρ is the material density, and $\mathbf{F}^{\mathrm{L}}$ is the electro-magneto-mechanical coupling body force per unit mass.

This coupling body force is known as the Lorentz ponderomotive force, which for a general case is defined by (Sedov 1971)

$$\mathbf{F}^{L} = \rho_e \mathbf{E} + (D_\alpha \nabla E_\alpha - E_\alpha \nabla D_\alpha + \boldsymbol{B}_\alpha \nabla H_\alpha - H_\alpha \nabla \boldsymbol{B}_\alpha) + (\mathbf{J} \times \boldsymbol{B}), \tag{1.4}$$

where $\mathbf{J}$ is the external electric current density, and Einstein's summation convention is adopted with respect to the index α. Here the solid body may possess properties of polarization and magnetization or anisotropy in electric and/or magnetic properties. In the case of CFRP matrix composites where the solid is linear and electrically anisotropic but magnetically isotropic, the Lorentz force (Equation 1.4) can be rewritten in the form (Zhupanska and Sierakowski 2007)

$$\begin{aligned} \mathbf{F}^{L} = \rho_e \left(\mathbf{E} + \frac{\partial u}{\partial t} \times \boldsymbol{B} \right) + \left(\boldsymbol{\sigma} \left(\mathbf{E} + \frac{\partial u}{\partial t} \times \boldsymbol{B} \right) \right) \\ \times \boldsymbol{B} + \nabla \left(\frac{\partial u}{\partial t} \right) (((\varepsilon - \varepsilon_0 \cdot \mathbf{1})\mathbf{E}) \times \boldsymbol{B}) + (\mathbf{J} \times \boldsymbol{B}). \end{aligned} \tag{1.5}$$

Equation 1.5 shows that the Lorentz ponderomotive force depends on electromagnetic parameters as well as the rate of deformations in the medium. The third term vanishes if the solid is electrically isotropic and the last term is the part of the body force that is caused by an external electric current in the solid body (e.g., passes through the conductive carbon fibers in a current-carrying transversely isotropic composite material).

Therefore, the set of Equations 1.1 and 1.3 together with Equations 1.2 and 1.5 form the general governing equations of a dynamic electro-magneto-mechanical coupling problem. Further development and simplification of the governing equations for the case of a thin current-carrying transversely isotropic laminated plate are presented in the following sections, divided into governing mechanical and electromagnetic equations.

1.2.1 Governing Mechanical Equations for the Laminated Plate

Consider an electrically conductive fiber-reinforced laminated plate with the thickness H that consists of N_L number of unidirectional layers of thickness h. The laminate coordinate system (x,y,z) (i.e., the global coordinate system) is chosen in such a way that plane $x-y$ coincides with the middle plane, and axis z is perpendicular to the middle plane (Figure 1.1). The orientation of the fibers may be different in each lamina layer.

Assume that each lamina in the laminate is transversely isotropic. Therefore, the stress–strain relations in the principal material directions (i.e.,

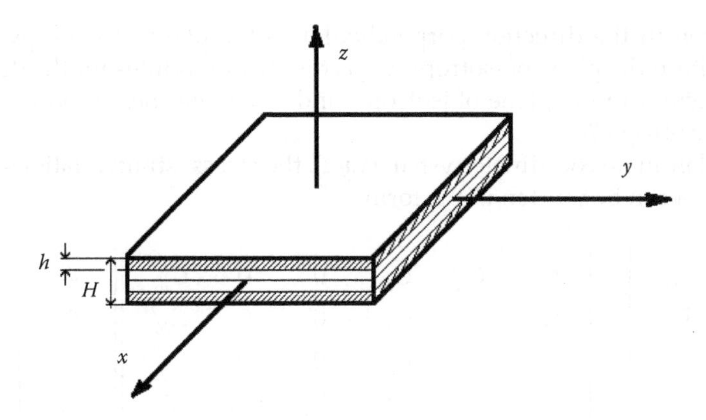

FIGURE 1.1
Laminated composite plate with laminate coordinate axes.

directions that are parallel to the planes of the material symmetry) for the i-th lamina are

$$\begin{bmatrix} \tau_{11} \\ \tau_{22} \\ \tau_{33} \\ \tau_{23} \\ \tau_{13} \\ \tau_{12} \end{bmatrix}_i = \begin{bmatrix} C_{11} & C_{12} & C_{12} & 0 & 0 & 0 \\ C_{12} & C_{22} & C_{23} & 0 & 0 & 0 \\ C_{12} & C_{23} & C_{22} & 0 & 0 & 0 \\ 0 & 0 & 0 & C_{44} & 0 & 0 \\ 0 & 0 & 0 & 0 & C_{55} & 0 \\ 0 & 0 & 0 & 0 & 0 & C_{55} \end{bmatrix} \begin{bmatrix} e_1 \\ e_2 \\ e_3 \\ \gamma_{23} \\ \gamma_{13} \\ \gamma_{12} \end{bmatrix}, \tag{1.6}$$

where C_{ij} are the components of the stiffness matrix C, which are defined in terms of the material properties as follows:

$$C_{11} = \frac{\left(1-\nu_{23}^2\right)E_1}{1-\upsilon}, \quad C_{12} = \frac{(\nu_{12}+\nu_{23}\nu_{12})E_2}{1-\upsilon}, \quad C_{23} = \frac{(\nu_{23}+\nu_{21}\nu_{12})E_2}{1-\upsilon},$$
$$C_{22} = \frac{(1-\nu_{12}\nu_{21})E_1}{1-\upsilon}, \quad C_{44} = G_{23}, \quad C_{55} = G_{12} \tag{1.7}$$

where

$$\nu = \nu_{12}\nu_{21} + \nu_{23}\nu_{32} + \nu_{31}\nu_{13} + 2\nu_{21}\nu_{32}\nu_{13}. \tag{1.8}$$

Here E_1 is Young's modulus for the fiber direction, E_2 is Young's modulus for the isotropy plane, ν_{23} is Poisson's ratio characterizing the contraction within the plane of isotropy for forces applied within the same plane, ν_{12} is Poisson's ratio characterizing contraction in the plane of isotropy due to forces in the direction perpendicular to it, ν_{21} is Poisson's ratio characterizing contraction in the direction perpendicular to the plane of isotropy due to forces within the plane of isotropy, G_{12} is the shear modulus for the direction perpendicular to the plane of isotropy, and G_{23} is the shear modulus in the plane of isotropy 2–3.

In the laminate coordinate system (x,y,z), the stress–strain relations for the i-th lamina can be written in the form

$$\begin{bmatrix} \tau_{xx} \\ \tau_{yy} \\ \tau_{zz} \\ \tau_{yz} \\ \tau_{xz} \\ \tau_{xy} \end{bmatrix}_i = \begin{bmatrix} \bar{C}_{11} & \bar{C}_{12} & \bar{C}_{13} & 0 & 0 & \bar{C}_{16} \\ \bar{C}_{12} & \bar{C}_{22} & \bar{C}_{23} & 0 & 0 & \bar{C}_{26} \\ \bar{C}_{13} & \bar{C}_{23} & \bar{C}_{33} & 0 & 0 & \bar{C}_{36} \\ 0 & 0 & 0 & \bar{C}_{44} & \bar{C}_{45} & 0 \\ 0 & 0 & 0 & \bar{C}_{45} & \bar{C}_{55} & 0 \\ \bar{C}_{16} & \bar{C}_{26} & \bar{C}_{36} & 0 & 0 & \bar{C}_{66} \end{bmatrix} \begin{bmatrix} e_x \\ e_y \\ e_z \\ \gamma_{yz} \\ \gamma_{xz} \\ \gamma_{xy} \end{bmatrix}_i . \tag{1.9}$$

Here $\bar{C}_{ij}$ represents the components of the transformed stiffness matrix $\bar{\mathbf{C}}$ defined as

$$\bar{\mathbf{C}} = \mathbf{T}^{-1}\mathbf{C}\mathbf{T}^{-\mathrm{T}}, \tag{1.10}$$

where $\boldsymbol{T}$ is the transformation matrix, the superscript –1 denotes the matrix inverse, and the superscript $\boldsymbol{T}$ denotes the matrix transpose. The transformation matrix $\boldsymbol{T}$ is defined as

$$\boldsymbol{T} = \begin{bmatrix} \cos^2\theta & \sin^2\theta & 0 & 0 & 0 & 2\cos\theta\sin\theta \\ \sin^2\theta & \cos^2\theta & 0 & 0 & 0 & -2\cos\theta\sin\theta \\ 0 & 0 & 1 & 0 & 0 & 0 \\ 0 & 0 & 0 & \cos\theta & -\sin\theta & 0 \\ 0 & 0 & 0 & \sin\theta & \cos\theta & 0 \\ -\cos\theta\sin\theta & \cos\theta\sin\theta & 0 & 0 & 0 & \cos^2\theta - \sin^2\theta \end{bmatrix}, \tag{1.11}$$

where θ is the angle from the x-axis to the 1-axis as shown in Figure 1.2.

It is also assumed that all laminae in the laminate are perfectly bonded, the laminate is thin, and a normal to the middle plane is assumed to remain straight and perpendicular to the middle plane when the laminate is deformed. This allows us to employ the classic Kirchhoff hypothesis of nondeformable normals, which suggests the following displacement field (Reddy 1999):

$$u_x = u(x,y,t) - z\frac{\partial w(x,y,t)}{\partial x}, \quad u_y = v(x,y,t) - z\frac{\partial w(x,y,t)}{\partial y}, \quad u_z = w(x,y,t), \tag{1.12}$$

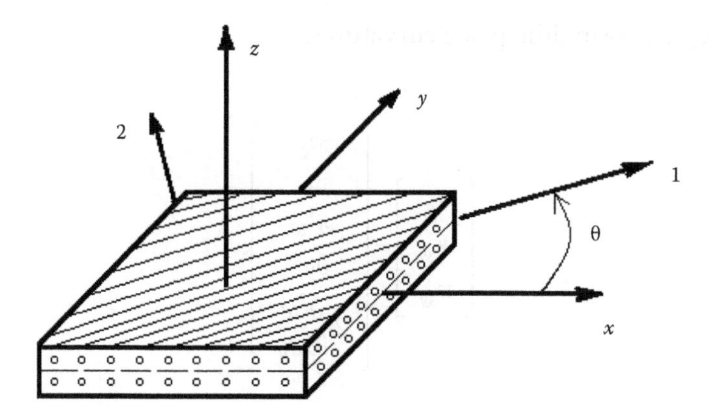

FIGURE 1.2
Laminate coordinate system (*x,y,z*) and in-plane principal material directions 1–2.

and strain field:

$$e_x = \frac{\partial u}{\partial x} - z\frac{\partial^2 w}{\partial x^2}, \quad e_y = \frac{\partial v}{\partial y} - z\frac{\partial^2 w}{\partial y^2}, \quad e_z = \frac{\partial w}{\partial z} = 0,$$

$$\gamma_{xy} = \frac{\partial u}{\partial y} + \frac{\partial v}{\partial x} - 2z\frac{\partial^2 w}{\partial x \partial y}, \quad \gamma_{yz} = \gamma_{xz} = 0, \tag{1.13}$$

or

$$\begin{bmatrix} e_x \\ e_y \\ \gamma_{xy} \end{bmatrix} = \begin{bmatrix} e_x^o \\ e_y^o \\ \gamma_{xy}^o \end{bmatrix} + z\begin{bmatrix} \kappa_x \\ \kappa_y \\ \kappa_{xy} \end{bmatrix} \tag{1.14}$$

where $u(x, y, t)$, $v(x, y, t)$, $w(x, y, t)$ are the corresponding middle-plane displacement components; $e_x^o, e_y^o, \gamma_{xy}^o$ are the middle-plane strains:

$$\begin{bmatrix} e_x^o \\ e_y^o \\ \gamma_{xy}^o \end{bmatrix} = \begin{bmatrix} \frac{\partial u}{\partial x} \\ \frac{\partial v}{\partial y} \\ \frac{\partial u}{\partial y} + \frac{\partial v}{\partial x} \end{bmatrix}; \tag{1.15}$$

and κ_x, κ_y, κ_{xy} are middle-plane curvatures:

$$\begin{bmatrix} \kappa_x \\ \kappa_y \\ \kappa_{xy} \end{bmatrix} = \begin{bmatrix} \dfrac{\partial^2 w}{\partial x^2} \\ \dfrac{\partial^2 w}{\partial y^2} \\ 2\dfrac{\partial^2 w}{\partial x \partial y} \end{bmatrix}. \tag{1.16}$$

The Kirchhoff hypothesis implies a linear variation of strain through the laminate thickness, whereas the stress variation through the thickness of the laminate is piecewise linear. In other words, the stress variation is linear through each lamina layer, but discontinues at lamina boundaries. Note that although shear strains are assumed to be zero, transverse shear stresses are not regarded as zeros but calculated from the equations of motion or equilibrium.

The stress and moment resultants are obtained by integration of the stresses through the thickness of the laminate:

$$\begin{bmatrix} N_{xx} \\ N_{yy} \\ N_{yz} \\ N_{xz} \\ N_{xy} \end{bmatrix} = \int_{-H/2}^{H/2} \begin{bmatrix} \tau_{xx} \\ \tau_{yy} \\ \tau_{yz} \\ \tau_{xz} \\ \tau_{xy} \end{bmatrix} dz = \sum_{i=1}^{N_L+1} \int_{z_i}^{z_i+1} \begin{bmatrix} \tau_{xx} \\ \tau_{yy} \\ \tau_{yz} \\ \tau_{xz} \\ \tau_{xy} \end{bmatrix}_i dz, \tag{1.17}$$

$$\begin{bmatrix} M_{xx} \\ M_{yy} \\ M_{xy} \end{bmatrix} = \int_{-H/2}^{H/2} \begin{bmatrix} \tau_{xx} \\ \tau_{yy} \\ \tau_{xy} \end{bmatrix} z dz = \sum_{i=1}^{N_L+1} \int_{z_i}^{z_i+1} \begin{bmatrix} \tau_{xx} \\ \tau_{yy} \\ \tau_{xy} \end{bmatrix}_i z dz, \tag{1.18}$$

where z_{i+1} is the distance to the top of the i-th layer in the laminate and z_i is the distance to the bottom of the i-th layer as shown in Figure 1.3.

Using stress–strain relations (Equation 1.9) and strain–middle-plane displacement relations (Equation 1.14), the stress and moment resultants

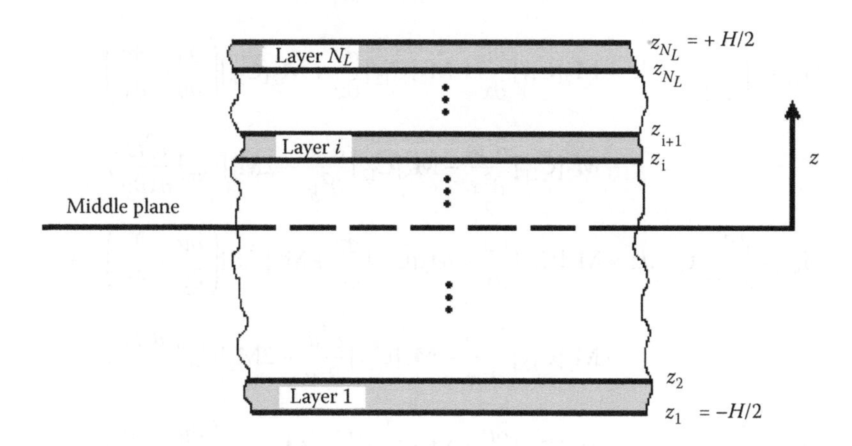

FIGURE 1.3
Coordinates of each layer in the laminate with N_L layers.

can be rewritten with respect to the derivatives of the middle-plane displacements:

$$
\begin{aligned}
N_{xx} &= \int_{-H/2}^{H/2} \tau_{xx}\,\mathrm{d}z = \mathbf{M}_1[\bar{C}_{11}]\frac{\partial u}{\partial x} + \mathbf{M}_1[\bar{C}_{12}]\frac{\partial v}{\partial y} + \mathbf{M}_1[\bar{C}_{16}]\left(\frac{\partial u}{\partial y} + \frac{\partial v}{\partial x}\right) \\
&\quad - \mathbf{M}_2[\bar{C}_{11}]\frac{\partial^2 w}{\partial^2 x} - \mathbf{M}_2[\bar{C}_{12}]\frac{\partial^2 w}{\partial^2 y} - 2\mathbf{M}_2[\bar{C}_{16}]\frac{\partial^2 w}{\partial x \partial y}, \\
N_{yy} &= \int_{-H/2}^{H/2} \tau_{yy}\,\mathrm{d}z = \mathbf{M}_1[\bar{C}_{12}]\frac{\partial u}{\partial x} + \mathbf{M}_1[\bar{C}_{22}]\frac{\partial v}{\partial y} + \mathbf{M}_1[\bar{C}_{26}]\left(\frac{\partial u}{\partial y} + \frac{\partial v}{\partial x}\right) \\
&\quad - \mathbf{M}_2[\bar{C}_{12}]\frac{\partial^2 w}{\partial^2 x} - \mathbf{M}_2[\bar{C}_{22}]\frac{\partial^2 w}{\partial^2 y} - 2\mathbf{M}_2[\bar{C}_{26}]\frac{\partial^2 w}{\partial x \partial y}, \\
N_{xy} &= \int_{-H/2}^{H/2} \tau_{xy}\,\mathrm{d}z = \mathbf{M}_1[\bar{C}_{16}]\frac{\partial u}{\partial x} + \mathbf{M}_1[\bar{C}_{26}]\frac{\partial v}{\partial y} + \mathbf{M}_1[\bar{C}_{66}]\left(\frac{\partial u}{\partial y} + \frac{\partial v}{\partial x}\right) \\
&\quad - \mathbf{M}_2[\bar{C}_{16}]\frac{\partial^2 w}{\partial^2 x} - \mathbf{M}_2[\bar{C}_{26}]\frac{\partial^2 w}{\partial^2 y} - 2\mathbf{M}_2[\bar{C}_{66}]\frac{\partial^2 w}{\partial x \partial y}, \\
N_{xz} &= \int_{-H/2}^{H/2} \tau_{xz}\,\mathrm{d}z, \\
N_{yz} &= \int_{-H/2}^{H/2} \tau_{yz}\,\mathrm{d}z,
\end{aligned}
\tag{1.19}
$$

$$M_{xx} = \int_{-H/2}^{H/2} \tau_{xx} z\,\mathrm{d}z = \mathbf{M}_2[\bar{C}_{11}]\frac{\partial u}{\partial x} + \mathbf{M}_2[\bar{C}_{12}]\frac{\partial v}{\partial y} + \mathbf{M}_2[\bar{C}_{16}]\left(\frac{\partial u}{\partial y} + \frac{\partial v}{\partial x}\right)$$

$$-\mathbf{M}_3[\bar{C}_{11}]\frac{\partial^2 w}{\partial^2 x} - \mathbf{M}_3[\bar{C}_{12}]\frac{\partial^2 w}{\partial^2 y} - 2\mathbf{M}_3[\bar{C}_{16}]\frac{\partial^2 w}{\partial x \partial y},$$

$$M_{yy} = \int_{-H/2}^{H/2} \tau_{xy} z\,dz = \mathbf{M}_2[\bar{C}_{12}]\frac{\partial u}{\partial x} + \mathbf{M}_2[\bar{C}_{22}]\frac{\partial v}{\partial y} + \mathbf{M}_2[\bar{C}_{26}]\left(\frac{\partial u}{\partial y} + \frac{\partial v}{\partial x}\right)$$

$$-\mathbf{M}_3[\bar{C}_{12}]\frac{\partial^2 w}{\partial^2 x} - \mathbf{M}_3[\bar{C}_{22}]\frac{\partial^2 w}{\partial^2 y} - 2\mathbf{M}_3[\bar{C}_{26}]\frac{\partial^2 w}{\partial x \partial y},$$

$$M_{xy} = \int_{-H/2}^{H/2} \tau_{xy} z\,dz = \mathbf{M}_2[\bar{C}_{16}]\frac{\partial u}{\partial x} + \mathbf{M}_2[\bar{C}_{26}]\frac{\partial v}{\partial y} + \mathbf{M}_2[\bar{C}_{66}]\left(\frac{\partial u}{\partial y} + \frac{\partial v}{\partial x}\right)$$

$$-\mathbf{M}_3[\bar{C}_{16}]\frac{\partial^2 w}{\partial^2 x} - \mathbf{M}_3[\bar{C}_{26}]\frac{\partial^2 w}{\partial^2 y} - 2\mathbf{M}_3[\bar{C}_{66}]\frac{\partial^2 w}{\partial x \partial y}$$

where the operator **M** is defined as

$$\mathbf{M}_k[f] = \int_{-H/2}^{H/2} f z^{k-1}\,\mathrm{d}z = \frac{1}{k}\sum_{i=1}^{N_L} f^{(i)}\left(z_{i+1}^k - z_i^k\right). \tag{1.20}$$

It should be noted that in Equation 1.19, N_{xz} and N_{yz} cannot be computed in the same manner in which the other resultants are found since they are neglected in the Kirchhoff hypothesis. Therefore, other methods should be employed to calculate them in this case.

The in-plane stress and moment resultants can also be rewritten in terms of the middle-plane strains and curvatures as

$$\begin{bmatrix} N_{xx} \\ N_{yy} \\ N_{xy} \end{bmatrix} = \begin{bmatrix} A_{11} & A_{12} & A_{16} \\ A_{12} & A_{22} & A_{26} \\ A_{16} & A_{26} & A_{66} \end{bmatrix} \begin{bmatrix} e_x^o \\ e_y^o \\ \gamma_{xy}^o \end{bmatrix} + \begin{bmatrix} B_{11} & B_{12} & B_{16} \\ B_{12} & B_{22} & B_{26} \\ B_{16} & B_{26} & B_{66} \end{bmatrix} \begin{bmatrix} \kappa_x \\ \kappa_y \\ \kappa_{xy} \end{bmatrix}, \tag{1.21}$$

$$\begin{bmatrix} M_{xx} \\ M_{yy} \\ M_{xy} \end{bmatrix} = \begin{bmatrix} B_{11} & B_{12} & B_{16} \\ B_{12} & B_{22} & B_{26} \\ B_{16} & B_{26} & B_{66} \end{bmatrix} \begin{bmatrix} e_x^o \\ e_y^o \\ \gamma_{xy}^o \end{bmatrix} + \begin{bmatrix} D_{11} & D_{12} & D_{16} \\ D_{12} & D_{22} & D_{26} \\ D_{16} & D_{26} & D_{66} \end{bmatrix} \begin{bmatrix} \kappa_x \\ \kappa_y \\ \kappa_{xy} \end{bmatrix}, \tag{1.22}$$

where

$$A_{ij} = \sum_{i=1}^{N_L+1} \left(\bar{Q}_{ij}\right)_i (z_{i+1} - z_i) = \mathbf{M}_1[\bar{Q}_{ij}],$$

$$B_{ij} = \frac{1}{2} \sum_{i=1}^{N_L+1} \left(\bar{Q}_{ij}\right)_i \left(z_{i+1}^2 - z_i^2\right) = \mathbf{M}_2[\bar{Q}_{ij}], \tag{1.23}$$

$$D_{ij} = \frac{1}{3} \sum_{i=1}^{N_L+1} \left(\bar{Q}_{ij}\right)_i \left(z_{i+1}^3 - z_i^3\right) = \mathbf{M}_3[\bar{Q}_{ij}].$$

Here A_{ij} are the components of the so-called extensional stiffnesses, B_{ij} are bending-extension coupling stiffnesses, D_{ij} are bending stiffnesses (Jones 1998), and Q_{ij} are the components of the reduced stiffness matrix $\bar{\mathbf{Q}}$:

$$\bar{\mathbf{Q}} = \begin{bmatrix} \bar{Q}_{11} & \bar{Q}_{12} & \bar{Q}_{16} \\ \bar{Q}_{12} & \bar{Q}_{22} & \bar{Q}_{26} \\ \bar{Q}_{16} & \bar{Q}_{26} & \bar{Q}_{66} \end{bmatrix} = \begin{bmatrix} \bar{C}_{11} & \bar{C}_{12} & \bar{C}_{16} \\ \bar{C}_{12} & \bar{C}_{22} & \bar{C}_{26} \\ \bar{C}_{16} & \bar{C}_{26} & \bar{C}_{66} \end{bmatrix}, \tag{1.24}$$

in which

$$\begin{aligned}
\bar{Q}_{11} &= \bar{C}_{11} = C_{11}\cos^4\theta + 2\left(C_{12} + 2C_{66}\right)\sin^2\theta\cos^2\theta + C_{22}\sin^4\theta, \\
\bar{Q}_{12} &= \bar{C}_{12} = \left(C_{11} + C_{22} - 4C_{66}\right)\sin^2\theta\cos^2\theta + C_{12}(\cos^4\theta + \sin^4\theta), \\
\bar{Q}_{22} &= \bar{C}_{22} = C_{11}\sin^4\theta + 2\left(C_{12} + 2C_{66}\right)\sin^2\theta\cos^2\theta + C_{22}\cos^4\theta, \\
\bar{Q}_{16} &= \bar{C}_{16} = \left(C_{11} - C_{12} - 2C_{66}\right)\sin\theta\cos^3\theta + \left(C_{12} - C_{22} + 2C_{66}\right)\sin^3\theta\cos\theta, \\
\bar{Q}_{26} &= \bar{C}_{26} = \left(C_{11} - C_{12} - 2C_{66}\right)\sin^3\theta\cos\theta + \left(C_{12} - C_{22} + 2C_{66}\right)\sin\theta\cos^3\theta, \\
\bar{Q}_{66} &= \bar{C}_{66} = \left(C_{11} + C_{22} - 2C_{12} - 2C_{66}\right)\sin^2\theta\cos^2\theta + C_{66}(\cos^4\theta + \sin^4\theta),
\end{aligned} \tag{1.25}$$

where θ is the angle between the x-axis and the 1-axis.

Therefore, the stress and moment resultants can be written in terms of extensional, bending-extension coupling, and bending stiffnesses as

$$
\begin{aligned}
N_{xx} &= A_{11}\frac{\partial u}{\partial x} + A_{12}\frac{\partial v}{\partial y} + A_{16}\left(\frac{\partial u}{\partial y} + \frac{\partial v}{\partial x}\right) - B_{11}\frac{\partial^2 w}{\partial^2 x} - B_{12}\frac{\partial^2 w}{\partial^2 y} - 2B_{16}\frac{\partial^2 w}{\partial x \partial y}, \\
N_{yy} &= A_{12}\frac{\partial u}{\partial x} + A_{22}\frac{\partial v}{\partial y} + A_{26}\left(\frac{\partial u}{\partial y} + \frac{\partial v}{\partial x}\right) - B_{12}\frac{\partial^2 w}{\partial^2 x} - B_{22}\frac{\partial^2 w}{\partial^2 y} - 2B_{26}\frac{\partial^2 w}{\partial x \partial y}, \\
N_{xy} &= A_{16}\frac{\partial u}{\partial x} + A_{26}\frac{\partial v}{\partial y} + A_{26}\left(\frac{\partial u}{\partial y} + \frac{\partial v}{\partial x}\right) - B_{16}\frac{\partial^2 w}{\partial^2 x} - B_{26}\frac{\partial^2 w}{\partial^2 y} - 2B_{66}\frac{\partial^2 w}{\partial x \partial y}, \\
N_{xz} &= \int_{-H/2}^{H/2} \tau_{xz}\, dz, \\
N_{yz} &= \int_{-H/2}^{H/2} \tau_{yz}\, dz, \\
M_{xx} &= B_{11}\frac{\partial u}{\partial x} + B_{12}\frac{\partial v}{\partial y} + B_{16}\left(\frac{\partial u}{\partial y} + \frac{\partial v}{\partial x}\right) - D_{11}\frac{\partial^2 w}{\partial^2 x} - D_{12}\frac{\partial^2 w}{\partial^2 y} - 2D_{16}\frac{\partial^2 w}{\partial x \partial y}, \\
M_{yy} &= B_{12}\frac{\partial u}{\partial x} + B_{22}\frac{\partial v}{\partial y} + B_{26}\left(\frac{\partial u}{\partial y} + \frac{\partial v}{\partial x}\right) - D_{12}\frac{\partial^2 w}{\partial^2 x} - D_{22}\frac{\partial^2 w}{\partial^2 y} - 2D_{26}\frac{\partial^2 w}{\partial x \partial y}, \\
M_{xy} &= B_{16}\frac{\partial u}{\partial x} + B_{26}\frac{\partial v}{\partial y} + B_{66}\left(\frac{\partial u}{\partial y} + \frac{\partial v}{\partial x}\right) - D_{16}\frac{\partial^2 w}{\partial^2 x} - D_{26}\frac{\partial^2 w}{\partial^2 y} - 2D_{66}\frac{\partial^2 w}{\partial x \partial y}.
\end{aligned}
\tag{1.26}
$$

Two-dimensional equations of motion for laminated plates are obtained by integration of equations of motion across the thickness of the plate (–H/2, H/2). In the presence of the electromagnetic loads and surface mechanical loads, this procedure yields

$$
\begin{aligned}
\frac{\partial N_{xx}}{\partial x} + \frac{\partial N_{xy}}{\partial y} + X_2 + \rho \int_{-H/2}^{H/2} F_x^L\, dz &= \rho H \frac{\partial^2 u}{\partial t^2}, \\
\frac{\partial N_{yy}}{\partial y} + \frac{\partial N_{xy}}{\partial x} + Y_2 + \rho \int_{-H/2}^{H/2} F_y^L\, dz &= \rho H \frac{\partial^2 v}{\partial t^2}, \\
\frac{\partial N_{xz}}{\partial x} + \frac{\partial N_{yz}}{\partial y} + Z_2 + \rho \int_{-H/2}^{H/2} F_z^L\, dz &= \rho H \frac{\partial^2 w}{\partial t^2},
\end{aligned}
$$

$$\frac{\partial M_{xx}}{\partial x}+\frac{\partial M_{xy}}{\partial y}+HX_1+\rho\int_{-H/2}^{H/2}F_x^L z\,dz=N_{xz}-\rho\frac{H^3}{12}\frac{\partial^3 w}{\partial t^2\partial x}, \tag{1.27}$$

$$\frac{\partial M_{yy}}{\partial y}+\frac{\partial M_{xy}}{\partial x}+HY_1+\rho\int_{-H/2}^{H/2}F_y^L z\,dz=N_{yz}-\rho\frac{H^3}{12}\frac{\partial^3 w}{\partial t^2\partial y}.$$

Here, ρ is the material density of the laminate; t is time; $\mathbf{F}^L=\left(F_x^L,F_y^L,F_z^L\right)$ is the Lorentz ponderomotive force per unit mass vector; and X_k, Y_k, and Z_k are combinations of tractions at the external surfaces of the laminate

$$X_1=\frac{1}{2}\left(\tau_{xz}\Big|_{z=\frac{H}{2}}+\tau_{xz}\Big|_{z=-\frac{H}{2}}\right),\quad X_2=\tau_{xz}\Big|_{z=\frac{H}{2}}-\tau_{xz}\Big|_{z=-\frac{H}{2}},$$

$$Y_1=\frac{1}{2}\left(\tau_{yz}\Big|_{z=\frac{H}{2}}+\tau_{yz}\Big|_{z=-\frac{H}{2}}\right),\quad Y_2=\tau_{yz}\Big|_{z=\frac{H}{2}}-\tau_{yz}\Big|_{z=-\frac{H}{2}}, \tag{1.28}$$

$$Z_2=\tau_{zz}\Big|_{z=\frac{H}{2}}-\tau_{zz}\Big|_{z=-\frac{H}{2}}.$$

Thus, the equations of motion are solved with respect to the middle-plane displacements $u(x, y, t)$, $v(x, y, t)$, $w(x, y, t)$. After that, the in-plane stresses τ_{xx}, τ_{yy}, τ_{xy} are found using stress–strain relations (Equation 1.9), and transverse shear stresses τ_{xz} and τ_{yz} and normal stress τ_{zz} are found by integrating equations of motion

$$\frac{\partial \tau_{xx}}{\partial x}+\frac{\partial \tau_{xy}}{\partial y}+\frac{\partial \tau_{xz}}{\partial z}+\rho\int_{-H/2}^{H/2}F_x^L\,dz=\rho\frac{\partial^2 u_x}{\partial t^2},$$

$$\frac{\partial \tau_{xy}}{\partial x}+\frac{\partial \tau_{yy}}{\partial y}+\frac{\partial \tau_{yz}}{\partial z}+\rho\int_{-H/2}^{H/2}F_y^L\,dz=\rho\frac{\partial^2 u_y}{\partial t^2}, \tag{1.29}$$

$$\frac{\partial \tau_{xy}}{\partial x}+\frac{\partial \tau_{yz}}{\partial y}+\frac{\partial \tau_{zz}}{\partial z}+\rho\int_{-H/2}^{H/2}F_z^L\,dz=\rho\frac{\partial^2 u_z}{\partial t^2}.$$

Furthermore, in the classic theory, the effect of transverse normal stress τ_{zz} on the stresses and the deformed state of the laminate is disregarded because this stress is considered to be small in comparison to in-plane stresses.

1.2.2 Governing Electromagnetic Equations for the Laminated Plate

Since generally the fibers in a lamina may not be in the direction of the global axes, the conductivity of layer i of a laminate in the laminate coordinate system (x,y,z) is defined as

$$\boldsymbol{\sigma}^{(i)} = \begin{bmatrix} \sigma_{11}^{(i)} & \sigma_{12}^{(i)} & 0 \\ \sigma_{12}^{(i)} & \sigma_{22}^{(i)} & 0 \\ 0 & 0 & \sigma_{33}^{(i)} \end{bmatrix} \tag{1.30}$$

where

$$\begin{aligned} \sigma_{11}^{(i)} &= \sigma_1^{(i)} \cos^2\theta + \sigma_2^{(i)} \sin^2\theta, \\ \sigma_{22}^{(i)} &= \sigma_2^{(i)} \cos^2\theta + \sigma_1^{(i)} \sin^2\theta, \\ \sigma_{12}^{(i)} &= (\sigma_1^{(i)} - \sigma_2^{(i)}) \sin\theta \cos\theta, \\ \sigma_{33}^{(i)} &= \sigma_3^{(i)}. \end{aligned} \tag{1.31}$$

Here $\sigma_1^{(i)}$, $\sigma_2^{(i)}$, and $\sigma_3^{(i)}$ are the conductivities along the principal material directions of the lamina i (see Figure 1.2).

In the coupled problems with mechanical and electromagnetic fields present, employing the Kirchhoff hypothesis is not sufficient to reduce the equations of motion to a 2D form without introducing additional hypotheses regarding the behavior of an electromagnetic field in thin plates and reducing the expression for the Lorentz force to a 2D form. The electromagnetic hypotheses are presented next.

It is assumed that the tangential components of the electric field vector and the normal component of the magnetic field vector do not change across the thickness of the plate:

$$E_x = E_x(x, y, t), \quad E_y = E_y(x, y, t), \quad H_z = H_z(x, y, t) \tag{1.32}$$

This set of hypotheses was obtained by Ambartsumyan et al. (1977) using asymptotic integration of 3D Maxwell's equations. It is important to note that the electromagnetic hypotheses (Equation 1.32) are valid only together with the hypothesis of nondeformable normals.

Furthermore, applying the electromagnetic hypotheses (Equation 1.32) and taking into account the constitutive relations (Equation 1.2), the second and fourth of Maxwell's equations (Equation 1.1) can be rewritten in the form

$$-\frac{\partial B_x}{\partial t}=\frac{\partial E_z}{\partial y}-\frac{\partial E_y}{\partial z},\quad -\frac{\partial B_y}{\partial t}=\frac{\partial E_x}{\partial z}-\frac{\partial E_z}{\partial x},\quad -\frac{\partial B_z}{\partial t}=\frac{\partial E_y}{\partial x}-\frac{\partial E_x}{\partial y},\tag{1.33}$$

$$J_x=\frac{\partial H_z}{\partial y}-\frac{\partial H_y}{\partial z},\quad J_y=\frac{\partial H_x}{\partial z}-\frac{\partial H_z}{\partial x},\quad J_z=\frac{\partial H_y}{\partial x}-\frac{\partial H_x}{\partial y},\tag{1.34}$$

where the components of the induced current density in the layer i are determined as

$$\begin{aligned}
J_x^{(i)}&=\sigma_{11}^{(i)}\left(E_x+\frac{\partial u_y}{\partial t}B_z-\frac{\partial u_z}{\partial t}B_y^{(i)}\right)+\sigma_{12}^{(i)}\left(E_y+\frac{\partial u_z}{\partial t}B_x^{(i)}-\frac{\partial u_x}{\partial t}B_z\right),\\
J_y^{(i)}&=\sigma_{12}^{(i)}\left(E_x+\frac{\partial u_y}{\partial t}B_z-\frac{\partial u_z}{\partial t}B_y^{(i)}\right)+\sigma_{22}^{(i)}\left(E_y+\frac{\partial u_z}{\partial t}B_x^{(i)}-\frac{\partial u_x}{\partial t}B_z\right),\\
J_z^{(i)}&=\sigma_{33}^{(i)}\left(E_z+\frac{\partial u_x}{\partial t}B_y^{(i)}-\frac{\partial u_y}{\partial t}B_x^{(i)}\right).
\end{aligned}\tag{1.35}$$

Relationships (Equation 1.35) are obtained from the third constitutive equation (Equation 1.2). To obtain Equation 1.34 from the Maxwell's equations (Equation 1.1), the term $\partial\mathbf{D}/\partial t$ is disregarded because it is small compared with the term $\boldsymbol{\sigma}(\boldsymbol{E}+\partial\boldsymbol{u}/\partial t\times\boldsymbol{B})$. This invokes the so-called quasistatic approximation to Maxwell's equations.

A linear approximation for the tangential components of the magnetic field (Zhupanska and Sierakowski 2007; Barakati and Zhupanska 2012a) can be assumed for each layer of the laminated plate, and, therefore, in-plane components of magnetic induction can be written in each layer as

$$\begin{aligned}
B_x^{(i)}&=\frac{1}{2}\left(B_x^{(i)+}+B_x^{(i)-}\right)+\frac{z}{h}\left(B_x^{(i)+}-B_x^{(i)-}\right)=\frac{1}{2}B_{x1}^{(i)}+\frac{z}{h}B_{x2}^{(i)},\\
B_y^{(i)}&=\frac{1}{2}\left(B_y^{(i)+}+B_y^{(i)-}\right)+\frac{z}{h}\left(B_y^{(i)+}-B_y^{(i)-}\right)=\frac{1}{2}B_{y1}^{(i)}+\frac{z}{h}B_{y2}^{(i)}.
\end{aligned}\tag{1.36}$$

Moreover, we assume that the distribution of the in-plane components of magnetic induction along the thickness of the laminate is linear and continuous so that the result of the integration of Equation 1.35 across the thickness of the plate depends only on the surface values of the induction on the top and bottom surfaces of the laminated plate.

By substituting Equations 1.35 and 1.36 into Equation 1.34, we have

$$
\begin{aligned}
&\sigma_{11}^{(i)}\left(E_x+\frac{\partial u_y}{\partial t}B_z-\frac{\partial u_z}{\partial t}B_y^{(i)}\right)+\sigma_{12}^{(i)}\left(E_y+\frac{\partial u_z}{\partial t}B_x^{(i)}-\frac{\partial u_x}{\partial t}B_z\right)\\
&\quad=\frac{\partial H_z}{\partial y}-\frac{1}{h}\left(H_y^{(i)+}-H_y^{(i)-}\right),\\
&\sigma_{12}^{(i)}\left(E_x+\frac{\partial u_y}{\partial t}B_z-\frac{\partial u_z}{\partial t}B_y^{(i)}\right)+\sigma_{22}^{(i)}\left(E_y+\frac{\partial u_z}{\partial t}B_x^{(i)}-\frac{\partial u_x}{\partial t}B_z\right)\\
&\quad=-\frac{\partial H_z}{\partial x}+\frac{1}{h}\left(H_x^{(i)+}-H_x^{(i)-}\right).
\end{aligned}
\tag{1.37}
$$

Application of the Kirchhoff hypothesis (Equation 1.12) to Equation 1.37 and integration of the resulting equations across the thickness of the laminate leads to the following electromagnetic governing equations in a laminate:

$$
\begin{aligned}
&\mathbf{M}_1[\sigma_{11}]E_x+\mathbf{M}_1[\sigma_{11}]\frac{\partial v}{\partial t}B_z-\mathbf{M}_2[\sigma_{11}]\frac{\partial^2 w}{\partial y\partial t}B_z-\frac{1}{2}\mathbf{M}_1[\sigma_{11}]\frac{\partial w}{\partial t}B_{y1}\\
&-\frac{1}{2}\mathbf{M}_2[\sigma_{11}]\frac{\partial w}{\partial t}B_{y2}+\mathbf{M}_1[\sigma_{12}]E_y+\frac{1}{2}\mathbf{M}_1[\sigma_{12}]\frac{\partial w}{\partial t}B_{x1}+\frac{1}{H}\mathbf{M}_2[\sigma_{12}]\frac{\partial w}{\partial t}B_{x2}\\
&-\mathbf{M}_1[\sigma_{12}]\frac{\partial u}{\partial t}B_z+\mathbf{M}_2[\sigma_{12}]\frac{\partial^2 w}{\partial x\partial t}B_z=H\frac{\partial H_z}{\partial y}-\left(H_y^+-H_y^-\right),\\
&\mathbf{M}_1[\sigma_{12}]E_x+\mathbf{M}_1[\sigma_{12}]\frac{\partial v}{\partial t}B_z-\mathbf{M}_2[\sigma_{12}]\frac{\partial^2 w}{\partial y\partial t}B_z-\frac{1}{2}\mathbf{M}_1[\sigma_{12}]\frac{\partial w}{\partial t}B_{y1}\\
&-\frac{1}{H}\mathbf{M}_2[\sigma_{12}]\frac{\partial w}{\partial t}B_{y2}+\mathbf{M}_1[\sigma_{22}]E_y+\frac{1}{2}\mathbf{M}_1[\sigma_{22}]\frac{\partial w}{\partial t}B_{x1}+\frac{1}{H}\mathbf{M}_2[\sigma_{22}]\frac{\partial w}{\partial t}B_{x2}\\
&-\mathbf{M}_1[\sigma_{22}]\frac{\partial u}{\partial t}B_z+\mathbf{M}_2[\sigma_{22}]\frac{\partial^2 w}{\partial x\partial t}B_z=-H\frac{\partial H_z}{\partial x}+\left(H_x^+-H_x^-\right).
\end{aligned}
\tag{1.38}
$$

It should be noted again that the normal component of the magnetic induction vector (i.e., H_z) and tangential components of the electric field vector are continuous across the thickness of the laminate. This can be concluded from the electromagnetic boundary conditions, which suggest that across any boundary of the discontinuity, the normal component of the magnetic induction vector **B** and tangential components of the electric field *E* are continuous.

From the second equation of Equation 1.38, an expression for E_y can be derived:

$$\begin{aligned} E_y = &-\frac{H}{\mu \mathbf{M}_1[\sigma_{22}]}\frac{\partial B_z}{\partial x} + \frac{1}{\mu \mathbf{M}_1[\sigma_{22}]} B_{x2} - \frac{\mathbf{M}_1[\sigma_{12}]}{\mathbf{M}_1[\sigma_{22}]} E_x - \frac{\mathbf{M}_1[\sigma_{12}]}{\mathbf{M}_1[\sigma_{22}]}\frac{\partial v}{\partial t} B_z \\ &+ \frac{\mathbf{M}_2[\sigma_{12}]}{\mathbf{M}_1[\sigma_{22}]}\frac{\partial W}{\partial t} B_z + \frac{1}{2}\frac{\mathbf{M}_1[\sigma_{12}]}{\mathbf{M}_1[\sigma_{22}]}\frac{\partial w}{\partial t} B_{y1} + \frac{1}{H}\frac{\mathbf{M}_2[\sigma_{12}]}{\mathbf{M}_1[\sigma_{22}]}\frac{\partial w}{\partial t} B_{y2} \\ &- \frac{1}{2}\frac{\mathbf{M}_1[\sigma_{22}]}{\mathbf{M}_1[\sigma_{22}]}\frac{\partial w}{\partial t} B_{x1} - \frac{1}{H}\frac{\mathbf{M}_2[\sigma_{22}]}{\mathbf{M}_1[\sigma_{22}]}\frac{\partial w}{\partial t} B_{x2} + \frac{\mathbf{M}_1[\sigma_{22}]}{\mathbf{M}_1[\sigma_{22}]}\frac{\partial u}{\partial t} B_z \\ &- \frac{\mathbf{M}_2[\sigma_{22}]}{\mathbf{M}_1[\sigma_{22}]}\frac{\partial^2 w}{\partial x \partial t} B_z. \end{aligned} \tag{1.39}$$

Furthermore, after applying the Kirchhoff hypothesis (Equation 1.12) and the electromagnetic hypotheses (Equations 1.32 and 1.36), the components of the Lorentz force for the lamina *i* read as

$$\begin{aligned} \rho F_x^{L(i)} = &\ \sigma_{12}^{(i)} E_x B_z + \sigma_{12}^{(i)} B_z^2 \left(\frac{\partial v}{\partial t} - z\frac{\partial^2 w}{\partial y \partial t}\right) - \sigma_{12}^{(i)} B_z B_y^{(i)} \frac{\partial w}{\partial t} + \sigma_{22}^{(i)} E_y B_z \\ &+ \sigma_{22}^{(i)} B_z B_x^{(i)} \frac{\partial w}{\partial t} - \left(\sigma_{22}^{(i)} B_z^2 + \sigma_{33}^{(i)} B_y^{(i)2}\right)\left(\frac{\partial u}{\partial t} - z\frac{\partial^2 w}{\partial x \partial t}\right) \\ &+ \sigma_{33}^{(i)} B_x^{(i)} B_y^{(i)} \left(\frac{\partial v}{\partial t} - z\frac{\partial^2 w}{\partial y \partial t}\right) + J_y^{*(i)} B_z, \end{aligned} \tag{1.40}$$

$$\begin{aligned} \rho F_y^{L(i)} = &-\sigma_{12}^{(i)} E_y B_z - \sigma_{12}^{(i)} B_z B_x^{(i)} \frac{\partial w}{\partial t} + \sigma_{12}^{(i)} B_z^2 \left(\frac{\partial u}{\partial t} - z\frac{\partial^2 w}{\partial x \partial t}\right) - \sigma_{11}^{(i)} E_x B_z \\ &+ \sigma_{33}^{(i)} B_x^{(i)} \left(\frac{1}{2} B_{y1} + \frac{z}{h} B_{y2}\right)\left(\frac{\partial u}{\partial t} - z\frac{\partial^2 w}{\partial x \partial t}\right) - (\sigma_{11}^{(i)} B_z^2 + \sigma_z B_x^{(i)2})\left(\frac{\partial v}{\partial t} - z\frac{\partial^2 w}{\partial y \partial t}\right) \\ &+ \sigma_{11}^{(i)} B_z B_y^{(i)} \frac{\partial w}{\partial t} - J_x^{*(i)} B_z. \end{aligned} \tag{1.41}$$

$$\rho F_z^{L(i)} = \sigma_{12}^{(i)} E_y B_y^{(i)} + \sigma_{12}^{(i)} B_x^{(i)} B_y^{(i)} \frac{\partial w}{\partial t} - \sigma_{12}^{(i)} B_z B_y^{(i)} \left(\frac{\partial u}{\partial t} - z \frac{\partial^2 w}{\partial x \partial t} \right)$$

$$- \sigma_{12}^{(i)} E_x B_x^{(i)} - \sigma_{12}^{(i)} B_z B_x^{(i)} \left(\frac{\partial v}{\partial t} - z \frac{\partial^2 w}{\partial y \partial t} \right) - \sigma_{11}^{(i)} E_x B_y^{(i)} - \sigma_{22}^{(i)} E_y B_x^{(i)}$$

$$+ \sigma_{11}^{(i)} B_z B_y^{(i)} \left(\frac{\partial v}{\partial t} - z \frac{\partial^2 w}{\partial y \partial t} \right) + \sigma_{22}^{(i)} B_z B_x^{(i)} \left(\frac{\partial u}{\partial t} - z \frac{\partial^2 w}{\partial x \partial t} \right)$$

$$- \left(\sigma_{11}^{(i)} B_y^{(i)2} + \sigma_{22}^{(i)} B_x^{(i)2} \right) \frac{\partial w}{\partial t} + J_x^{*(i)} B_y^{(i)} - J_y^{*(i)} B_x^{(i)} \tag{1.42}$$

where $\boldsymbol{\varepsilon} \approx \varepsilon_0 \mathbf{1}$, and $J_x^{*(i)}$ and $J_y^{*(i)}$ are components of the external electric current, which depend on the fiber orientation in the lamina.

Note that 2D approximation of the coupled mechanical and electromagnetic field equations presented in this work is different from the previous studies (Ambartsumyan et al. 1977; Hasanyan and Piliposyan 2001; Librescu et al. 2003; Hasanyan et al. 2005), where the small disturbance concept was used to simplify the nonlinear magnetoelastic problems for anisotropic and laminated composite plates. The approach adopted in the present work is not limited to the small disturbance problems and enables one to treat highly dynamic coupled problems.

1.2.3 Coupled System of Governing Equations for the Laminated Plate

The 2D system of equations of motion (Equation 1.27) and Maxwell's equations (Equation 1.38) constitutes a mathematical framework within which coupled mechanical and electromagnetic response of electrically conductive laminated plates is studied. From the mathematical standpoint, the system of Equations 1.27 and 1.38 is a nonlinear mixed system of parabolic and hyperbolic PDEs. This system can be solved using the numerical solution procedure described in the next section.

From Equation 1.26, the derivatives of the middle-plane displacements with respect to the y-direction can be found in terms of the derivative of field variables with respect to the x-direction as

$$\begin{aligned}
\frac{\partial w}{\partial y} &= W, \\
\frac{\partial u}{\partial y} &= q_{11} \frac{\partial u}{\partial x} + q_{12} \frac{\partial v}{\partial x} + q_{13} N_{xy} + q_{14} N_{yy} + q_{15} \frac{\partial^2 w}{\partial x^2} + q_{16} \frac{\partial W}{\partial x} + q_{17} M_{yy}, \\
\frac{\partial v}{\partial y} &= q_{21} \frac{\partial u}{\partial x} + q_{22} \frac{\partial v}{\partial x} + q_{23} N_{xy} + q_{24} N_{yy} + q_{25} \frac{\partial^2 w}{\partial x^2} + q_{26} \frac{\partial W}{\partial x} + q_{27} M_{yy}, \\
\frac{\partial W}{\partial y} &= q_{61} \frac{\partial u}{\partial x} + q_{62} \frac{\partial v}{\partial x} + q_{63} N_{xy} + q_{64} N_{yy} + q_{65} \frac{\partial^2 w}{\partial x^2} + q_{66} \frac{\partial W}{\partial x} + q_{67} M_{yy},
\end{aligned} \tag{1.43}$$

Here, the coefficients q_{1i}, q_{2i}, and q_{6i} are defined as below:

$$q_{11} = \frac{A_{26}B_{12}B_{22} - A_{12}A_{26}D_{22} - A_{16}B_{22}^2 - A_{22}B_{12}B_{26} + A_{12}B_{22}B_{26} + A_{22}A_{16}D_{22}}{Q},$$

$$q_{12} = \frac{-D_{22}A_{26}^2 + 2A_{26}B_{22}B_{26} - A_{66}B_{22}^2 - A_{22}B_{26}^2 + A_{22}A_{66}D_{22}}{Q},$$

$$q_{13} = \frac{B_{22}^2 - A_{22}D_{22}}{Q}, \quad q_{14} = \frac{A_{26}D_{22} - B_{22}B_{26}}{Q},$$

$$q_{15} = \frac{-A_{26}B_{22}D_{12} + A_{26}B_{12}D_{22} + B_{16}B_{22}^2 - B_{22}B_{12}B_{26} + A_{22}B_{26}D_{12} - A_{22}B_{16}D_{22}}{Q},$$

$$q_{16} = \frac{2\left(-A_{26}B_{22}D_{26} + A_{26}B_{26}D_{22} + A_{22}B_{26}D_{26} - B_{22}B_{26}^2 + B_{66}B_{22}^2 - A_{22}B_{66}D_{22}\right)}{Q},$$

$$q_{17} = \frac{A_{22}B_{26} - A_{26}B_{22}}{Q}, \tag{1.44}$$

$$q_{21} = \frac{-A_{12}B_{26}^2 + A_{26}B_{12}B_{26} + A_{16}B_{22}B_{26} - A_{66}B_{22}B_{12} - A_{16}B_{26}D_{22} + A_{12}A_{66}D_{22}}{Q},$$

$$q_{22} = 0, \quad q_{23} = \frac{-B_{22}B_{26} + A_{26}D_{22}}{Q}, \quad q_{24} = \frac{B_{26}^2 = A_{66}D_{22}}{Q}$$

$$q_{25} = \frac{B_{12}B_{26}^2 - B_{22}B_{16}B_{26} - A_{26}B_{12}B_{26} + A_{66}B_{22}D_{12} - A_{66}B_{12}D_{22} + A_{26}A_{16}D_{22}}{Q},$$

$$q_{26} = \frac{2\left(B_{26}^3 - A_{26}B_{26}D_{26} - B_{22}B_{66}B_{26} - A_{66}B_{26}D_{22} + A_{66}B_{22}D_{26} + A_{26}B_{66}D_{22}\right)}{Q},$$

$$q_{27} = \frac{-A_{26}B_{26} + A_{66}B_{22}}{Q}, \tag{1.45}$$

$$q_{61} = \frac{B_{12}A_{26}^2 - A_{16}A_{26}B_{16} - A_{12}A_{26}B_{26} - A_{22}A_{66}B_{12} + A_{12}A_{66}B_{22} + A_{22}A_{16}B_{26}}{Q},$$

$$q_{62} = 0, \quad q_{63} = \frac{B_{22}A_{26} - A_{22}B_{26}}{Q}, \quad q_{64} = \frac{A_{26}B_{26} - A_{66}B_{22}}{Q},$$

$$q_{65} = \frac{-D_{12}A_{26}^2 + A_{26}B_{22}B_{16} + A_{26}B_{12}B_{26} - A_{66}B_{22}B_{12} - A_{22}B_{16}B_{26} + A_{22}A_{66}D_{12}}{Q},$$

$$q_{66} = \frac{2\left(-D_{26}A_{26}^2 + A_{26}B_{26}^2 + A_{26}B_{22}B_{66} - A_{66}B_{22}B_{26} - A_{22}B_{26}B_{66} + A_{22}A_{66}D_{26}\right)}{Q},$$

$$q_{67} = \frac{A_{22}A_{66} - A_{26}^2}{Q}, \tag{1.46}$$

where

$$Q = A_{26}^2 D_{22} - 2A_{26} B_{22} B_{26} + B_{26}^2 B_{22} + A_{66}\left(B_{22}^2 - A_{22} D_{22}\right). \tag{1.47}$$

Moreover, from Equation 1.26, the following resultants can be rewritten as

$$\begin{aligned}
N_{xx} &= s_1 M_{yy} + s_2 N_{xy} + s_3 \frac{\partial v}{\partial x} + s_4 \frac{\partial u}{\partial x} + s_5 N_{yy} + s_6 \frac{\partial^2 w}{\partial x^2} + s_7 \frac{\partial W}{\partial x}, \\
M_{xx} &= f_1 M_{yy} + f_2 N_{xy} + f_3 \frac{\partial v}{\partial x} + f_4 \frac{\partial u}{\partial x} + f_5 N_{yy} + f_6 \frac{\partial^2 w}{\partial x^2} + f_7 \frac{\partial W}{\partial x}, \\
M_{xy} &= l_1 M_{yy} + l_2 N_{xy} + l_3 \frac{\partial v}{\partial x} + l_4 \frac{\partial u}{\partial x} + l_5 N_{yy} + l_6 \frac{\partial^2 w}{\partial x^2} + l_7 \frac{\partial W}{\partial x},
\end{aligned} \tag{1.48}$$

where

$$\begin{aligned}
s_1 &= A_{16} q_{17} + A_{12} q_{27} - B_{12} q_{67}, & s_2 &= A_{16} q_{13} + A_{12} q_{23} - B_{12} q_{63}, \\
s_3 &= A_{16}(q_{12} + 1) + A_{12} q_{22} - B_{12} q_{62}, & s_4 &= A_{16} q_{11} + A_{12} q_{21} - B_{12} q_{61} + A_{11}, \\
s_5 &= A_{16} q_{14} + A_{12} q_{24} - B_{12} q_{64}, & s_6 &= A_{16} q_{15} + A_{12} q_{25} - B_{12} q_{65} - B_{11}, \\
s_7 &= A_{16} q_{16} + A_{12} q_{26} - B_{12} q_{66} - 2B_{16},
\end{aligned} \tag{1.49}$$

$$\begin{aligned}
f_1 &= B_{16} q_{17} + B_{12} q_{27} - D_{12} q_{67} & f_2 &= B_{16} q_{13} + B_{12} q_{23} - D_{12} q_{63}, \\
f_3 &= B_{16}(q_{12} + 1) + B_{12} q_{22} - D_{12} q_{62}, & f_4 &= B_{16} q_{11} + B_{12} q_{21} - D_{12} q_{61} + B_{11}, \\
f_5 &= B_{16} q_{14} + B_{12} q_{24} - D_{12} q_{64}, & f_6 &= B_{16} q_{15} + B_{12} q_{25} - D_{12} q_{65} - D_{11}, \\
f_7 &= B_{16} q_{16} + B_{12} q_{26} - D_{12} q_{66} - 2D_{12},
\end{aligned} \tag{1.50}$$

$$\begin{aligned}
l_1 &= B_{66} q_{17} + B_{26} q_{27} - D_{26} q_{67}, & l_2 &= B_{66} q_{13} + B_{26} q_{23} - D_{26} q_{63}, \\
l_3 &= B_{66}(q_{12} + 1) + B_{26} q_{22} - D_{26} q_{62}, & l_4 &= B_{66} q_{11} + B_{26} q_{21} - D_{26} q_{61} - B_{16}, \\
l_5 &= B_{66} q_{14} + B_{26} q_{24} - D_{26} q_{64}, & l_6 &= B_{66} q_{15} + B_{26} q_{25} - D_{26} q_{65} - D_{16}, \\
l_7 &= B_{66} q_{16} + B_{26} q_{26} - D_{26} q_{66} - 2D_{66}
\end{aligned} \tag{1.51}$$

Using Equations 1.26, 1.39, and 1.48 together with the Lorentz force equations (Equations 1.40 through 1.42), the following first derivatives of resultants with respect to the y-direction can be derived from the equations of motion (Equation 1.27):

$$\frac{\partial N_{xy}}{\partial y} = \rho H \frac{\partial^2 u}{\partial t^2} - s_1 \frac{\partial M_{yy}}{\partial x} - s_2 \frac{\partial N_{xy}}{\partial x} - s_3 \frac{\partial^2 v}{\partial x^2} - s_4 \frac{\partial^2 u}{\partial x^2} - s_5 \frac{\partial N_{yy}}{\partial x} - s_6 \frac{\partial^3 w}{\partial x^3}$$

$$- s_7 \frac{\partial^2 W}{\partial x^2} + \frac{H}{\mu} B_z \frac{\partial B_z}{\partial x} + \frac{1}{4}\mathbf{M}_1[\sigma_{33}] B_{y1}^2 \frac{\partial u}{\partial t} - \frac{1}{4}\mathbf{M}_2[\sigma_{33}] B_{y1}^2 \frac{\partial^2 w}{\partial x \partial t} - \mathbf{M}_1\left[J_y^*(t)\right] B_z$$

$$\frac{\partial N_{yy}}{\partial y} = \rho H \frac{\partial^2 v}{\partial t^2} - \frac{\partial N_{xy}}{\partial x} + q_{42} E_x B_z + q_{42} B_z^2 \frac{\partial v}{\partial t} + q_{46} B_z^2 \frac{\partial W}{\partial t}$$

$$+ q_{45} B_{y1} B_z \frac{\partial w}{\partial t} + q'_{45} B_z^2 \frac{\partial w}{\partial t} + q_{40} B_z \frac{\partial B_z}{\partial x} + \mathbf{M}_1\left[J_y^*(t)\right] B_z,$$

$$\frac{\partial N_{yz}}{\partial y} = \rho H \frac{\partial^2 w}{\partial t^2} + p(y,t) - \frac{\partial N_{xz}}{\partial x} + q_{82} B_{y1} E_x + q_{82} B_{y1} B_z \frac{\partial v}{\partial t} + q_{86} B_{y1} B_z \frac{\partial W}{\partial t}$$

$$+ q_{85} B_{y1}^2 \frac{\partial w}{\partial t} + q_{80} B_{y1} \frac{\partial B_z}{\partial x} + q'_{85} B_{y1} B_z \frac{\partial^2 w}{\partial x \partial t} + \frac{1}{2} B_{y1} \mathbf{M}_1\left[J_x^*(t)\right],$$

$$\frac{\partial M_{yy}}{\partial y} = -\frac{\rho H^3}{12} \frac{\partial^2 W}{\partial t^2} + N_{yz} - l_1 \frac{\partial M_{yy}}{\partial x} + l_2 \frac{\partial N_{xy}}{\partial x} + l_3 \frac{\partial^2 v}{\partial x^2} + l_4 \frac{\partial^2 u}{\partial x^2} + l_5 \frac{\partial N_{yy}}{\partial x}$$

$$+ l_6 \frac{\partial^3 w}{\partial x^3} + l_7 \frac{\partial^2 W}{\partial x^2} - + q_{72} E_x B_z + q_{72} B_z^2 \frac{\partial v}{\partial t} + q_{76} B_z^2 \frac{\partial W}{\partial t} + q_{75} B_{y1} B_z \frac{\partial w}{\partial t}$$

$$+ q'_{75} B_z^2 \frac{\partial^2 w}{\partial x \partial t} + q_{70} B_z \frac{\partial B_z}{\partial x} + \mathbf{M}_2\left[J_x^*(t)\right] B_z. \tag{1.52}$$

in which B_{y2} and B_x are considered to be zero ($B_{y2} = B_{x1} = B_{x2} = 0$). The coefficients q_{ij} are

$$q_{42} = \mathbf{M}_1[\sigma_{11}] - \frac{\mathbf{M}_1^2[\sigma_{12}]}{\mathbf{M}_1[\sigma_{22}]}, \qquad q_{46} = \frac{\mathbf{M}_1[\sigma_{12}]\mathbf{M}_2[\sigma_{12}]}{\mathbf{M}_1[\sigma_{22}]} - \mathbf{M}_2[\sigma_{11}],$$

$$q_{45} = \frac{1}{2}\left(\frac{\mathbf{M}_1^2[\sigma_{12}]}{\mathbf{M}_1[\sigma_{22}]} - \mathbf{M}_1[\sigma_{11}]\right), \qquad q'_{45} = \mathbf{M}_2[\sigma_{12}] - \frac{\mathbf{M}_1[\sigma_{12}]\mathbf{M}_2[\sigma_{22}]}{\mathbf{M}_1[\sigma_{22}]},$$

$$q_{40} = -\frac{H}{\mu}\frac{\mathbf{M}_1[\sigma_{12}]}{\mathbf{M}_1[\sigma_{22}]}$$

$$q_{72} = \mathbf{M}_2[\sigma_{11}] - \frac{\mathbf{M}_1[\sigma_{12}]\mathbf{M}_2[\sigma_{12}]}{\mathbf{M}_1[\sigma_{22}]}, \qquad q_{76} = \frac{\mathbf{M}_2^2[\sigma_{12}]}{\mathbf{M}_1[\sigma_{22}]} - \mathbf{M}_3[\sigma_{11}],$$

$$q_{75} = \frac{1}{2}\left(\frac{\mathbf{M}_1[\sigma_{12}]\mathbf{M}_2[\sigma_{12}]}{\mathbf{M}_1[\sigma_{22}]} - \mathbf{M}_2[\sigma_{11}]\right), \qquad q'_{75} = \mathbf{M}_3[\sigma_{12}] - \frac{\mathbf{M}_2[\sigma_{12}]\mathbf{M}_2[\sigma_{22}]}{\mathbf{M}_1[\sigma_{22}]}$$

$$q_{70} = -\frac{H}{\mu}\frac{\mathbf{M}_2[\sigma_{12}]}{\mathbf{M}_1[\sigma_{22}]},$$

$$q_{82}=\frac{1}{2}\left(\frac{\mathbf{M}_1^2[\sigma_{12}]}{\mathbf{M}_1[\sigma_{22}]}-\mathbf{M}_1[\sigma_{11}]\right),\qquad q_{86}=-\frac{1}{2}\left(\frac{\mathbf{M}_1[\sigma_{12}]\mathbf{M}_2[\sigma_{12}]}{\mathbf{M}_1[\sigma_{22}]}-\mathbf{M}_2[\sigma_{11}]\right),$$

$$q_{85}=\frac{1}{4}\left(\mathbf{M}_1[\sigma_{11}]-\frac{\mathbf{M}_1^2[\sigma_{12}]}{\mathbf{M}_1[\sigma_{22}]}\right),\qquad q'_{85}=\frac{1}{2}\left(\frac{\mathbf{M}_1[\sigma_{12}]\mathbf{M}_2[\sigma_{22}]}{\mathbf{M}_1[\sigma_{22}]}-\mathbf{M}_2[\sigma_{12}]\right).$$

$$q_{80}=\frac{1}{2}\frac{H}{\mu}\frac{\mathbf{M}_1[\sigma_{12}]}{\mathbf{M}_1[\sigma_{22}]},\tag{1.53}$$

The stress resultant N_{xz} in the equation of N_{yz} in Equation 1.52 can be found from the equations of motion in the form

$$\begin{aligned}N_{xz}=&\frac{\rho H^3}{12}\frac{\partial^3 w}{\partial x\partial t^2}-\frac{\rho H^3}{12}l_1\frac{\partial^2 W}{\partial t^2}+\rho Hl_2\frac{\partial^2 u}{\partial t^2}+\rho Hl_5\frac{\partial^2 v}{\partial t^2}+l_1N_{yz}\\&+q_{31}\frac{\partial^2 u}{\partial x^2}+q_{32}\frac{\partial^2 v}{\partial x^2}+q_{33}\frac{\partial N_{xy}}{\partial x}+q_{34}\frac{\partial N_{yy}}{\partial x}+q_{35}\frac{\partial^3 w}{\partial x^3}+q_{36}\frac{\partial^2 W}{\partial x^2}+q_{37}\frac{\partial M_{yy}}{\partial x}\\&+q_{51}\left(E_xB_z+B_z^2\frac{\partial v}{\partial t}\right)+q_{52}B_z^2\frac{\partial W}{\partial t}+q_{53}B_{y1}B_z\frac{\partial w}{\partial t}+q_{54}B_z^2\frac{\partial^2 w}{\partial x\partial t}+q_{50}B_z\frac{\partial B_z}{\partial x}\\&+q'_{51}B_{yl}^2\frac{\partial v}{\partial t}q'_{52}B_{yl}^2\frac{\partial^2 w}{\partial x\partial t}+\mathbf{M}_1\left[J_x^*(t)\right]l_5B_z-\mathbf{M}_1\left[J_y^*(t)\right]l_2B_z+\mathbf{M}_2\left[J_y^*(t)\right]B_z\end{aligned}\tag{1.54}$$

where

$$q_{31}=f_4-l_1l_4-s_4l_2+q_{21}l_3+q_{11}l_4+q_{61}l_7,$$

$$q_{32}=f_3-l_1l_3-s_3l_2+q_{22}l_3+q_{12}l_4+q_{62}l_7,$$

$$q_{33}=f_2-l_1l_2-s_2l_2+q_{23}l_3+q_{13}l_4+q_{63}l_7,$$

$$q_{34}=f_5-l_1l_5-s_5l_2+q_{24}l_3+q_{14}l_4+q_{64}l_7,$$

$$q_{35}=f_6-l_1l_6-s_6l_2+q_{25}l_3+q_{15}l_4+q_{65}l_7,$$

$$q_{36}=f_7-l_1l_7-s_7l_2+q_{26}l_3+q_{16}l_4+q_{66}l_7+l_6,$$

$$q_{37}=f_1-l_1^2-s_1l_2+q_{27}l_3+q_{17}l_4+q_{67}l_7,$$

$$q_{51}=q_{72}l_1+q_{42}l_5+q_{451},\qquad q_{52}=q_{76}l_1+q_{46}l_5-q_{751},$$

$$q_{53}=q_{75}l_1+q_{45}l_5+q_{851},\qquad q_{54}=q_{751}l_1+q_{451}l_5+\mathbf{M}_3[\boldsymbol{\sigma}_{22}]-\frac{\mathbf{M}_2^2[\boldsymbol{\sigma}_{22}]}{\mathbf{M}_1[\boldsymbol{\sigma}_{22}]}$$

$$q_{50}=q_{70}l_1+l_2\frac{H}{\mu}+2q_{80}l_5+\frac{\mathbf{M}_2[\sigma_{22}]}{\mathbf{M}_1[\sigma_{22}]}\frac{H}{\mu}$$

$$q'_{51}=\frac{1}{4}(\mathbf{M}_1[\sigma_{33}]l_1-\mathbf{M}_2[\sigma_{33}]),\qquad q'_{52}=\frac{1}{4}\left(\mathbf{M}_3[\sigma_{33}]-\mathbf{M}_2[\sigma_{33}]l_2\right),\tag{1.55}$$

For the electromagnetic governing equations, considering $B_x = B_{y2} = 0$, from Equation 1.38, we can write

$$\frac{\partial B_z}{\partial y} = q_{02}E_x + q_{02}\frac{\partial v}{\partial t}B_z + q_{06}\frac{\partial W}{\partial t}B_z + q_{05}B_{y1}\frac{\partial w}{\partial t} + q'_{05}\frac{\partial^2 w}{\partial x \partial t}B_z + q_{00}\frac{\partial B_z}{\partial x} \tag{1.56}$$

where

$$q_{02} = \frac{\mu}{H}\left(\mathbf{M}_1[\sigma_{11}] - \frac{\mathbf{M}_1^2[\sigma_{12}]}{\mathbf{M}_1[\sigma_{22}]}\right), \qquad q_{06} = \frac{\mu}{H}\left(\frac{\mathbf{M}_1[\sigma_{12}]\mathbf{M}_2[\sigma_{12}]}{\mathbf{M}_1[\sigma_{22}]} - \mathbf{M}_2[\sigma_{11}]\right),$$

$$q_{05} = \frac{\mu}{2H}\left(\frac{\mathbf{M}_1^2[\sigma_{12}]}{\mathbf{M}_1[\sigma_{22}]} - \mathbf{M}_1[\sigma_{11}]\right), \qquad q'_{05} = \frac{\mu}{H}\left(\mathbf{M}_2[\sigma_{12}] - \frac{\mathbf{M}_1[\sigma_{12}]\mathbf{M}_2[\sigma_{22}]}{\mathbf{M}_1[\sigma_{22}]}\right),$$

$$q_{00} = -\frac{\mathbf{M}_1[\sigma_{12}]}{\mathbf{M}_1[\sigma_{22}]}. \tag{1.57}$$

The last governing equation can be obtained from the third equations of Equations 1.38 and 1.39 as

$$\begin{aligned}\frac{\partial E_x}{\partial y} = \frac{\partial B_z}{\partial x} &- \frac{H}{\mu \mathbf{M}_1[\sigma_{22}]}\frac{\partial^2 B_z}{\partial x^2} - \frac{\mathbf{M}_1[\sigma_{12}]}{\mathbf{M}_1[\sigma_{22}]}\frac{\partial E_x}{\partial x} + \frac{\mathbf{M}_1[\sigma_{12}]}{2\mathbf{M}_1[\sigma_{22}]}B_{y1}\frac{\partial^2 w}{\partial t \partial x} \\ &+ \frac{\mathbf{M}_2[\sigma_{12}]}{\mathbf{M}_1[\sigma_{22}]}\left(\frac{\partial^2 W}{\partial t \partial x}B_z + \frac{\partial W}{\partial t}\frac{\partial B_z}{\partial x}\right) - \frac{\mathbf{M}_1[\sigma_{12}]}{\mathbf{M}_1[\sigma_{22}]}\left(\frac{\partial^2 v}{\partial t \partial x}B_z + \frac{\partial v}{\partial t}\frac{\partial B_z}{\partial x}\right) \\ &+ \frac{\mathbf{M}_1[\sigma_{22}]}{\mathbf{M}_1[\sigma_{22}]}\left(\frac{\partial^2 u}{\partial t \partial x}B_z + \frac{\partial u}{\partial t}\frac{\partial B_z}{\partial x}\right) - \frac{\mathbf{M}_2[\sigma_{22}]}{\mathbf{M}_1[\sigma_{22}]}\left(\frac{\partial^3 w}{\partial t \partial x^2}B_z + \frac{\partial^2 w}{\partial x \partial t}\frac{\partial B_z}{\partial x}\right).\end{aligned} \tag{1.58}$$

Finally, the tenth-order system of governing equations for a laminated plate includes the four equations of Equation 1.43, the four equations of Equation 1.52, and the two electromagnetic equations of Equations 1.56 and 1.58. There is no known analytical solution for such a system; therefore, a numerical solution procedure is proposed to solve the developed system of governing equations in the following section.

1.3 Numerical Solution Procedure

There are different approaches to solving the governing system of PDEs developed in Section 1.2. As the problem is coupled and highly dynamic, the numerical solution procedure needs to deal with an ill-conditioned system. Among all possible numerical solution methods (Kubíček and Hlaváček 1983, Atkinson et al. 2009, Roberts and Shipman 1972, Scott and Watts 1977), such as shooting techniques, the finite element method, and quasilinearization, in this work a sequential application of finite difference (FD) time and spatial (with respect to one coordinate) integration schemes, method of lines (MOL), quasilinearization of the resulting system of the nonlinear ordinary differential equations (ODEs), an FD spatial integration of the obtained two-point boundary-value problem is employed. The final solution is obtained by the application of the superposition method followed by orthonormalization. A discussion of the details of the suggested numerical solution procedure is presented next. The numerical solution procedure is the extension of the procedure developed in earlier work (Barakati and Zhupanska 2012a).

1.3.1 Time Integration

The first step of the numerical solution procedure is the time integration. For this purpose, Newmark's scheme (Newmark 1959) is employed in this work because of its wide use in dynamic problems due to simplicity. In this method, the derivatives of any function f with respect to time can be written in the form

$$\left.\frac{\partial^2 f}{\partial t^2}\right|_{t+\Delta t} = \frac{1}{\beta(\Delta t)^2}\left(f\big|_{t+\Delta t} - f\big|_{t}\right) - \frac{1}{\beta}\left(\frac{1}{\Delta t}\left.\frac{\partial f}{\partial t}\right|_{t} + \left(\frac{1}{2}-\beta\right)\left.\frac{\partial^2 f}{\partial t^2}\right|_{t}\right),$$

$$\left.\frac{\partial f}{\partial t}\right|_{t+\Delta t} = \left.\frac{\partial f}{\partial t}\right|_{t} + \Delta t\left((1-\gamma)\left.\frac{\partial^2 f}{\partial t^2}\right|_{t} + \gamma\left.\frac{\partial^2 f}{\partial t^2}\right|_{t+\Delta t}\right), \tag{1.59}$$

where β and γ are the scheme parameters and Δt is the time integration step. The parameters β and γ are considered to be 0.25 and 0.5, respectively, as these values yield unconditional stability in linear problems. However, the size of the time step is also very critical in the stability of nonlinear problems.

1.3.2 Method of Lines

After the time integration, the next step is the spatial integration. For the numerical solution procedure that we employ in this work, we need to reduce the system of PDEs to a system of ODEs. To this end, we employ the MOL, which is a well-established numerical (or semianalytical) technique that has been widely used to solve the governing PDEs of physical boundary-value problems (Sadiku and Obiozor 2000). The basic idea of the MOL is to approximate the original PDE by discretizing all but one of the independent variables in order to obtain a set of ODEs. This is done by replacing the derivatives with respect to one independent variable with algebraic approximations such as FD, spline, or weighted residual techniques. Therefore, the PDE can be reduced to an initial-value ODE system, which can be easily solved by employing a numerical integration algorithm (Schiesser and Griffiths 2009). The popular algebraic approximation used in most MOL solutions is the FD scheme.

In this work, the governing PDEs of the 2D problem have three independent variables: x, y, and t. As mentioned earlier, the Newmark's scheme is used for the time integration. The numerical procedure can be followed by the application of the method of lines to eliminate the explicit presence of one spatial independent variable in the governing equations, converting the system of the PDEs into a system of ODEs. For this purpose, the plate domain is divided using straight lines perpendicular to the x-direction, and the central FD is employed to approximate the derivatives with respect to x:

$$\begin{aligned}
&\frac{\partial g_j^i}{\partial x} \approx \frac{g_j^{i+1} - g_j^{i-1}}{2\Delta x}, \\
&\frac{\partial^2 g_j^i}{\partial x^2} \approx \frac{g_j^{i+1} - 2g_j^i + g_j^{i-1}}{\Delta x^2}, \\
&\frac{\partial^3 g_j^i}{\partial x^3} \approx \frac{g_j^{i+2} - 2g_j^{i+1} + 2g_j^{i-1} - g_j^{i-2}}{2\Delta x^3}, \\
&\frac{\partial^4 g_j^i}{\partial x^4} \approx \frac{g_j^{i+2} - 4g_j^{i+1} + 6g_j^i - 4g_j^{i-1} + g_j^{i-2}}{\Delta x^4},
\end{aligned} \tag{1.60}$$

where the index j represents any of the N variables in the vector of unknowns g, index i designates a position along the grid in the x-direction, and Δx is the spacing in x. Thus, the system of ODEs approximates the solution of the original system of PDEs at the grid points i = 1, 2,…, n_x. The final form of the vector of unknowns g is now n_x times larger:

$$g = \left[g_1^1, g_2^1, \ldots, g_N^1, g_1^2, g_2^2, \ldots, g_N^2, \ldots, g_1^{n_x}, g_2^{n_x}, \ldots, g_N^{n_x} \right]^T, \tag{1.61}$$

where the vector g is of the size $(N \cdot n_x) \times 1$.

It is worth mentioning that in MOL, the system of equations is solved for the unknowns on the lines that are located inside the domain, while the known boundary conditions related to the discretized spatial dimension need to be applied manually to the system of equations. One important advantage of MOL is that it can be easily set aside from the solution procedure for 1D problems (e.g., a long plate). Conversely, if the numerical solution procedure for the 1D problem is already developed, MOL is the best option for extending the solution to the 2D case.

1.3.3 Quasilinearization and Superposition Method

Now that the nonlinear system of PDEs is reduced to an initial-value ODE system by the application of Newmark's scheme and MOL, it is time to linearize the system of equations. After employing the FD space integration with respect to one of the spatial coordinates (the x-coordinate, for instance), the system of equations can be written in the form

$$\frac{\partial g}{\partial y} = \mathbf{\Phi}\left(y, t, g, \frac{\partial g}{\partial t}, \frac{\partial^2 g}{\partial t^2}\right), \tag{1.62}$$

where the unknown $(N \cdot n_x) \times 1$-dimensional vector $g(y, t)$ includes the unknown middle-plane displacements and their first derivatives, stress and moment resultants, and electromagnetic components. Moreover, $\mathbf{\Phi}$ is a smooth and continuously differentiable function of g. It should be noted that in order to reduce the system of second-order governing equations in the form of the system of first-order ODEs (Equation 1.62), the second derivatives of the unknowns with respect to y are replaced with the first derivatives of new unknown functions, which themselves are the first derivatives of the unknowns of the system of equations with respect to the y-direction.

To solve the nonlinear system (Equation 1.62), the quasilinearization method proposed by Bellman and Kalaba (1969) is employed. In this method, a sequence vector $\{g^{k+1}\}$ is generated by the linear equations

$$\frac{\mathrm{d}g^{k+1}}{\mathrm{d}y} = \mathbf{\Phi}(g^k) + J(g^k)(g^{k+1} - g^k), \tag{1.63}$$

and the linearized boundary conditions

$$\begin{aligned} \mathbf{D}_1(g^k)g^{k+1}(y_0, t+\Delta t) &= d_1(g^k), \\ \mathbf{D}_2(g^k)g^{k+1}(y_N, t+\Delta t) &= d_2(g^k) \end{aligned} \tag{1.64}$$

with g^0 being an initial guess. Here g^k and g^{k+1} are the solutions at the k-th and $(k + 1)$-th iterations, matrices $\mathbf{D}_1(g^k)$ and $\mathbf{D}_2(g^k)$ together with vectors $\mathbf{d}_1(g^k)$ and $\mathbf{d}_2(g^k)$ are determined from the given boundary conditions at the edges of the plate (i.e., points y_0 and y_N, correspondingly), and $\mathbf{J}(g^k)$ is the Jacobian matrix defined as

$$\{\mathbf{J}_{ij}(g^k)\} = \left\{ \frac{\partial \Phi_i}{\partial g_j} \left(g_1^k, g_2^k, \ldots g_N^k \right) \right\}, \tag{1.65}$$

which needs to be calculated analytically. The sequence of solutions $\{g^{k+1}\}$ of the linear system (Equation 1.63) rapidly converges to the solution of the original nonlinear system (Equation 1.62). An initial approximation to the solution of the nonlinear problem is needed at the first time step, and for the next time steps, the nonlinear solution at the previous time step is used for the initial approximation. Finally, the iterative process is terminated when the desired accuracy of the solution is achieved

$$\left| \frac{g_i^{k+1} - g_i^k}{g_i^k} \right| \leq \delta, \tag{1.66}$$

where δ is the convergence parameter.

To solve the linear system of the two-point boundary-value problem in Equations 1.63 and 1.64, the superposition method along with the stable discrete orthonormalization technique (Zhupanska and Sierakowski 2005; Scott and Watts 1977; Godunov 1961; Conte 1966; Mol'chenko and Loos 1999) is employed here. In the superposition method, the solution of the boundary-value problem at the $(k + 1)$-th iteration can be obtained by the linear summation of J linearly independent general solutions (base solutions) and one particular solution as

$$g^{k+1}(y, t + \Delta t) = \sum_{j=1}^{J} c_j G^j(y, t + \Delta t) + G^{J+1}(y, t + \Delta t), \tag{1.67}$$

where $\mathbf{G}^j$, $j = 1, 2, 3, \ldots, J$, are solutions of the Cauchy problem for the homogeneous system (Equation 1.63) with homogeneous initial condition at the left endpoint, where the solution is sought; $\mathbf{G}^{J+1}$ is the solution of the Cauchy problem for the inhomogeneous system (Equation 1.63) with true initial condition at the left endpoint; and c_j, $j = 1, 2, 3, \ldots, J$ are the solution constants. If there are the same number of boundary conditions on both ends and they are separated, $N/2$ base solutions $\mathbf{G}^j$ are needed (Scott and Watts 1977), where $J = N/2$ here. Using straightforward integration to obtain a solution in the form of Equation 1.67 will not lead to satisfactory results since the matrix of the system (Equation 1.63) is "ill-conditioned." Therefore, straightforward integration will result in the loss of linear independency in the solution vectors $\mathbf{G}^j$, $j = 1, 2, 3, \ldots, J + 1$. See the literature (Scott and Watts 1977; Godunov 1961, Conte 1966) for further discussion of the loss of linear independence in the stiff boundary-value problems. The loss of linear independence in the solution vectors can be bypassed by applying an orthonormalization procedure.

1.3.4 Orthonormalization

To ensure that the solution vectors are properly linearly independent, the orthonormalization procedure is applied at each step of the integration. A modified Gram–Schmidt method is employed for this purpose because of its numerical stability and simplicity in computations. To show how the orthonormalization method can be included in the numerical solution procedure, we first write the solution (Equation 1.67) at iteration $(k + 1)$ in the form

$$g^{k+1}(y, t + \Delta t) = \boldsymbol{\Omega} \boldsymbol{c} + \boldsymbol{G}^{J+1}, \tag{1.68}$$

where matrix $\boldsymbol{\Omega}$ is the set of base solutions $\boldsymbol{G}^j$, $j = 1, 2, 3, \ldots, J$, and $\boldsymbol{c}$ is the vector of solution constants c_j. After the application of the orthonormalization process to the solution vectors, the matrix of the new orthonormal base solutions, $\boldsymbol{\Omega}_{\text{new}}$, can be written in terms of the matrix of the old orthonormal base solutions as

$$\boldsymbol{\Omega}_{\text{new}} = \boldsymbol{\Omega}_{\text{old}}\, \mathbf{P}, \tag{1.69}$$

where $\boldsymbol{P}$ is a nonsingular upper triangular matrix. This matrix is determined using a procedure described in the literature (Conte 1966). The particular solution is then calculated as

$$\boldsymbol{G}_{\text{new}}^{J+1} = \boldsymbol{G}_{\text{old}}^{J+1} - \boldsymbol{\Omega}_{\text{new}} \boldsymbol{\eta}_{\text{new}}, \tag{1.70}$$

where the elements of the vector $\boldsymbol{\eta}_{\text{new}}$ are the inner products of $\boldsymbol{G}_{\text{old}}^{J+1}$ and the new base solution vectors of $\boldsymbol{\Omega}_{\text{new}}$. The particular solution is orthogonal to the new set of the orthonormal base solutions.

Starting from the left end of the plate and performing orthonormalization, the solution of the boundary-value problem (Equations 1.63 and 1.64) can be continued to the last integration point on the right side where the boundary conditions (Equation 1.64) give the unknown solution constants. The solution after orthonormalization (or reorthonormalization) is

$$\begin{aligned} g_m^{k+1}(y_m, t + \Delta t) &= \boldsymbol{\Omega}_m(y_m)\mathbf{c}_m + \boldsymbol{G}_m^{J+1}(y_m), \\ \boldsymbol{G}_m^{J+1}(y_m) &= \boldsymbol{G}_{m-1}^{J+1}(y_m) - \boldsymbol{\Omega}_m(y_m)\boldsymbol{\eta}_m(y_m), \end{aligned} \tag{1.71}$$

where $y = y_m$ is the end point. The continuity in the solution is preserved by requiring

$$g_{m-1}^{k+1}(y_m, t + \Delta t) = g_m^{k+1}(y_m, t + \Delta t). \tag{1.72}$$

The solution constants are obtained by substituting Equation 1.71 into Equation 1.72 and using Equation 1.69:

$$\mathbf{c}_{m-1} = \mathbf{P}_m\,(\mathbf{c}_m - \boldsymbol{\eta}_m) \tag{1.73}$$

This enables obtaining the solution at all integration points without performing a complete reintegration.

Furthermore, when this orthonormalization process is inadequate and the orthonormalized vectors are still linearly dependent to some extent, *K*-criterion reorthonormalization is performed with $K=\sqrt{2}$ based on the Euclidean norms of the solution vectors (Ruhe 1983).

1.3.5 Spatial Integration and Final Solution

The last step of the numerical solution procedure is the spatial integration and solving the resulting linear system of equations. For the spatial integration, explicit fourth-order Runge–Kutta's FD procedure is applied to the system of ODEs (Equation 1.62) as

$$\begin{aligned}
g_{i+1} &= g_i + \sum_{m=1}^{4} b_m f_m,\\
f_1 &= \Delta x \boldsymbol{\Phi}(y_i, g_i),\\
f_2 &= \Delta x \boldsymbol{\Phi}(y_i + c_2\Delta y, g_i + \beta_{21} f_2),\\
f_3 &= \Delta x \boldsymbol{\Phi}(y_i + c_3\Delta y, g_i + \beta_{31} f_1 + \beta_{32} f_2),\\
f_4 &= \Delta x \boldsymbol{\Phi}(y_i + c_4\Delta y, g_i + \beta_{41} f_1 + \beta_{42} f_2 + \beta_{43} f_3),
\end{aligned} \tag{1.74}$$

where

$$\begin{aligned}
&c_1 = 0,\quad c_4 = 1,\quad c_2 = u,\quad c_3 = v,\\
&u = 0.3,\quad v = 0.6,\\
&b_2 = \frac{2v-1}{12u(v-u)(1-u)},\quad b_3 = \frac{1-2u}{12v(v-u)(1-v)},\\
&b_4 = \frac{6uv-4u-4v+3}{12(1-u)(1-v)},\quad b_1 = 1-b_2-b_3-b_4,\\
&\beta_{21} = u,\quad \beta_{32} = \frac{1}{24b_3u(1-v)},\quad \beta_{31} = v-\beta_{32},\\
&\beta_{43} = \frac{b_3(1-v)}{b_4} = \frac{1-2u}{12b_4v(u-v)},\quad \beta_{42} = -\frac{v(4v-5)-u+2}{24b_4u(v-u)(1-v)},\\
&\beta_{41} = 1-\beta_{42}-\beta_{43}.
\end{aligned} \tag{1.75}$$

Other suggested integration techniques are multistep methods such as the fourth-order Adams–Bashforth defined as (Atkinson et al. 2009):

$$g_{i+1} = g_i + \frac{\Delta y}{24}[55\mathbf{\Phi}_i - 59\mathbf{\Phi}_{i-1} + 37\mathbf{\Phi}_{i-2} - 9\mathbf{\Phi}_{i-3}], \tag{1.76}$$

and the fourth-order Adams–Moulton method, which reads as (Atkinson et al. 2009)

$$g_{i+1} = g_i + \frac{\Delta y}{24}[9\mathbf{\Phi}_{i+1} + 19\mathbf{\Phi}_i - 5\mathbf{\Phi}_{i-1} + \mathbf{\Phi}_{i-2}]. \tag{1.77}$$

It should be noted that since the Adams–Bashforth and Adams–Moulton methods are not self-starting, the fourth-order Runge–Kutta method is used for the first four steps.

The spatial integration along with the orthonormalization of the solution vectors at each nodal point is performed starting from the first node on the left side of the plate ($y = -a/2$) until reaching the final node on the right ($y = +a/2$). At this point, using the boundary conditions at $y = +a/2$, a linear system of equations can be formed to find the unknown solution constants in Equation 1.67. To solve this linear system of equations, say $\boldsymbol{Ax} = \boldsymbol{b}$, the Cholesky decomposition method (Kincaid and Cheney 2002) is employed in which the $n \times n$ matrix of coefficients $\boldsymbol{A}$ is decomposed into a lower triangular matrix $\boldsymbol{L}$ and an upper triangular matrix U:

$$A = LU \tag{1.78}$$

where

$$\begin{aligned} U_{ij} &= A_{ij} - \sum_{k=1}^{i-1} L_{ik}U_{kj}, \quad j = i,\ldots,n \\ L_{ij} &= \frac{1}{U_{jj}}\left(A_{ij} - \sum_{k=1}^{j-1} L_{ik}U_{kj}\right), \quad j = 1,\ldots,i-1. \end{aligned} \tag{1.79}$$

To overcome the rounding errors in computing the vector of unknowns $\boldsymbol{x}$, the following iterative refinement is used: after solving the system $\mathbf{A}\boldsymbol{x}^{(i)} = \boldsymbol{b}$, the residual vector $\mathbf{r}^{(i)} = \mathbf{A}\boldsymbol{x}^{(i)} - \boldsymbol{b}$ is computed. This follows by solving the new system $\mathbf{A}\boldsymbol{dx}^{(i+1)} = \boldsymbol{r}^{(i)}$ and updating the solution $\boldsymbol{x}^{(i+1)} = \boldsymbol{x}^{(i)} + \boldsymbol{dx}^{(i+1)}$. This procedure is repeated until an accurate enough solution is achieved. Finally, by solving for the solution constants at the final node, the constants at other nodes can easily be found by the recursive formulation (Equation 1.73), which leads to the final solution of unknowns all over the plate using the superposition method (Equation 1.67). A FORTRAN code has been developed to implement the described numerical procedure for the solution of the boundary-value problem (Equations 1.63 and 1.64). The next section presents the results of the solution of a nonlinear coupling problem for the unidirectional and cross-ply composite plates.

1.4 Mechanical Response of the Composite Plate Subjected to Impact and Electromagnetic Loads

1.4.1 Problem Statement

Consider a thin fiber-reinforced electrically conductive laminated composite plate of width a, length l, and thickness H subjected to the transverse short duration load p, pulsed electric current of density $\mathbf{J}^*$, and immersed in the magnetic field with the induction $\mathbf{B}^*$ (Figure 1.4). The density of the applied pulsed electric current is

$$\mathbf{J}^* = \left(J_x^*, 0, 0\right),$$
$$J_x^* = J_x^*(t) = J_0 e^{-t/\tau_c} \sin\frac{\pi t}{\tau_c}, \quad t \geq 0. \tag{1.80}$$

where τ_c is the characteristic time of the electric current. A pulsed current is considered in this study because it has been proven that it produces considerably less heat in the composite plate compared with other types of electric

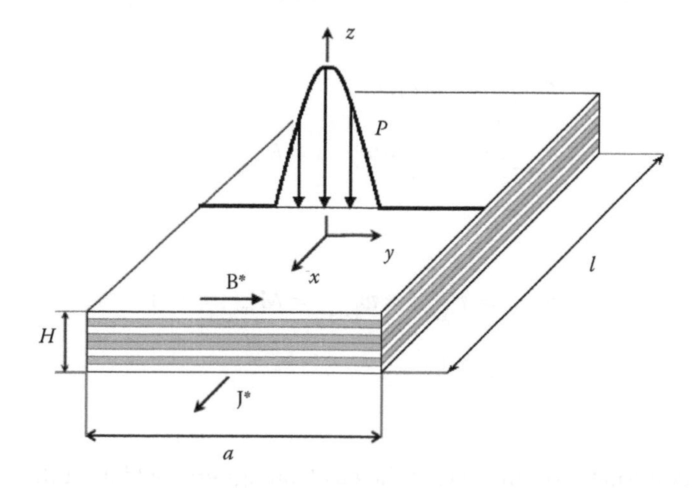

FIGURE 1.4
Composite plate subjected to pulsed electric current and transverse impact load and immersed in magnetic field.

currents, e.g., DC and AC (Barakati and Zhupanska 2012b). Therefore, the effect of thermal stresses can be neglected by the application of a pulsed current.

The plate is also immersed in the constant in-plane magnetic field

$$\mathbf{B}^* = \left(0, B_y^*, 0\right), \qquad B_y^* = \text{const}. \tag{1.81}$$

In addition, it is assumed that the plate is subjected to a short-duration impact load applied transversely to the plate, and this load results in the time-varying compressive pressure distribution, $p(y,t)$, given by

$$p(y,t) = \begin{cases} p_0\sqrt{1-\left(\dfrac{y}{b}\right)^2}\sin\dfrac{\pi t}{\tau_p}, & |y| \le b, \quad 0 < t \le \tau_p, \\ 0, \quad b < |y| \le \dfrac{a}{2}, & t > \tau_p. \end{cases} \tag{1.82}$$

Here p_0 is the maximum contact pressure, b is the half-size of the contact zone, and τ_p is the characteristic time parameter, which determines the duration of the applied pressure. Moreover, the load is assumed to result only in elastic deformation, and the plate is assumed to be initially at rest.

As for the boundary conditions, the plate is simply supported:

$$\tau_{zz}\Big|_{z=\frac{H}{2}} = -p(y,t), \qquad u\Big|_{y=\pm\frac{a}{2}} = v\Big|_{y=\pm\frac{a}{2}} = w\Big|_{y=\pm\frac{a}{2}} = M_{yy}\Big|_{y=\pm\frac{a}{2}} = 0, \tag{1.83}$$

$$u\Big|_{x=\pm\frac{l}{2}} = v\Big|_{x=\pm\frac{l}{2}} = w\Big|_{x=\pm\frac{l}{2}} = M_{xx}\Big|_{x=\pm\frac{l}{2}} = 0, \qquad E_y\Big|_{x=\pm\frac{l}{2}} = 0, \tag{1.84}$$

and the boundary conditions for the electromagnetic field are taken as

$$\left(E_x - \frac{\partial w}{\partial t}B_y^* + \frac{\partial v}{\partial t}B_z\right)\Bigg|_{y=-\frac{a}{2}} = 0,$$

$$E_x\Big|_{y=\frac{a}{2}} = 0, \tag{1.85}$$

$$E_y\Big|_{x=\pm\frac{l}{2}} = 0.$$

The following plate parameters are considered in the work. The width of the plate is $a = 0.1524$ m, and the thickness is $H = 0.0021$ m. The plate is assumed to be made of the AS4/3501-6 CFRP matrix composite with 60% fiber volume fraction. The material properties of the composite are as follows: density $\rho = 1594$ kg/m^3; Young's moduli in the fiber and transverse directions are $E_x = 102.97$ GPa and $E_y = 7.55$ GPa, respectively; Poisson's ratios, $\nu_{yx} = \nu_{xz} = 0.3$; and electric conductivity in the fiber direction, $\sigma_x = 39{,}000$ S/m. The half-size of the contact zone is $b = H/100$.

1.4.2 Numerical Results for the Unidirectional Composite Plate

In this section, the results of the numerical studies of the unidirectional rectangular electrically conductive transversely isotropic plate subjected to the mechanical load in Equation 1.82 and the pulsed electromagnetic loads in Equations 1.80 and 1.81 are presented.

First we briefly discuss how to bring the system of governing equations to the vector form (Equation 1.62). Considering the type of loading on the plate and ignoring the small terms that contain $(\varepsilon_y - \varepsilon_0)$, the system of equations for a 2D plate reads as

$$\frac{\partial u}{\partial y} = \frac{1}{HB_{66}}N_{xy} - \frac{\partial v}{\partial x},$$

$$\frac{\partial v}{\partial y} = \frac{1}{HB_{22}}N_{yy} - \frac{B_{12}}{B_{22}}\frac{\partial u}{\partial x},$$

$$\frac{\partial N_{xy}}{\partial y} = \rho H\frac{\partial^2 u}{\partial t^2} - \frac{B_{12}}{B_{22}}\frac{\partial N_{yy}}{\partial x} - \left(B_{11} - \frac{B_{12}^2}{B_{22}}\right)H\frac{\partial^2 u}{\partial x^2} + \frac{H}{\mu}\frac{\partial B_z}{\partial x}B_z$$

$$+\frac{H}{4}\sigma_z\left(B_{y1}^2 + \frac{1}{3}B_{y2}^2\right)\frac{\partial u}{\partial t} - \frac{H^2}{12}\sigma_z B_{y1}B_{y2}\frac{\partial^2 w}{\partial x \partial t} + (\varepsilon_x - \varepsilon_0)HE_xB_z\frac{\partial^2 v}{\partial x \partial t}$$

$$-\frac{H}{2}(\varepsilon_x - \varepsilon_0)B_{y1}E_x\frac{\partial^2 w}{\partial x \partial t},$$

$$\frac{\partial N_{yy}}{\partial y} = \rho H \frac{\partial^2 v}{\partial t^2} - \frac{\partial N_{xy}}{\partial x} + \sigma_x H B_z^2 \frac{\partial v}{\partial t} - \frac{H}{2}\sigma_x B_{y1} B_z \frac{\partial w}{\partial t} + \frac{\varepsilon_x - \varepsilon_0}{B_{22}} E_x B_z \frac{\partial N_{yy}}{\partial t}$$

$$-(\varepsilon_x - \varepsilon_0) H \frac{B_{12}}{B_{22}} E_x B_z \frac{\partial^2 u}{\partial x \partial t} - \frac{H}{2}(\varepsilon_x - \varepsilon_0) B_{y1} E_x \frac{\partial W}{\partial t} + \sigma_x H E_x B_z + H B_z J_x^*(t),$$

$$\frac{\partial w}{\partial y} = W,$$

$$\frac{\partial W}{\partial y} = -\frac{12}{H^3 B_{22}} M_{yy} - \frac{B_{12}}{B_{22}} \frac{\partial^2 w}{\partial x^2},$$

$$\frac{\partial M_{yy}}{\partial y} = -\frac{\rho H^3}{12}\frac{\partial^2 W}{\partial t^2} + N_{yz} + \frac{H^3}{6} B_{66} \frac{\partial^2 W}{\partial x^2} - \frac{H^2}{12}\sigma_x B_z B_{y2} \frac{\partial w}{\partial t} - \frac{H^3}{12}\sigma_x B_z^2 \frac{\partial W}{\partial t}$$

$$-\frac{1}{12}(\varepsilon_x - \varepsilon_0) H^2 E_x B_{y2} \frac{\partial W}{\partial t} + \frac{(\varepsilon_x - \varepsilon_0)}{B_{22}} E_x B_z \frac{\partial M_{yy}}{\partial t}$$

$$+\frac{H^3}{12}\frac{B_{12}}{B_{22}}(\varepsilon_x - \varepsilon_0) E_x B_z \frac{\partial^3 w}{\partial x^2 \partial t},$$

$$\frac{\partial N_{yz}}{\partial y} = \rho H \frac{\partial^2 w}{\partial t^2} + p(y,t) - \frac{H}{2}\sigma_x E_x B_{y1} - \frac{H}{2}\sigma_x B_{y1} B_z \frac{\partial v}{\partial t} + \frac{H}{4}\sigma_x \left(B_{y1}^2 + \frac{1}{3}B_{y2}^2\right)\frac{\partial w}{\partial t}$$

$$+\frac{H^2}{12}\sigma_x B_z B_{y2} \frac{\partial W}{\partial t} - (\varepsilon_x - \varepsilon_0) H E_x B_z \frac{\partial W}{\partial t} - \frac{H}{2} B_{y1} J_x^*(t) - \frac{\rho H^3}{12}\frac{\partial^4 w}{\partial x^2 \partial t^2}$$

$$-\left(\frac{B_{12}}{B_{22}} + \frac{2B_{66}}{B_{22}}\right)\frac{\partial^2 M_{yy}}{\partial x^2} - \frac{H^3}{12}\left(\frac{2B_{66}B_{12}}{B_{22}} - B_{11} + \frac{B_{12}^2}{B_{22}}\right)\frac{\partial^4 w}{\partial x^4}$$

$$-\frac{H^3}{12}\left(\sigma_y B_z^2 + \frac{1}{4}\sigma_z B_{y1}^2\right)\frac{\partial^3 w}{\partial x^2 \partial t} - \frac{H^3}{2}\left(\frac{1}{3}\sigma_y B_z \frac{\partial B_z}{\partial x} + \frac{1}{40}\sigma_z B_{y2}^2\right)\frac{\partial^2 w}{\partial x \partial t}$$

$$+\frac{H^2}{12}\sigma_z B_{y1} B_{y2} \frac{\partial^2 u}{\partial x \partial t} - \frac{H^3}{12}(\varepsilon_x - \varepsilon_0)\left(B_z \frac{\partial E_x}{\partial x} + E_x \frac{\partial B_z}{\partial x}\right)\frac{\partial^2 W}{\partial x \partial t}$$

$$-\frac{H^3}{12}(\varepsilon_x - \varepsilon_0) E_x B_z \frac{\partial^3 W}{\partial x^2 \partial t} - \frac{H^2}{12}(\varepsilon_x - \varepsilon_0) B_{y2}\left(\frac{\partial E_x}{\partial x}\frac{\partial^2 w}{\partial x \partial t} + E_x \frac{\partial^3 w}{\partial x^2 \partial t}\right),$$

$$\frac{\partial E_x}{\partial y} = \frac{\partial B_z}{\partial t} + \frac{\partial u}{\partial t}\frac{\partial B_z}{\partial x} + \frac{\partial^2 u}{\partial t \partial x} B_z - \frac{1}{\sigma_y \mu}\frac{\partial^2 B_z}{\partial x^2},$$

$$\frac{\partial B_z}{\partial y} = \sigma_x \mu E_x + \sigma_x \mu \frac{\partial v}{\partial t} B_z - \sigma_x \mu \frac{B_{y1}}{2}\frac{\partial w}{\partial t} + \frac{B_{y2}}{H}, \tag{1.86}$$

which includes eight mechanical and two electromagnetic variables. This system can be rewritten in the vector form (Equation 1.62), where the unknown vector g stands for

$$g = [u, v, N_{xy}, N_{yy}, w, W, M_{yy}, N_{yz}, E_x, B_z]^T, \tag{1.87}$$

in which the order of the variables was selected such that the resulting matrix of coefficients of the system is close to a band matrix: the four in-plane displacements and resultants first, followed by the four out-of-plane unknowns and, finally, the two electromagnetic variables. This is helpful for yielding a less ill-conditioned matrix of coefficients.

The MOL discussed in Section 1.2.4 is now applied to the system in Equation 1.86 to discretize one of the spatial independent variables (x). To do so, all the derivatives with respect to x are replaced with the corresponding FD approximations. The plate is partitioned in the x-direction by n_x number of lines where $x = \pm l/2$ are the boundary lines, the line $x = -l/2 + \Delta x$ is the first line $i = 1$, and the line $x = +l/2 - \Delta x$ is the last line $i = n_x$. After applying MOL, the final vector of unknowns reads as

$$\begin{aligned} g = \Big[& u^1, v^1, N_{xy}^1, N_{yy}^1, w^1, W^1, M_{yy}^1, N_{yz}^1, E_x^1, B_z^1, \\ & u^2, v^2, N_{xy}^2, N_{yy}^2, w^2, W^2, M_{yy}^2, N_{yz}^2, E_x^2, B_z^2, \dots, \\ & u^{n_x}, v^{n_x}, N_{xy}^{n_x}, N_{yy}^{n_x}, w^{n_x}, W^{n_x}, M_{yy}^{n_x}, N_{yz}^{n_x}, E_x^{n_x}, B_z^{n_x} \Big]^T \end{aligned} \tag{1.88}$$

The boundary conditions for $x = \pm l/2$ need to be applied manually to the adjacent lines in the system of equations. Therefore, the PDE system of Equation 1.86 is now reduced to a system of ODEs that can be solved by the same numerical procedure used for the 1D case: MOL is followed by the Newmark's time integration and quasilinearization. Then the resulting linear system of ODEs is integrated in the y-direction while orthonormalization is applied, which yields the final solution over the plate.

The $5n_x$ homogeneous vectors and one nonhomogenous initial vector for the 2D problem are

$$\mathbf{N}_{1,i}^{\text{hom}} = \Big[\overbrace{0,\dots,0}^{10(i-1)},0,0,1,0,0,0,0,0,0,0,\overbrace{0,\dots,0}^{10(n_x-i)}\Big]^T,$$

$$\mathbf{N}_{2,i}^{\text{hom}} = \Big[\overbrace{0,\dots,0}^{10(i-1)},0,0,0,1,0,0,0,0,0,0,\overbrace{0,\dots,0}^{10(n_x-i)}\Big]^T,$$

$$\mathbf{N}_{3,i}^{\text{hom}} = \Big[\overbrace{0,\dots,0}^{10(i-1)},0,0,0,0,0,1,0,0,0,0,\overbrace{0,\dots,0}^{10(n_x-i)}\Big]^T,$$

$$\mathbf{N}_{4,i}^{\text{hom}} = \left[\overbrace{0,\ldots,0}^{10(i-1)},0,0,0,0,0,0,0,1,0,0,\overbrace{0,\ldots,0}^{10(n_x-i)}\right]^T,$$

$$\mathbf{N}_{5,i}^{\text{hom}} = \left[\overbrace{0,\ldots,0}^{10(i-1)},0,0,0,0,0,0,0,\frac{g_2^k}{2\beta\Delta t}+\zeta_2,-1,\overbrace{0,\ldots,0}^{10(n_x-i)}\right]^T,$$

$$\mathbf{N}^{\text{non hom}} = \left[\overbrace{0,\ldots,0}^{10(i-1)},0,0,0,0,0,0,0,\zeta_2 g_{10}^k - \zeta_5 B_y^*,-g_{10}^k,\overbrace{0,\ldots,0}^{10(n_x-i)}\right]^T. \tag{1.89}$$

where ζ_j includes all the terms related to g_j^k after the combination of the two Newmark's equations (Zhupanska and Sierakowski 2005).

The boundary conditions for the plate at $y = \pm a/2$, as introduced in Equation 1.83, can be used to define the matrices $\mathbf{D}_1$ and $\mathbf{D}_2$ and vectors $\boldsymbol{d}_1$ and $\boldsymbol{d}_2$ in Equation 1.64 as

$$\mathbf{D}_1 = \begin{bmatrix} \mathrm{A}_1, & 0, & \cdots, & 0 \\ 0, & \mathrm{A}_1, & \cdots, & 0 \\ \vdots & & \ddots & \vdots \\ 0, & \cdots, & 0, & \mathrm{A}_1 \end{bmatrix}, \quad \mathbf{D}_2 = \begin{bmatrix} \mathrm{A}_2, & 0, & \cdots, & 0 \\ 0, & \mathrm{A}_2, & \cdots, & 0 \\ \vdots & & \ddots & \vdots \\ 0, & \cdots, & 0, & \mathrm{A}_2 \end{bmatrix},$$

$$\boldsymbol{d}_1 = [\ \mathbf{a}_1,\ \ \mathbf{a}_1,\ \ \cdots\ \ \mathbf{a}_1\ ]^T,\quad \boldsymbol{d}_2 = [\ \mathbf{a}_2,\ \ \mathbf{a}_2,\ \ \cdots\ \ \mathbf{a}_2\ ]^T, \tag{1.90}$$

where $\mathbf{D}_1$ and $\mathbf{D}_2$ are matrices of the size $5n_x \times 5n_x$ and $\boldsymbol{d}_1$ and $\boldsymbol{d}_2$ are vectors of the size $5n_x$, in which

$$\mathbf{A}_1 = \begin{bmatrix} 1, & 0, & 0, & 0, & 0, & 0, & 0, & 0, & 0, & 0 \\ 0, & 1, & 0, & 0, & 0, & 0, & 0, & 0, & 0, & 0 \\ 0, & 0, & 0, & 0, & 1, & 0, & 0, & 0, & 0, & 0 \\ 0, & 0, & 0, & 0, & 0, & 0, & 1, & 0, & 0, & 0 \\ 0, & \dfrac{g_{10}^k}{2\beta\Delta t} & 0, & 0, & \dfrac{-B_y^*}{2\beta\Delta t} & 0, & 0, & 0, & 1, & \dfrac{g_1^k}{2\beta\Delta t}+\zeta_1 \end{bmatrix},$$

$$\mathbf{A}_2 = \begin{bmatrix} 1, & 0, & 0, & 0, & 0, & 0, & 0, & 0 \\ 0, & 1, & 0, & 0, & 0, & 0, & 0, & 0 \\ 0, & 0, & 0, & 0, & 0, & 1, & 0, & 0 \\ 0, & 0, & 0, & 0, & 0, & 0, & 1, & 0 \end{bmatrix},$$

$$\mathbf{a}_1 = \left[\ 0,\ 0,\ 0,\ 0,\ -B_y^*\zeta_2\ \ -\frac{g_1^k g_8^k}{2\beta\Delta t}\ \right]^T,\quad \mathbf{a}_2 = [\ 0,\ 0,\ 0,\ 0,\ 0\ ]^T. \tag{1.91}$$

Moreover, the mechanical and electromagnetic boundary conditions at $x = \pm l/2$ in Equation 1.84 result in the following values of nonzero variables at $x = \pm l/2$:

$$
\begin{aligned}
&N_{xy}\Big|_{x=-\frac{l}{2}} = HB_{66}\frac{v^1}{\Delta x}, \quad N_{xy}\Big|_{x=+\frac{l}{2}} = -HB_{66}\frac{v^{n_x}}{\Delta x},\\
&N_{yy}\Big|_{x=-\frac{l}{2}} = HB_{12}\frac{u^1}{\Delta x}, \quad N_{yy}\Big|_{x=+\frac{l}{2}} = -HB_{12}\frac{u^{n_x}}{\Delta x},\\
&M_{yy}\Big|_{x=-\frac{l}{2}} = \frac{H^3}{12}\left(\frac{B_{11}B_{22}}{B_{12}} - B_{12}\right)\frac{1}{\Delta x^2}(w^2 - 2w^1),\\
&M_{yy}\Big|_{x=+\frac{l}{2}} = -\frac{H^3}{12}\left(\frac{B_{11}B_{22}}{B_{12}} - B_{12}\right)\frac{1}{\Delta x^2}(w^{n_x-2} - 2w^{n_x-1}),\\
&W\Big|_{x=\pm\frac{l}{2}} = 0,\\
&M_{yy}\Big|_{x=+\frac{l}{2}} = -\frac{H^3}{12}\left(\frac{B_{11}B_{22}}{B_{12}} - B_{12}\right)\frac{1}{\Delta x^2}(w^{n_x-1} - 2w^{n_x}),\\
&B_z\Big|_{x=-\frac{l}{2}} = B_z^1, \quad B_z\Big|_{x=+\frac{l}{2}} = B_z^{n_x}.
\end{aligned}
\qquad (1.92)
$$

Next, the effects of the application of various electromagnetic loads on the mechanical response of the plate have been studied.

The results reported below were obtained for the following parameters. The plate was assumed to be rectangular with $l = 2a$, and the mechanical load (Equation 1.82) was such that $p_0 = 1$ MPa, $\tau_p = 10$ ms. The characteristic time of the pulsed electric current was $\tau_c = \tau_p = 10$ ms. Moreover, in all simulations the time step was $dt = 10^{-4}$ s, the number of lines was five, $n_x = 5$, and $n_y = 10^5$.

Figure 1.5 shows the effect of the magnitude of the external magnetic induction, B_y^*, on the plate's deflection. In this figure, the current density of the pulsed current is fixed at $J_0 = 10^5$ A/m^2 for all cases. It can be seen that an increase in the magnitude of the magnetic field leads to a decrease in the amplitude of the deflection and a more rapid decay in vibrations.

The effect of the magnitude of the electric current density is presented in Figure 1.6. Here, the magnetic induction is $B_y^* = 0.1$ T for all cases. Since the magnetic field is small, a noticeable change in the vibration amplitudes is seen only when the current density is large enough, here as large as $J_0 = 10^7$ A/m^2. Furthermore, the damping effect can be ignored when the magnetic field is small.

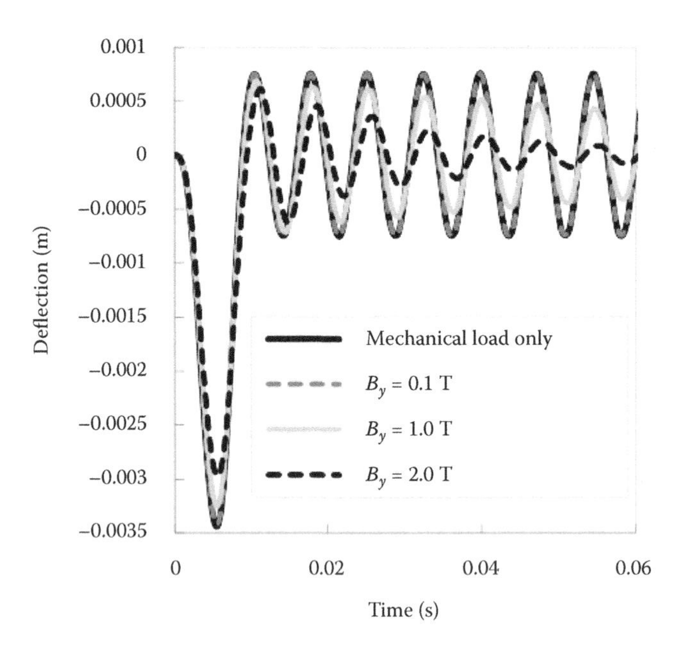

FIGURE 1.5
Deflection of unidirectional composite plate: effect of magnitude of magnetic induction when $J_0 = 10^5$ A/m^2.

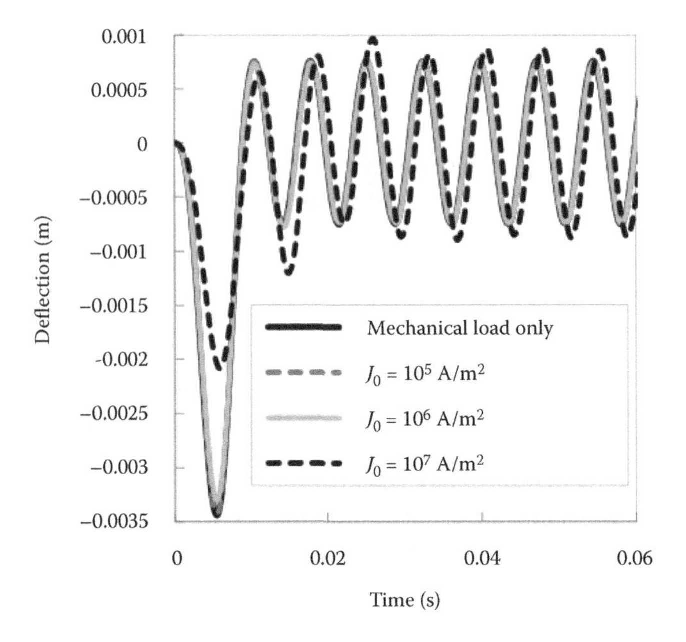

FIGURE 1.6
Deflection of unidirectional composite plate: effect of magnitude of current density when $B_y^* = 0.1$ T (three curves coincide).

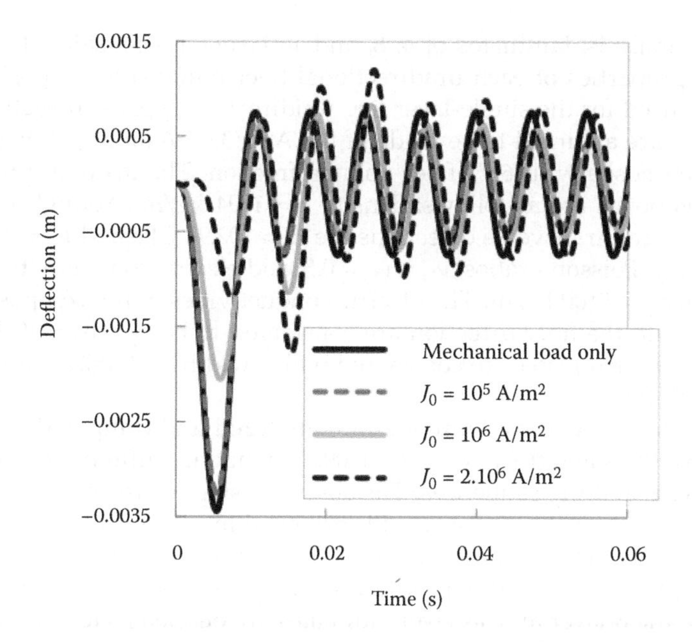

FIGURE 1.7
Deflection of unidirectional composite plate: effect of magnitude of current density when $B_y^* = 1.0$ T.

The mechanical response of the unidirectional composite plate in the presence of a larger magnetic induction $B_y^* = 1.0$ T is shown in Figure 1.7. Now that the magnetic field is relatively large, not only is the damping effect noticeable but also the deflection of the plate is considerably reduced, at least during the application of the impact load.

1.4.3 Numerical Results for the Laminated Plate

To investigate the response of a laminated plate subjected to the mechanical loads in Equation 1.82 and the electromagnetic loads in Equations 1.80 and 1.81, the system of governing equations developed in Section 1.2 is solved by the numerical solution procedure introduced in Section 1.3. For the sake of simplicity, only symmetric cross-ply laminates with layers of equal thickness are considered here. In a symmetric cross-ply laminate, the geometry and material properties of the layers are symmetric with respect to the middle plane of the laminate, and the fiber orientations of the layers are either $\theta = 0°$ or $\theta = 90°$. In such laminates, there is no bending-extension coupling (B_{ij} are zero in Equations 1.21 and 1.22). Moreover, we have $A_{16} = A_{26} = 0$. Thus, the equations of the resultants (Equation 1.26) are significantly simplified for the case of symmetric cross-ply laminates. Due to the ease of manufacturing and analysis, these types of laminates are widely used in civil and aerospace industries.

In this analysis, laminates of 4, 8, and 16 layers are considered, and the material properties of each unidirectional fiber reinforced composite layer are those used for the single-layer (i.e., unidirectional) plate in Section 1.4.1. Laminates are assumed to be made of the AS4/3501-6 unidirectional CFRP matrix composite with 60% fiber volume fraction. The material properties of the composite are as follows: density $\rho = 1594$ kg/m^3; Young's modulus in the fiber and transverse directions are $E_1 = 102.97$ GPa and $E_2 = 7.55$ GPa, respectively; Poisson's ratios, $\nu_{21} = \nu_{13} = 0.3$; and electric conductivity in fiber direction, $\sigma_1 = 39{,}000$ S/m. The electric conductivities of the composite perpendicular to the fiber direction are considered to be $\sigma_2 = \sigma_3 = 10^{-4}\sigma_1$. The square laminated plates are considered with a width $a = 0.1524$ m and thickness $H = 0.0021$ m.

Four different types of laminates are analyzed and compared. All laminates have the same thickness, $H = 0.0021$ m, but are different in the number of layers and ply sequences. The so-called single-layer plate consists of one transversely isotropic layer with principal material directions coinciding with the laminate coordinate axes. The four-layer laminate is laid up in the form [0/90/90/0] or $[0/90]_s$, where subscript "s" stands for "symmetric." Similarly, the 8-layer and 16-layer laminates are defined as $[0/90/0/90]_s$ and $[0/90/0/90/0/90/0/90]_s$, respectively.

Simply supported boundary conditions are assumed as in Equations 1.83 through 1.85, and a laminated plate is assumed to be subjected to a transient mechanical load (Equation 1.82) with the characteristic time $\tau_p = 10$ ms and maximum pressure $p_0 = 1$ MPa, constant in-plane magnetic field (Equation 1.81), and pulsed electric current (Equation 1.80), where $\tau_c = \tau_p = 10$ ms. Moreover, in all numerical studies, the time step was $dt = 10^{-4}$ s, and $n_x = 5$, while $n_y = 6000$. The half-size of the contact zone is $b = H/10$.

Figure 1.8 shows middle-plane transverse deflection, w, in the center of the plate ($x = 0$, $y = 0$). It also shows that adding layers with fiber orientations of $\theta = 90°$ to those of $\theta = 0°$ significantly reduces deflection of the laminate. Moreover, the frequent use of the layers of $\theta = 90°$ in between the plies of $\theta = 0°$ will result in further increase in the impact resistance of the laminated plate.

A similar trend can be observed when, in addition to the mechanical load, an electromagnetic load is applied to the laminated plate. In Figure 1.9, a pulsed electric current and an external magnetic induction $\left(B_y^* = 1.0\text{ T}\right)$ are applied together with the mechanical load. It can be seen that the addition of the magnetic induction leads to decay in the plate vibrations.

Next, the response of the laminated plates under various electromagnetic loads is discussed. Figures 1.10 through 1.15 show the effect of the magnitude of the external magnetic induction, B_y^*. Each figure shows the results for 1-layer, 4-layer, 8-layer, and 16-layer plates subjected to the same mechanical and electromagnetic loads. Thus, the difference in the number of layers on the response of the plate is emphasized. Figure 1.10 shows deflection of the different laminated plates subjected to the pulsed electric current, $J_0 = 10^5$ A/m^2, $\tau_c = \tau_p = 10$ ms, and low magnetic induction $B_y^* = 0.1\text{ T}$. As noted earlier, the

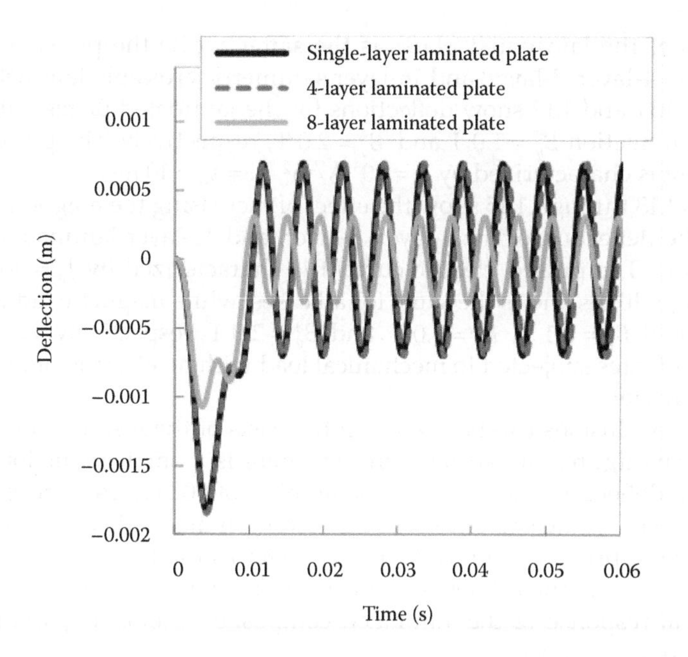

FIGURE 1.8
Deflection of laminated plate: effect of ply sequence with no electromagnetic load applied.

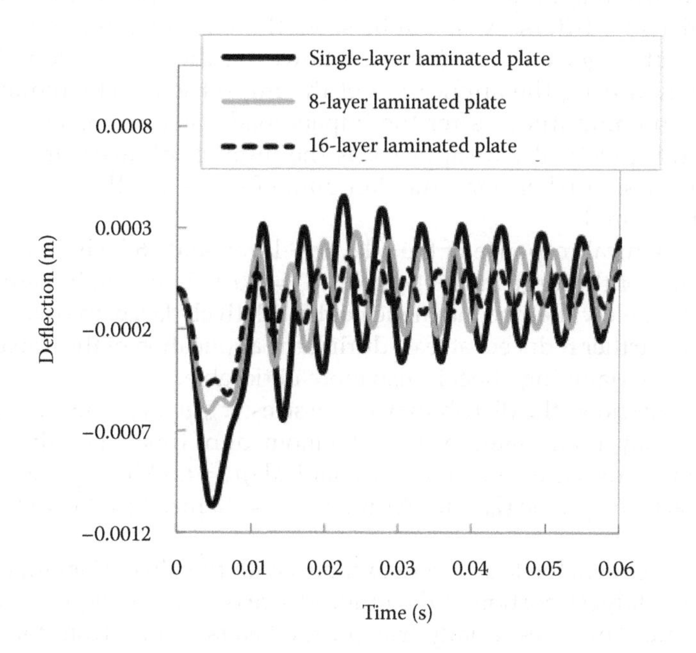

FIGURE 1.9
Deflection of laminated plate: effect of ply sequence in the presence of electromagnetic field.

thickness of the laminated plates is the same, while the ply sequence is different in 4-layer, 8-layer, and 16-layer symmetric cross-ply laminates.

Figures 1.11 and 1.12 show deflections for the laminated plates with large magnetic induction $B_y^* = 1.0$ T and $B_y^* = 2.0$ T, respectively. The pulsed electric current is characterized by $J_0 = 10^5$ A/m^2, $\tau_c = \tau_p = 10$ ms.

Figures 1.13 through 1.15 show the effect of increasing the magnetic induction on the deflection of the 4-layer, 8-layer, and 16-layer laminated plates, respectively. The pulsed electric current is characterized by $J_0 = 10^5$ A/m^2 and $\tau_c = \tau_p = 10$ ms and is the same for all cases, while magnetic induction is different and $B_y^* = 0.1$ T, $B_y^* = 1.0$ T, and $B_y^* = 2.0$ T, respectively. The results for the laminates subjected to mechanical load with no electromagnetic load are also present.

Several conclusions can be drawn on the basis of the results presented in the previous figures. It can be seen that there is a small reduction in the maximum deflection and stress as the number of 90° layers increases. This stays true even in the presence of a high-strength magnetic field. It can also be seen that vibration magnitude decays faster as the number of 90° layers decreases. Overall, the influence of an electromagnetic field on the dynamic mechanical response of the laminated composites is most apparent in the unidirectional composites.

The effect of the magnitude of the electric current density on the deflection of the 4-layer, 8-layer, and 16-layer laminates is presented in Figures 1.16 through 1.18. Here, the magnetic induction is $B_y^* = 0.1$ T for all cases, while the current density differs. As it can be seen, there is some reduction in the deflection at larger current densities; however, the noticeable reduction occurs only during the application of the impact load. The reduction in the vibration amplitudes after the impact load has diminished is small. Moreover, it practically disappears as the number of layers in the laminate increases. Furthermore, the damping effect is small when the magnetic field is small.

The mechanical response of the 4-layer, 8-layer, and 16-layer laminates in the presence of the large magnetic field, $B_y^* = 1.0$ T, is shown in Figures 1.19 through 1.21. Now that the magnetic field is relatively large, the deflection of the plate is further reduced, at least during the application of the impact load. Moreover, the damping effect is also more noticeable.

Figure 1.22 shows the distribution of the stress τ_{yy}/p_0 over the cross-section of the eight-layer laminated plate at the moment of time when the stress is maximum for the case when both mechanical ($p_0 = 1.0$ MPa, $\tau_p = 10$ ms) and pulsed electromagnetic ($J_0 = 10^6$ A/m^2, $\tau_c = \tau_p = 10$ ms, $B_y^* = 1.0$ T) loads are applied.

As it can be seen, the four-layer laminates with the fiber orientations of $\theta = 90°$ bear the largest portion of the induced stress τ_{yy} in an eight-layer cross-ply laminate. This is especially true for the layers farther from the middle plane of the plate. The magnitude of the stress caused in the laminated plate can be compared with the case when the plate is subjected to the mechanical

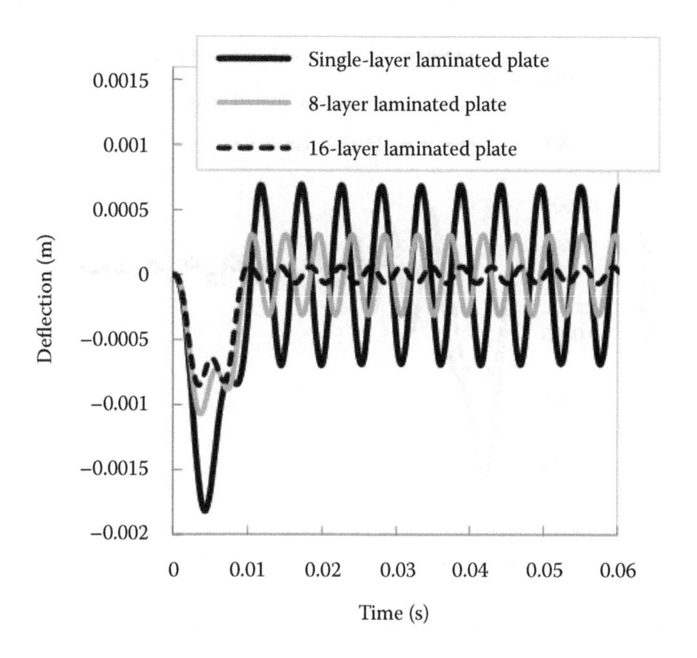

FIGURE 1.10
Deflection of laminated plate: effect of number of layers and low magnetic induction, $B_y^* = 0.1$ T.

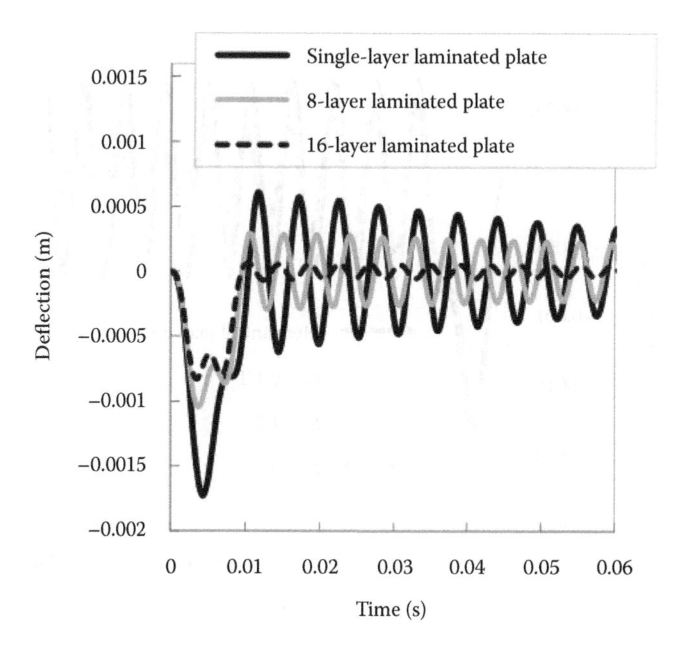

FIGURE 1.11
Deflection of laminated plate: effect of number of layers and magnetic induction, $B_y^* = 1.0$ T.

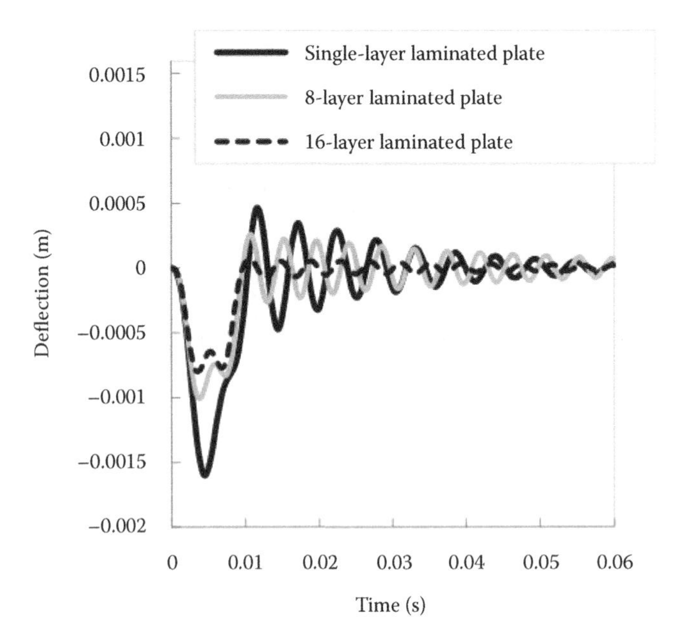

FIGURE 1.12
Deflection of laminated plate: effect of number of layers and magnetic induction, $B_y^* = 2.0$ T.

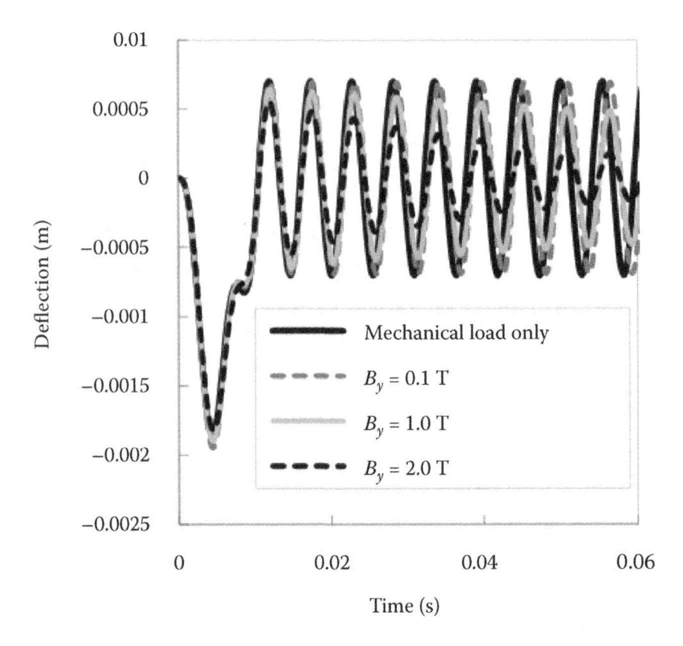

FIGURE 1.13
Deflection of four-layer laminated plate: effect of increasing magnetic induction.

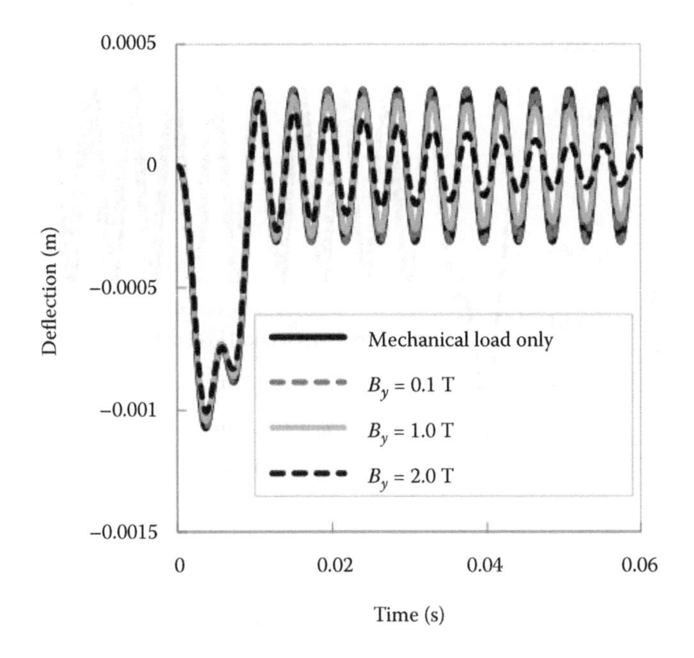

FIGURE 1.14
Deflection of eight-layer laminated plate: effect of increasing magnetic induction.

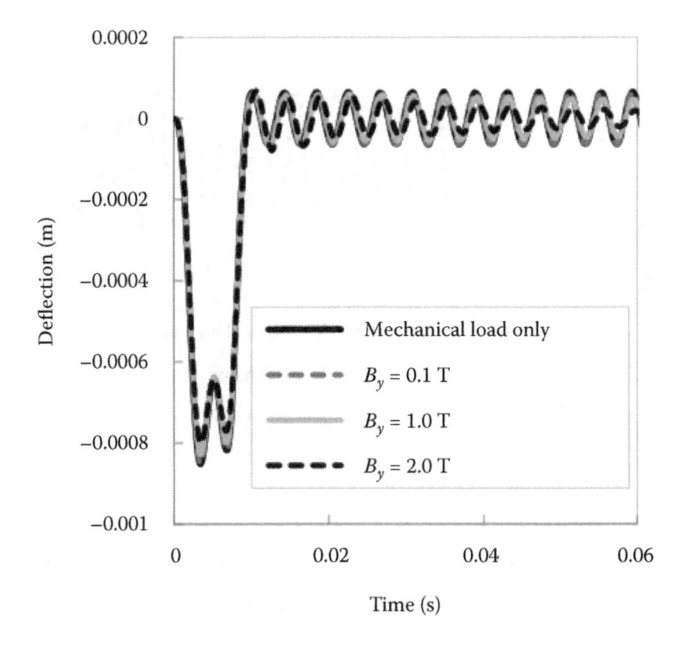

FIGURE 1.15
Deflection of 16-layer laminated plate: effect of increasing magnetic induction.

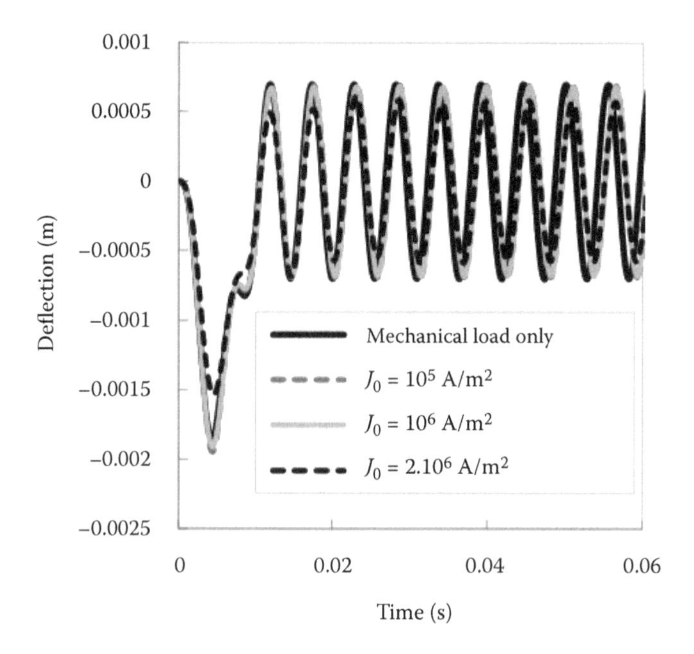

FIGURE 1.16
Deflection of four-layer laminated plate: effect of increasing electric current at $B_y^* = 0.1$ T.

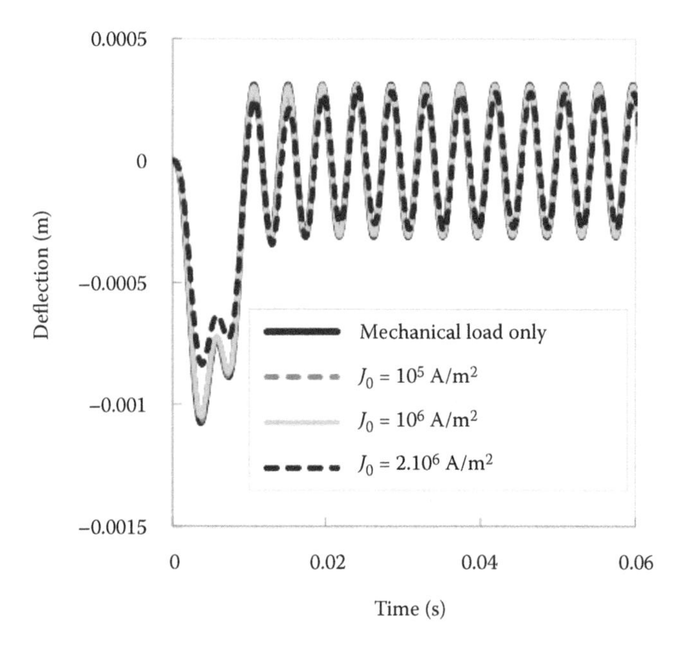

FIGURE 1.17
Deflection of eight-layer laminated plate: effect of increasing electric current at $B_y^* = 0.1$ T.

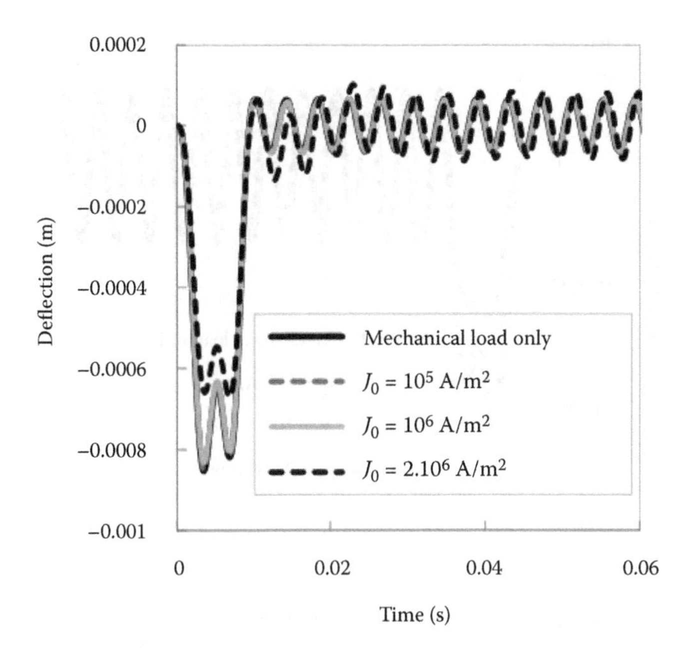

FIGURE 1.18
Deflection of 16-layer laminated plate: effect of increasing electric current at $B_y^* = 0.1$ T.

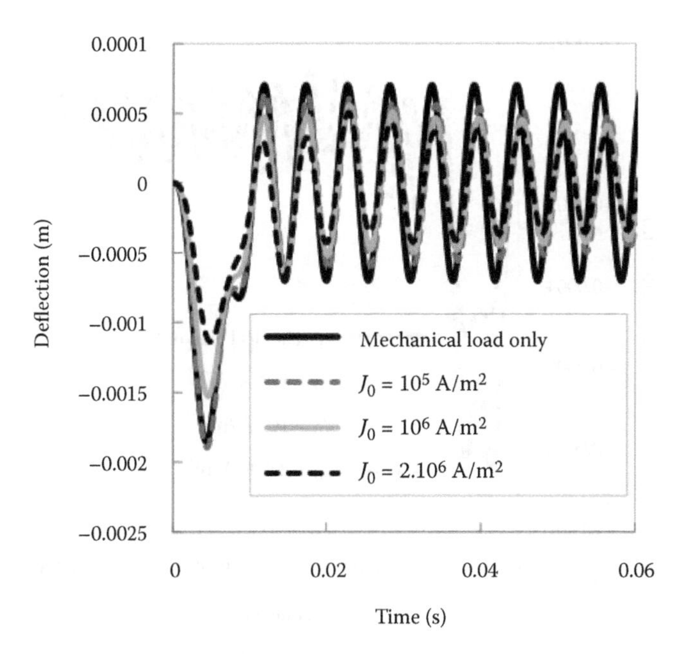

FIGURE 1.19
Deflection of four-layer laminated plate: effect of increasing electric current at $B_y^* = 1.0$ T.

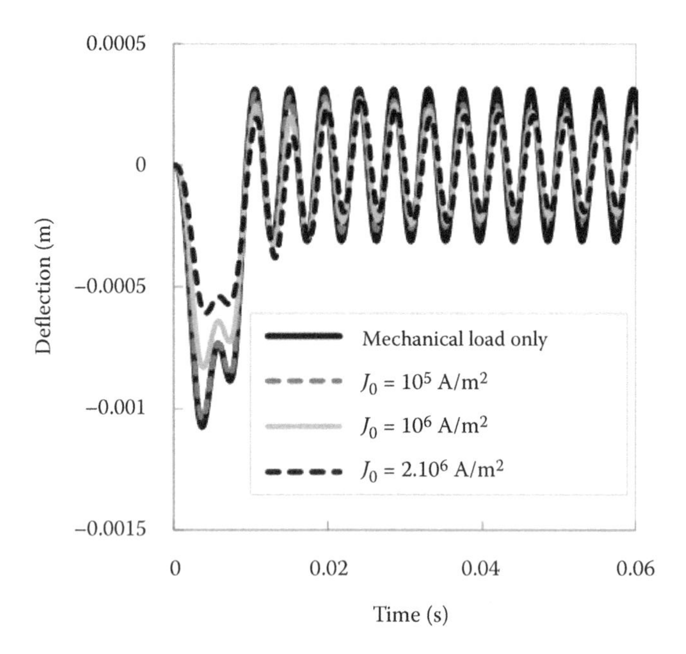

FIGURE 1.20
Deflection of eight-layer laminated plate: effect of increasing electric current at $B_y^* = 1.0$ T.

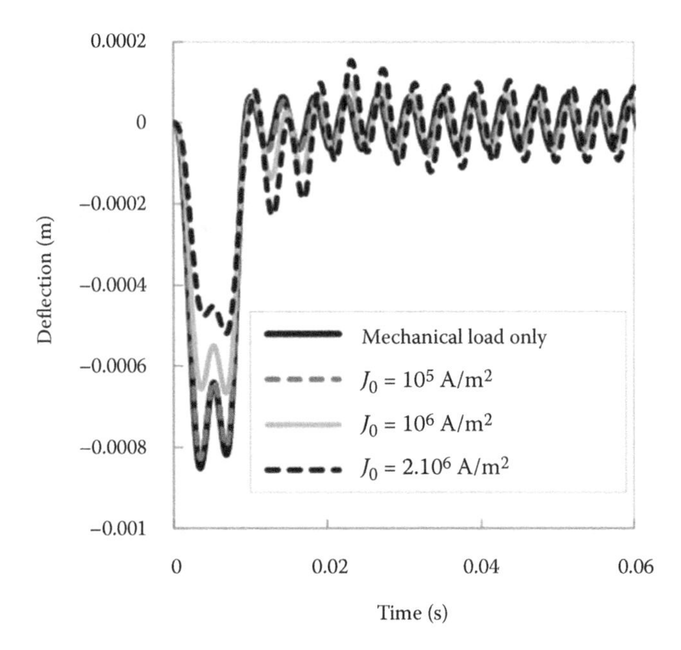

FIGURE 1.21
Deflection of 16-layer laminated plate: effect of increasing electric current at $B_y^* = 1.0$ T.

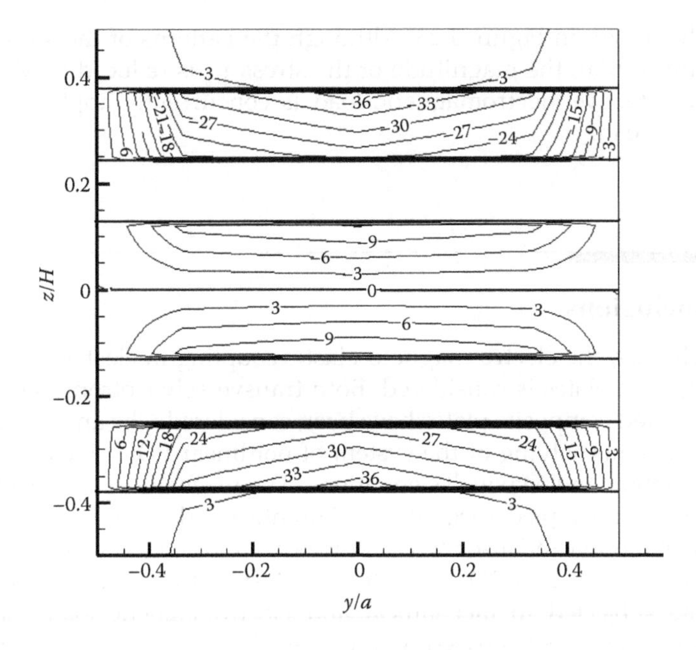

FIGURE 1.22
Contours of the stress τ_{yy}/p_0 at t = 3.8 ms in an eight-layer laminated plate subjected to both mechanical and electromagnetic loads (p_0 = 1.0 MPa, $J_0 = 10^6$ A/m^2, $\tau_c = \tau_p$ = 10 ms, $B_y^* = 1.0$ T).

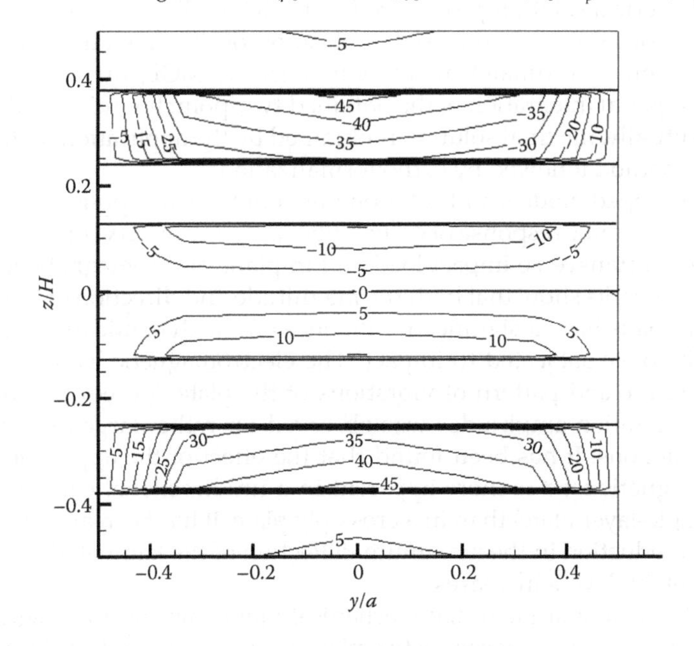

FIGURE 1.23
Contours of the stress τ_{yy}/p_0 at t = 3.6 ms in an eight-layer laminated plate subjected to mechanical load only (p_0 = 1.0 MPa, τ_p = 10 ms).

load only, shown in Figure 1.23. Although the patterns of the stress distribution are similar, the magnitude of the stress τ_{yy} is reduced by about 20% when an effective electromagnetic load is concurrently applied with the mechanical load.

1.5 Conclusions

In this chapter, the electro-magneto-elastic coupling in electrically conductive composite plates is considered. Both transversely isotropic single-layer and laminated composite plates have been considered. The analysis is based on simultaneous solving of the system of nonlinear PDEs, including equations of motion and Maxwell's equations. The mechanical and electromagnetic fields are coupled through the Lorentz ponderomotive force in the equation of motion and also the velocity terms in the electromagnetic constitutive equations. A new 2D model for the electrically conductive laminated composite subjected to mechanical and electromagnetic loads has been developed. The model is based on the extension of the 2D model for transversely isotropic electrically conductive plates and utilizes physics-based simplifying hypotheses for both mechanical and electromagnetic parts.

The numerical solution procedure for the 2D nonlinear system of governing PDEs consists of a sequential application of FD time and spatial (with respect to one coordinate) integration schemes, MOL, quasilinearization, and a FD spatial integration of the obtained two-point boundary-value problem. Eventually, the final solution is obtained by the application of the superposition method followed by orthonormalization.

The developed models and solution methodology are applied to the problem of the dynamic response of carbon fiber polymer matrix composite plates subjected to transverse impact load and in-plane electromagnetic load. The numerical results show that both the magnitude and direction of the electromagnetic loads have a significant influence on the dynamic response of the composite plate subjected to impact. The electromagnetic load can change the amplitude and pattern of vibrations of the plate. Moreover, a damping effect in vibration amplitudes was observed for rather large external magnetic inductions. It has been found that the effect of the application of the electromagnetic load is more apparent in a unidirectional composite plate (i.e., a single-layer plate) than in a cross-ply plate. It has been also shown that there is a reduction in the maximum deflection of the laminated plate as the number of 90° layers increases.

Overall, the amplitude of the mechanical vibrations and the magnitude of stresses in the electrically conductive plate can be significantly reduced by the application of an appropriate combination of a pulsed electric current and magnetic induction during the occurrence of the impact and also afterward.

Acknowledgments

The authors would like to acknowledge the support of AFOSR (FA9550-09-1-0359) and DARPA (N66001-11-1-4133).*

References

Ambartsumyan, S.A., Belubekyan, M.B. and Bagdasaryan, G.E. 1977. *Magnetoelasticity of Thin Shells and Plates*. Moscow: Nauka.

Atkinson, K.E., Han, W. and Stewart, D.E. 2009. *Numerical Solutions of Ordinary Differential Equations*. New Jersey: John Wiley & Sons, Inc.

Barakati, A. and Zhupanska, O.I. 2012a. Analysis of the effects of a pulsed electromagnetic field on the dynamic response of electrically conductive composites. *J. Appl. Math. Model.* 36: 6072–6089.

Barakati, A. and Zhupanska, O.I. 2012b. Thermal and mechanical response of a carbon fiber reinforced composite to a transverse impact and in-plane pulsed electromagnetic loads. *J. Eng. Mat. Tech.* 134: 031004.

Bellman, R.F. and Kalaba, R.E. 1969. *Quasilinearization and Nonlinear Boundary-Value Problems*. New York: American Elsevier Publishing Company.

Conte, S.D. 1966. The numerical solution of linear boundary value problems. *SIAM Rev.* 8: 309–321.

Godunov, S.K. 1961. On the numerical solution of boundary value problems for system of linear ordinary differential equations. *Uspekhi Mat. Nauk* 16: 171–174.

Hasanyan, D.J. and Piliposyan, G.T. 2001. Modelling and stability of magnetosoft ferromagnetic plates in a magnetic field. *Proc. R. Soc. A.* 457: 2063–2077.

Hasanyan, D.J., Librescu, L., Qin, Z. and Ambur, D.R. 2005. Magneto-thermoelastokinetics of geometrically nonlinear laminated composite plates. Part 1: Foundation of theory. *J. Sound Vib.* 287: 153–175.

Jones, R.M. 1998. *Mechanics of Composite Materials*, 2nd ed. Ann Arbor, MI: Taylor & Francis.

Kincaid, D.R. and Cheney, E.W. 2002. *Numerical Analysis: Mathematics of Scientific Computing*, 3rd ed. Providence, RI: American Mathematical Society.

Kubíček, M. and Hlaváček, V. 1983. *Numerical Solution of Nonlinear Boundary Value Problems with Applications*. Englewood Cliffs, NJ: Prentice-Hall.

Librescu, L., Hasanyan, D., Qin, Z. and Ambur, D.R. 2003. Nonlinear magnetothermoelasticity of anisotropic plates immersed in a magnetic field. *J. Therm. Stress* 26: 1277–1304.

Mol'chenko, L.V. and Loos, I.I. 1999. Magnetoelastic nonlinear deformation of a conical shell of variable stiffness. *Int. Appl. Mech.* 35: 1111–1116.

* Disclaimer: Any opinions, findings, and conclusions or recommendations expressed in this publication are those of the authors and do not necessarily reflect the views of AFOSR and DARPA.

Newmark, N.M. 1959. A method of computation for structural dynamics. *J. Eng. Mech. Div. Proc. ASCE* 85: 67–97.
Panofsky, W.K.H. and Phillips, M. 1962. *Classical Electricity and Magnetism*. New York: Addison-Wesley.
Reddy, J.N. 1999. *Theory and Analysis of Elastic Plates*. Philadelphia: Taylor & Francis.
Roberts, S.M. and Shipman, J.S. 1972. *Two-Point Boundary Value Problems: Shooting Methods*. New York: American Elsevier Publishing Company.
Ruhe, A. 1983. Numerical aspects of Gram–Schmidt orthogonalization of vectors. *Lin. Algebra Appl.* 52–53: 591–601.
Sadiku, M.N.O. and Obiozor C.N. 2000. A simple introduction to the method of lines. *Int. J. Elec. Eng. Educ.* 37: 282–296.
Schiesser, W.E. and Griffiths, G.W. 2009. *A Compendium of Partial Differential Equation Models: Method of Lines Analysis with MATLAB*. New York: Cambridge University Press.
Scott, M.R. and Watts, H.A. 1977. Computational solution of linear two-point boundary value problem via orthonormalization. *Siam J. Numer. Anal.* 14: 40–70.
Sedov, L.I. 1971. *A Course in Continuum Mechanics*, Vol. 1. Groningen: Wolters-Noordhoff.
Zhupanska, O.I. and Sierakowski, R.L. 2005. Mechanical response of composites in the presence of an electromagnetic field. In: *Proceeding of 46th AIAA/ASME/ASCE/AHS/ASC Structures, Structural Dynamics & Materials Conference*, Paper # AIAA 2005–1949.
Zhupanska, O.I. and Sierakowski, R.L. 2007. Effects of an electromagnetic field on the mechanical response of composites. *J. Compos. Mater.* 41: 633–652.
Zhupanska, O.I. and Sierakowski, R.L. 2011. Electro-thermo-mechanical coupling in carbon fiber polymer matrix composites. *Acta Mech.* 218: 219–232.

2

Design and Characterization of Magnetostrictive Composites

Rani Elhajjar, Chiu Law, and Anastasia Muliana

CONTENTS

2.1 Introduction

The term magnetoelasticity refers to the interaction of the elastic material and its magnetic state. Magnetoelasticity allows the analysis of magnetostriction or magnetic field–induced deformations. There are several magnetoelastic effects; these are volume magnetostriction, Joule magnetostriction, dipolar magnetostriction, direct Wiedemann effect, and changes in the elastic properties due to magnetoelastic contributions (Lacheisserie 1993). The most commonly used magnetostriction effect was observed by the renowned physicist James Joule by documenting changes in length in ferromagnetic materials in the presence of a magnetic field. Joule magnetostriction refers to a deformation that transforms a spherical sample into an ellipsoid whose symmetry axis lies along the magnetization direction. The strain in the material can also result from orientation changes in small magnetic domains within the material. The magnetostriction response is observed in most ferromagnetic materials and can range from zero to nearly 1% in rare-earth-based intermetallic compounds. Values ranging from 1000 to 2000 microstrain are observed in fields ranging from 50 to 200 kA/m (Verhoeven et al. 1989). These strains are amplified if the sensor is operated in the dynamic range under the correct conditions. Certain magnetostrictive underwater transducers can outperform lead zirconate titanate transducers in the low-frequency domain (Hartmut 2007).

2.2 Behavior of High Magnetostriction Materials

These unique properties of magnetostrictive materials allow constructing actuators that have small displacements and large forces operating at low voltages. These can be in the form of cantilevers, single elements, or embedded in laminates. Compared with piezoelectric devices, magnetostrictive-based devices offer several advantages related to the ability to obtain higher deformations and forces. In addition, the higher Young's modulus and lower operating voltage range offer important advantages especially when considering that electrical contact is not necessary. In addition, high magnetostrictive materials are capable of producing large amounts of force in a short response time. The delay time between applying a magnetic field and the occurrence of deformation is approximately 1 μs (Kondo 1997). Magnetostrictive materials (Figure 2.1) have been widely used in transducer applications. For example, Kim and Kim (2007) proposed an ultrasonic Terfenol-D transducer for

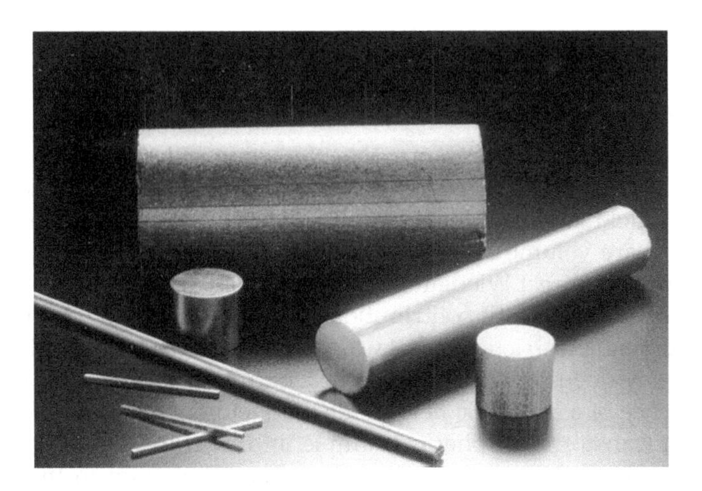

FIGURE 2.1
High magnetostriction Terfenol-D rods. (Courtesy of ETREMA Products Inc., USA.)

transmitting and receiving longitudinal ultrasonic waves in rotating shafts with the capability to transmit and receive ultrasonic-guided waves wirelessly. Dynamic assessments of Terfenol-D actuators have been performed to 100 Hz (Lovisolo et al. 2008). A significant advantage of this material (Figure 2.2) compared with ceramics is the high value of Young's modulus since that may allow higher actuation loads. A bias magnetic field is applied so that the material will work in the linear strain/magnetic field.

2.3 Magnetostrictive Composites

Combining high magnetostrictive materials within polymer matrices improves the moldability highly of such materials. An important advantage of embedding Terfenol-D in a polymer binder, having a high electric resistivity, is in extending the frequency response to 10–100 kHz, which is much higher than the 1-kHz frequency achieved in Terfenol-D rods (Lim et al. 1999). Terfenol–polymer composites isolate the particles from each other and reduce the eddy current losses at high frequencies. The effect of particle size has been found to favor large particle sizes in a narrow range compared with smaller ones that are not properly aligned in the magnetization process in manufacturing (Rodriguez et al. 2009).

FIGURE 2.2
Actuators with magnetostrictive materials. (Courtesy of ETREMA Products Inc., USA.)

Particle distribution and packing density have also been found to affect the response of Terfenol-D particulate composites in a polymer matrix (Duenas and Carman 2001). Particle distributions with a wide range of particles produced better results and reduced the demagnetization effects. By combining smaller and larger particles in the composite, larger packing efficiencies can be obtained with minor effects on the modulus. The elastic modulus as a function of the magnetic field in ferromagnetic materials typically shows a slight increase followed by decrease, then finally increases after reaching a critical threshold. The differences due to the applied load are attributed to the effect of the mechanical energy and its impact on the movement of the domain walls (Duenas and Carman 2001).

Smaller particle sizes have yielded composites with higher compressive strengths (Lim et al. 1999). The same study reported improvements in compressive strength with curing pressure; however, this often resulted in reduced magnetostrictive properties. Hudson et al. (2000) studied the dependence of the particle size and volume fraction on the dynamic magneto-mechanical properties of epoxy-bonded Terfenol-D at a high-frequency response. They found that the effects of eddy currents can be reduced so the frequency range can be extended to 200 kHz. Rodriguez et al. (2008) found that the composites with preferential alignment orientation of Terfenol-D powders exhibit a greater saturation magnetostriction value compared with non-oriented composites.

Some studies have suggested that for ferromagnetic fractions, >30% inadequate preload-induced compressive stresses are applied by the epoxy during cure, whereas for values ≤10% too large of a load is provided (Duenas and Carman 2000). For composites with higher volume fractions of ferromagnetic particles, larger preloads are required to increase the magnetostriction response. Interesting magneto-mechanical coupling properties can be obtained when using crystallographically aligned (112) magnetostrictive particle composites (Altin et al. 2007, McKnight and Carman 2001, Ho et al. 2006). The Terfenol-D material was cut into fibrils <1000 microns in diameter with an aspect ratio greater than 3:1. The longer dimension of the particles corresponds to the (112) direction. Their measurements indicate that these particulate composites achieve properties near that of monolithic Terfenol-D. The results also indicate that residual stresses play a role in determining the initial domain state in the material. The specimens with the highest volume fraction of 49% showed properties approaching that of the monolithic material, including a strain of 1600 ppm (nearly 90% of the monolithic value). Similarly, dynamic improvements in behavior have also been reported (Or et al. 2003). Shear lag and demagnetization are some of the issues to contend with in the magnetoelectric response of laminated Terfenol-D/piezoelectric composites. Chang and Carman (2007a,b) compared experimental measurements and theoretical predictions accounting for shear lag and demagnetization effects and obtained good agreement with experimental results.

2.4 Coupling between Mechanical and Magnetic Properties

The behavior of magnetostrictive materials is generally nonlinear; however, linearized analysis can be used in many instances depending on the operating magnetic fields and frequencies. Similar to the deformation of a material under mechanical stress, the presence of a magnetic field intensity, H, will induce a magnetic flux density, B, inside a medium. In a vacuum, these properties are related by μ_o, the vacuum permeability:

$$B = \mu_o H. \tag{2.1}$$

The relationship between the strain and magnetic field curve, $\lambda(H)$, is generally nonlinear and hysteresis effects are usually observed. In addition, it is also known that the material properties of Terfenol-D are dependent on the bias field and prestress conditions (Moffett et al. 1991). The effects of prestress on the magnetoelastic properties of Terfenol-D are shown in Table 2.1. Their results indicate that the magnetoelastic properties of Terfenol-D depend on the stress and magnetic field applied. The magnetostriction is also dependent on temperature and stress levels, although the saturation levels are generally not affected by temperature. The stress level on polycrystalline rods of Terfenol-D has shown larger sensitivity to the magnetic field in the presence of compressive stress. Under magnetic field, the movement of the magnetic moments inside the material causes the material to become anisotropic. Similarly, the application of mechanical stress (due to the magnetoelastic coupling) results in the magnetic moments becoming anisotropic: applying a compressive stress in Terfenol-D causes the moments to be oriented perpendicular to the stress direction.

To predict the magnetoelastic behavior of the composite material, one can assume an orthotropic composite material with a material displaying a general elastomagnetic response. The relationship that governs the linear behavior between the normal and shear strains (ε, γ), magnetic flux density

TABLE 2.1

Magnetoelastic Coefficients at 90 kA/m Bias versus Prestress

Prestress (MPa)	30	40	50
$\mu_{33}^{\sigma HT}/\mu_o$	3.7	3.8	3.0
d_{33} (nm/A)	8.0	9.7	5.0
k_{33} (%)	63.1	67.4	52.0

Source: Hartmut, J., *Adaptronics and Smart Structures, Basics, Materials, Design, and Applications*, vol. 2, rev. ed., 2007.

(B), normal and shear stresses (σ, τ), and magnetic field intensity (H) is given by

$$\begin{bmatrix} \varepsilon_{11} \\ \varepsilon_{22} \\ \varepsilon_{33} \\ \gamma_{23} \\ \gamma_{31} \\ \gamma_{12} \\ B_1 \\ B_2 \\ B_3 \end{bmatrix} = \begin{bmatrix} S_{11}^{\sigma HT} & S_{12}^{\sigma HT} & S_{13}^{\sigma HT} & 0 & 0 & 0 & d_{11} & d_{21} & d_{31} \\ S_{12}^{\sigma HT} & S_{22}^{\sigma HT} & S_{23}^{\sigma HT} & 0 & 0 & 0 & d_{12} & d_{22} & d_{32} \\ S_{13}^{\sigma HT} & S_{23}^{\sigma HT} & S_{33}^{\sigma HT} & 0 & 0 & 0 & d_{13} & d_{23} & d_{33} \\ 0 & 0 & 0 & S_{44}^{\sigma HT} & 0 & 0 & d_{14} & d_{24} & d_{34} \\ 0 & 0 & 0 & 0 & S_{55}^{\sigma HT} & 0 & d_{15} & d_{25} & d_{35} \\ 0 & 0 & 0 & 0 & 0 & S_{66}^{\sigma HT} & d_{16} & d_{26} & d_{36} \\ d_{11} & d_{12} & d_{13} & d_{14} & d_{15} & d_{16} & \mu_{11}^{\sigma HT} & \mu_{12}^{\sigma HT} & \mu_{13}^{\sigma HT} \\ d_{21} & d_{22} & d_{23} & d_{24} & d_{25} & d_{26} & \mu_{21}^{\sigma HT} & \mu_{22}^{\sigma HT} & \mu_{23}^{\sigma HT} \\ d_{31} & d_{32} & d_{33} & d_{34} & d_{35} & d_{31} & \mu_{31}^{\sigma HT} & \mu_{32}^{\sigma HT} & \mu_{33}^{\sigma HT} \end{bmatrix} \begin{bmatrix} \sigma_{11} \\ \sigma_{22} \\ \sigma_{33} \\ \tau_{23} \\ \tau_{31} \\ \tau_{12} \\ H_1 \\ H_2 \\ H_3 \end{bmatrix}. \tag{2.2}$$

In this equation, the strains and the magnetic flux density are coupled using the 9×9 matrix of the elastic compliances (S), the elastomagnetic coefficients (d), and the permeability (μ). Using the symmetry of a polarized polycrystalline specimen reduces the coefficients to two magnetic permeability and four piezomagnetic coefficients. Typically, three independent piezomagnetic coefficients, d_{31}, d_{33}, and d_{15}, are the most dominant of the elastomagnetic response; therefore, the coupling equations can be written as

$$\begin{bmatrix} \varepsilon_{11} \\ \varepsilon_{22} \\ \varepsilon_{33} \\ \gamma_{23} \\ \gamma_{31} \\ \gamma_{12} \\ B_1 \\ B_2 \\ B_3 \end{bmatrix} = \begin{bmatrix} S_{11}^{\sigma HT} & S_{12}^{\sigma HT} & S_{13}^{\sigma HT} & 0 & 0 & 0 & 0 & 0 & d_{31} \\ S_{12}^{\sigma HT} & S_{22}^{\sigma HT} & S_{23}^{\sigma HT} & 0 & 0 & 0 & 0 & 0 & d_{32} \\ S_{13}^{\sigma HT} & S_{23}^{\sigma HT} & S_{33}^{\sigma HT} & 0 & 0 & 0 & 0 & 0 & d_{33} \\ 0 & 0 & 0 & S_{44}^{\sigma HT} & 0 & 0 & 0 & d_{15} & 0 \\ 0 & 0 & 0 & 0 & S_{55}^{\sigma HT} & 0 & d_{15} & 0 & 0 \\ 0 & 0 & 0 & 0 & 0 & S_{66}^{\sigma HT} & 0 & 0 & 0 \\ 0 & 0 & 0 & 0 & d_{15} & 0 & \mu_{11}^{\sigma HT} & 0 & 0 \\ 0 & 0 & 0 & d_{15} & 0 & 0 & 0 & \mu_{11}^{\sigma HT} & 0 \\ d_{31} & d_{32} & d_{33} & 0 & 0 & 0 & 0 & 0 & \mu_{33}^{\sigma HT} \end{bmatrix} \begin{bmatrix} \sigma_{11} \\ \sigma_{22} \\ \sigma_{33} \\ \tau_{23} \\ \tau_{31} \\ \tau_{12} \\ H_1 \\ H_2 \\ H_3 \end{bmatrix}. \tag{2.3}$$

Under quasi-static loading conditions, in the absence of the prestress, the coupled constitutive stress–strain relationship can be linearized and the piezomagnetic linear coefficient d_{33}^{H} is related to the strain ε_{33} and the magnetic field H, so that (Gaudenzi 2009)

$$\varepsilon_{33} = d_{33}^{H} H_3. \tag{2.4}$$

The quasi-static magnetoelastic coupling factor k can be used to characterize the ability of a given material to convert magnetic energy into mechanical energy or the reverse potential. For a long cylindrical specimen magnetized along the axis of symmetry and subjected to a stress in the z-direction, the coupling factor is (Lacheisserie 1993)

$$k_{33} = \frac{d_{33}}{\sqrt{S_{33}^{\sigma HT} \mu_{33}^{\sigma HT}}}. \tag{2.5}$$

Dynamic loading of a Terfenol-D-based actuator (noncomposite) shows a sharp peak in the strain-versus-frequency response (Or et al. 2003). In contrast to static strains, the strains at resonance are magnified by a mechanical coupling factor, Q_m. This magnetic coupling factor is due to damping caused by internal mechanical losses and the effects of prestress. Incorporating this factor, the strain can be expressed as

$$\varepsilon_{33} = Q_m d_{33}^{H} H_3. \tag{2.6}$$

2.5 Micromechanical Analysis of Magnetostrictive Composites

To determine the overall performance of composites, several micromechanics models have been developed. The results of the micromechanics models are predictions of the effective properties and responses of heterogeneous materials based on the properties of the constituents and microstructural morphologies. The development of a micromechanics model is traditionally based on the assumption that a heterogeneous body is considered a statistically homogeneous medium so that the overall property of a heterogeneous body can be evaluated by taking volume average of the corresponding properties of all constituents in a representative volume element (RVE). The volume-averaging scheme has been extended for predicting responses of active composites with coupled mechanical and nonmechanical effects. The simplest micromechanics model, which is the rule of mixture (ROM), has been applied to determine the effective properties of piezoelectric and piezomagnetic composites. Altin et al. (2007) used upper-bound ROM to determine the effective properties of piezomagnetic composites. Experimental data suggest that the elastic modulus of the composite as a function of the volume fraction is more comparable to theoretical predictions indicating that the composites are closer to a 1–3 configuration rather than to a 0–3 (Nersessian et al. 2003). Aboudi (2001) used the method of cells (MOC) model, in which the composite microstructures are idealized with the periodically distributed arrays of cubic RVEs, to obtain the effective electro-magneto-mechanical properties of multiphase composites. The linear electro-magneto-mechanical coupling constitutive model was considered. Dunn and Taya (1993a,b) and Dunn (1994) applied well-known dilute distribution, Mori–Tanaka (MT), self-consistent (SC), and differential models to evaluate the effective properties of piezoelectric composites with linear electro-mechanical relations. Kim (2011) presented an exact solution to predict the effective properties of magneto-electro-thermo-elastic multilayer composites, and the results coincide with those evaluated by the MT model.

Owing to its simplicity, the ROM approach is commonly used to obtain effective piezomagnetic and/or magnetoelastic properties of active composites. However, ROM is limited in capturing the effect of detailed microstructural morphologies of the active composites such as the shape and size of the inclusions and the distribution of the inclusions in the matrix constituent. Refined micromechanics models, such as MT, SC, and MOC, attempt to incorporate some of the microstructural aspects in predicting the effective properties of composites. Here, we will briefly discuss the MT micromechanics model for predicting effective linear elastomagnetic properties of active composites.

Let the above constitutive model for elastomagnetic material be expressed as

$$Z_I = M_{IK}\Sigma_K, \quad \text{or} \quad \Sigma_I = L_{IK}Z_K, \quad \text{where} \quad I, K = 1, 2, \ldots 9 \tag{2.6}$$

where

$$\boldsymbol{\Sigma}^{\mathrm{T}} = \{\sigma_{11}, \sigma_{22}, \sigma_{33}, \sigma_{23}, \sigma_{13}, \sigma_{12}, H_1, H_2, H_3\}, \tag{2.7}$$

$$\mathbf{Z}^{\mathrm{T}} = \{\varepsilon_{11}, \varepsilon_{22}, \varepsilon_{33}, 2\varepsilon_{23}, 2\varepsilon_{13}, 2\varepsilon_{12}, B_1, B_2, B_3\}, \tag{2.8}$$

$$\mathbf{M} = \begin{bmatrix} \underset{(6\times6)}{\mathbf{S}} & \underset{(6\times3)}{\mathbf{d}^{\mathrm{T}}} \\ \underset{(3\times6)}{\mathbf{d}} & \underset{(3\times3)}{\boldsymbol{\mu}} \end{bmatrix}; \quad \mathbf{L} = \mathbf{M}^{-1}. \tag{2.9}$$

On the basis of a volume-averaging scheme, the generalized effective field variables of a composite consisting of N phases are written as

$$\bar{\mathbf{Z}} = \sum_{r=0}^{N} c_r \mathbf{Z}_r, \tag{2.10}$$

$$\bar{\Sigma} = \sum_{r=0}^{N} c_r \, \Sigma_r, \tag{2.11}$$

where the subscript r denotes the phase, $r = 0$ is the matrix phase and c_r is the volume fraction of phase r, which has the property $1 = \sum_{r=0}^{N} c_r$. An overbar denotes the effective (macroscopic) quantity by a volume-averaging scheme over a considered RVE. In an average sense, the constitutive relations of a composite can be expressed by the following relations:

$$\bar{\mathbf{\Sigma}} = \bar{\mathbf{L}}\bar{\mathbf{Z}}. \tag{2.12}$$

For the constituent of phase r within a composite, the constitutive relations is given as

$$\mathbf{\Sigma}_r = \mathbf{L}_r \mathbf{Z}_r. \tag{2.13}$$

To relate the generalized field variables for strains or electric fields between the micro- and macro-scales, the concentration tensor $\mathbf{A}_r$ ($r > 0$) for the r-th inhomogeneity is defined through

$$\mathbf{Z}_r = \mathbf{A}_r \bar{\mathbf{Z}}_r. \tag{2.14}$$

Once the concentration tensor $\mathbf{A}_r$ has been determined, the effective generalized stiffness tensor of active composite can be derived via Equations 2.10 through 2.14.

$$\bar{\mathbf{L}} = \mathbf{L}_0 + \sum_{r=1}^{N} c_r (\mathbf{L}_r - \mathbf{L}_0)\mathbf{A}_r. \tag{2.15}$$

Thus, Equation 2.15 is capable of evaluating the effective property tensor $\bar{\mathbf{L}}$ of a composite with the known inclusion volume fraction c_r and constituent properties $\mathbf{L}_0$ and $\mathbf{L}_r$ ($r \geq 1$). Several efforts have been made in formulating the concentration tensor, which can be seen in Dunn (1994).

The linear elastomagnetic constitutive model is only applicable when the material is subjected to relatively low external stimuli such as low stress and low magnetic field. Typical magnetostrictive materials, such as Terfenol-D, galfenol, and amorphous $Co_{77}B_{23}$ alloy, experience nonlinear response when subjected to high mechanical stress and high magnetic field. Furthermore, these materials are capable of dissipating energy as shown by their hysteretic response. Thus, to extend applications of magnetostrictive materials and their composites, modeling nonlinear and hysteretic electromagnetic response becomes necessary. Several constitutive models based on continuum mechanics and thermodynamics approaches have been developed for nonlinear hysteretic electromagnetic response (Miehe et al. 2011, Hauser et al. 2006, Linnemann et al. 2009, Sun and Zheng 2006). In an analogy to plasticity, the total strains and magnetic fields are additively decomposed into reversible and irreversible components. The threshold for the reversible and irreversible parts is defined by magnetic coercive field strength H_c; below this threshold, the response is reversible. Sun and Zheng (2006) proposed a nonlinear and coupling constitutive model that facilitated modeling the effects of compressive strain on the magnetostriction response. Substrate thickness effects on the magnetization and magnetostriction of Terfenol-D films were studied using a nonlinear constitutive model (Lu and Li 2010). Guan et al. (2009) used the Eshelby equivalent inclusion and MT method to estimate the average magnetostriction of the composites. The approximate models indicated higher volume fraction and lower modulus resulting in increased sensitivity. Note the heat is used to cure the extremely low viscosity polymer resin. The entire assembly is placed under a vacuum to ensure all voids are removed from the specimen.

On the basis of a phenomenological approach, the total magnetization can be additively decomposed into the reversible B_r and irreversible B_{irr} parts. The one-dimensional expression is

$$B[H,t] = B_r[H,t] + B_{irr}[H,t]; \quad \dot{B}[H,t] = \dot{B}_r[H,t] + \dot{B}_{irr}[H,t], \tag{2.16}$$

where overdot denotes the time derivative of the magnetization. The reversible magnetization is a general function of magnetic field that can also depend on time. We consider the following form for the rate of the irreversible magnetization, adopting the polarization switching model of ferroelectric material (Muliana 2011)

$$\frac{dB_{irr}}{dT} = \left\{ \begin{array}{lll} \lambda \left| \dfrac{H}{H_c} \right|^n & 0 \le H \le H_c, \dot{H} \ge 0.0 \quad \text{or} & -H_c \le H \le 0, \dot{H} \le 0.0 \\ \mu \exp\left[-\omega \left(\dfrac{|H|}{H_c} - 1 \right) \right] & H_c < H \le H_m, \dot{H} \ge 0.0 \quad \text{or} & -H_m \le H < -H_c, \dot{H} \le 0.0 \\ 0 & 0 \le H \le H_m, \dot{H} < 0.0 \quad \text{or} & -H_m \le H \le 0, \dot{H} > 0.0 \end{array} \right\}, \tag{2.17}$$

where λ, μ, ω, n are the material parameters that are calibrated from experiments and H_m is the maximum magnetic field intensity. A time integration algorithm (Muliana 2011) is used to obtain an incremental formulation of the magnetization in Equation 2.16. Figure 2.3 illustrates the normalized hysteretic response generated using Equations 2.16 and 2.17 due to the input $H_m \sin(2\pi f t)$ with a frequency of 1 Hz. The reversible magnetization is assumed to be a linear function of magnetic field $B_r = \mu_o H$. Upon removing the magnetic field from the maximum magnetic field H_m, the magnetization response

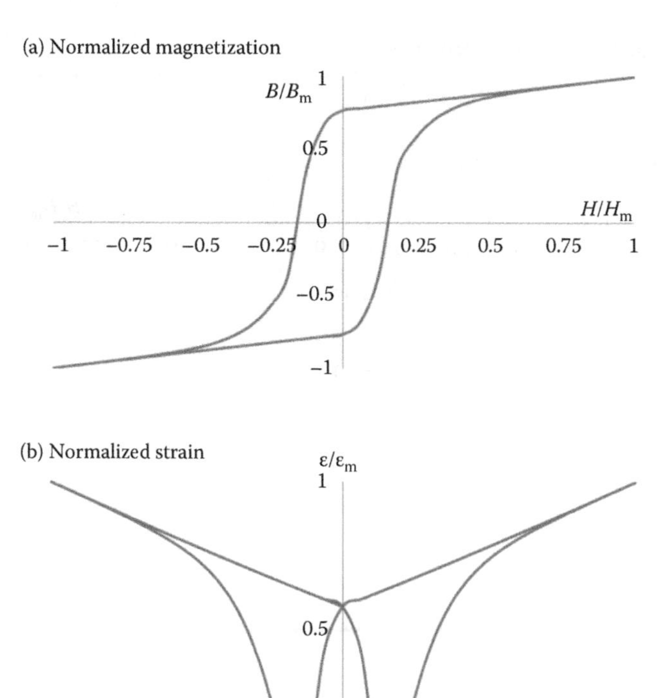

FIGURE 2.3
Hysteretic magnetization and butterfly strain responses.

is only due to the reversible part, which is shown by a linear response till zero magnetic field is reached. It is also possible to pick a nonlinear function of the magnetic field for the reversible magnetization, allowing the response to reach a saturated (steady) value. In absence of the reversible magnetization, upon removal of the magnetic field, constant (saturated) magnetization and strain are expected, as seen in Figure 2.4.

When the magnetostrictive materials are used as inclusions in active composites, there is also a need to obtain an effective nonlinear hysteretic response. When the active composites experience nonlinear (field-dependent) behaviors, it is necessary to quantify field variables in the constituents due to prescribed external stimuli. In such a case, the ROM method is incapable of predicting the nonlinear response of the materials since the effective properties obtained from the ROM rely only on the volume content and properties of the constituents. Refined micromechanics models, such as MT and MOC, in which field variables of each constituent are determined through the introduction of concentration tensors (e.g., Equation 2.14), are necessary to predict

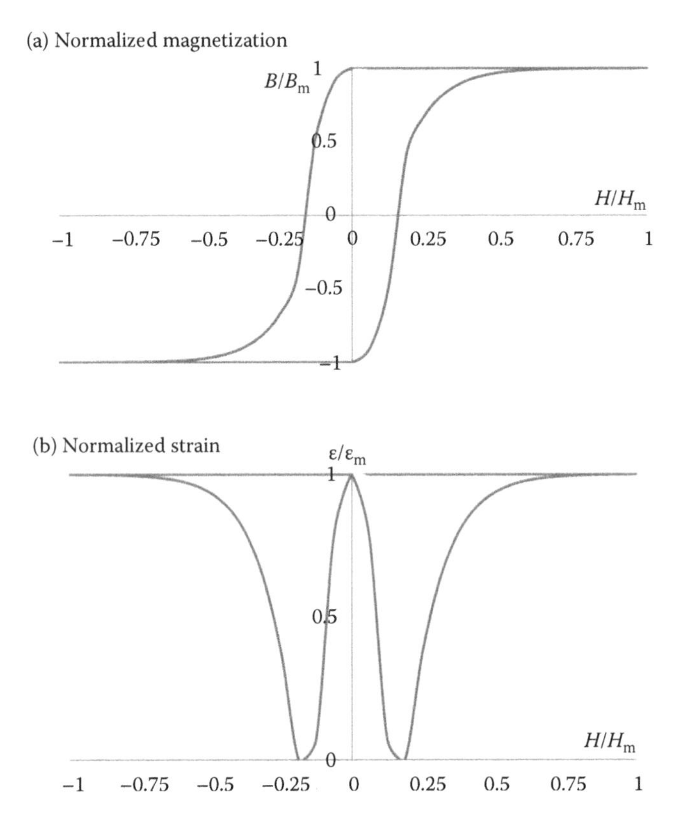

FIGURE 2.4
Hysteretic magnetization and butterfly strain responses from the irreversible part.

the nonlinear response of active composites. Aboudi (2005) and Muliana (2010) have presented micromechanics models for predicting nonlinear hysteretic response of active composites comprising ferroelectric constituents. The overall nonlinear hysteretic responses are obtained numerically, leading to approximate solutions.

2.6 Characterization of Terfenol-D Composite for Fiber Optic Current Sensor

Terfenol-D is a giant magnetostrictive material that produces strain on the order of 1000 ppm. However, its operation frequency is limited to the order of kHz and its brittleness hinders the use of conventional machining methods for device fabrication. Moreover, Terfenol-D is a rather rare and expensive alloy. To address these issues, researchers have investigated Terfenol-D composites (Or et al. 2003). In a composite with Terfenol-D particles surrounded by a nonmetallic binder (such as epoxy resin), the flow of eddy current is interrupted among particles by the increased electrical resistivity of the resin. As a result, the heat generation by eddy current losses is reduced and higher operational frequency can be reached (Or et al. 2003, Hudson et al. 1999). In general, Terfenol-D has a delayed deformation response in the order of microseconds after applying the magnetic field (Or et al. 2003, Kondo 1997) Another advantage of a Terfenol-D composite is its flexibility in manufacturing and machining. For instance, complex shapes can be produced with mold injection methods. Since the composite has a volume fraction of Terfenol-D of <1, the fiber optic current sensor (FOCS) based on composite Terfenol-D will incur less material cost. For the same reason, the strain induced by composite Terfenol-D is expected to be less than that of monolithic Terfenol-D. Moreover, the ability to control strain distribution by varying the Terfenol-D particle concentration along the magnetic sensing direction provides opportunities in designing a new type of FOCS that may be able to compensate for the reduction in maximum strain.

2.6.1 Fabrication of Terfenol-D Composite

Specimens were fabricated using a range of fiber volume fractions, $V_f = 0.3$–0.45 and a dimension of 25 × 25 × 6 mm. The following procedure was used for fabrication. The monolithic Terfenol-D bar was crushed into fine powders. The Terfenol-D particle range was from 100 to 300 microns. The powder was then poured into a mold and mixed with an epoxy that has a very low viscosity (cps = 65) and allows sufficient powder wetting and void reduction. Care has to be taken with lower volume fractions as the larger density of the particles will result in stratification. The mixture was degassed under a vacuum for 30 min to eliminate air bubbles. Following this step, the mold with the mixture was placed between a pair of rare earth magnets for alignment of Terfenol-D particles along the maximum magnetostriction direction. The whole assembly was placed inside a 70°C oven for 12 h to ensure full cure of the epoxy. The resulting microstructure of the composite is shown in Figure 2.5. The sample was demolded and sprayed with a couple layers of paint to encode a random speckle pattern on its surface. The resulting Terfenol-D composite with a speckle pattern is shown in Figure 2.6.

2.6.2 Surface Strain and Deformation Measurements

Three-dimensional digital image correlation (DIC) was used to capture the displacements and strains on the surface of the composite specimen. The DIC technique uses a random speckle pattern applied to the specimen that is captured using a couple-charged device (CCD) camera. These images are then processed using correlation algorithms to compare strained maps to a

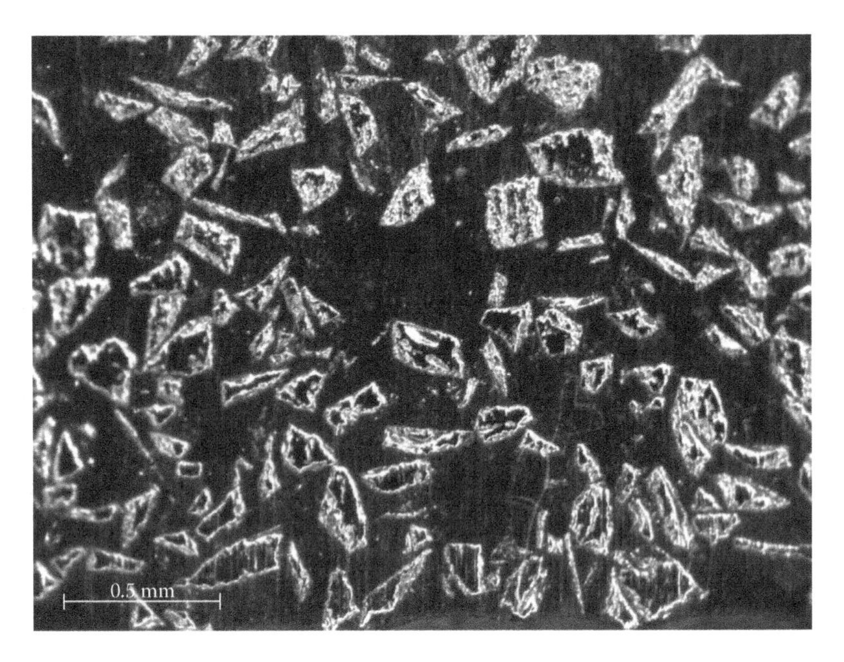

FIGURE 2.5
Microstructure of Terfenol-D/epoxy composite.

FIGURE 2.6
Speckle pattern on Terfenol-D/epoxy composite specimen.

reference image taken before the loading is applied. The procedure has been previously used for carbon/epoxy laminates (Elhajjar and Petersen 2011). The digital image correlation was performed using the Dantec Dynamics Q-400 system. A resolution of 5 megapixels is used for the CCD cameras together with a 50-mm Schneider Xenoplan lens. The DIC technique is used to visualize the surface strain of the sample that was in between a pair of N52 rare earth magnets (25 × 25 × 13 mm size). Various surface displacement components (see Figure 2.7) were calculated after the registered facets were tracked through the deformation process. With this digital technique, we are also able to examine strains (Figure 2.8). It is important to note that the strain variations were clearly affected by the facet sizes chosen for the analysis. However, the average strains over the area were indeed reflective of the actual deformation.

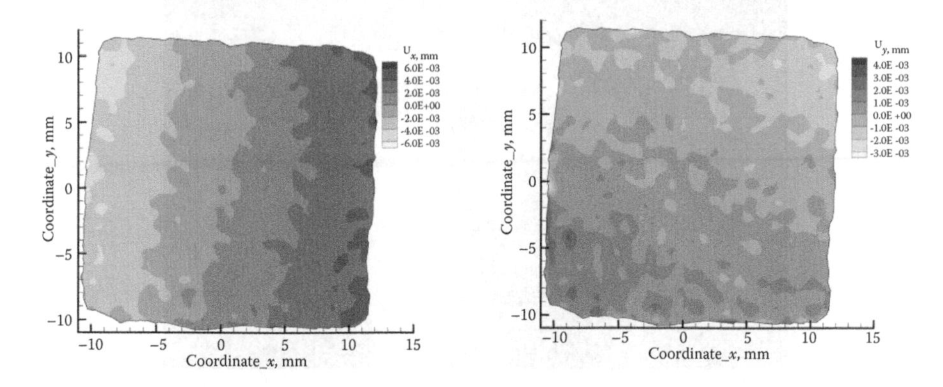

FIGURE 2.7
Displacement distribution of a Terfenol-D/epoxy composite between a pair of N52 rare earth magnets (160 kA/m).

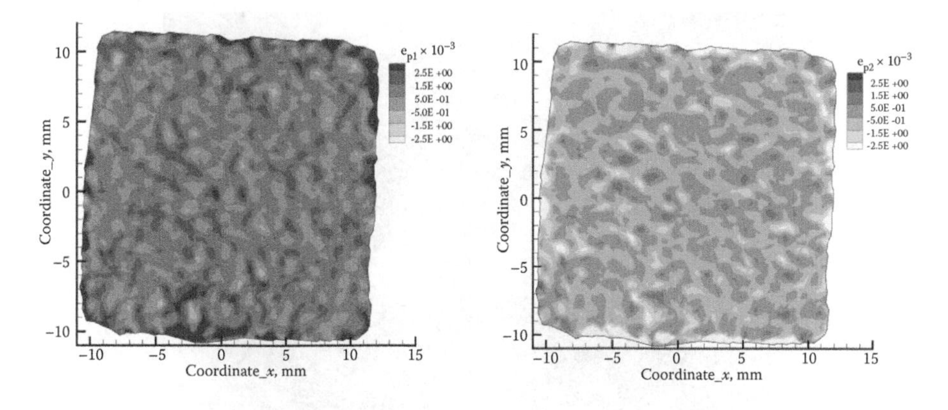

FIGURE 2.8
Strain distribution of a Terfenol-D/epoxy composite between a pair of N52 rare earth magnets (160 kA/m).

2.6.3 Magnetic Property Measurement of the Sample

We constructed a sample holder that secured the sample on two sides with plastic wedges. Then, we attached a fiber Bragg grating (FBG) along the magnetostriction axis at two points (see Figure 2.9) and used a 12-gauge copper wire laying across the top of the sample with adhesive tapes to prevent it from escaping from the holder (see Figures 2.10 and 2.11). The sample with the holder was clamped on one arm of a PC board holder and was placed in between two N52 rare earth magnets (1-in diameter and 0.5-in thickness) that were attached with duct tapes to the front and back jaws of a bench vice (see Figure 2.11). By turning the spindle of the vice, we can adjust the distance between the magnet and the sample, and hence, control the magnetic field

FIGURE 2.9
Sample was secured by a holder and an FBG was attached at two points with adhesive.

FIGURE 2.10
Measurement point 1 for field B_1.

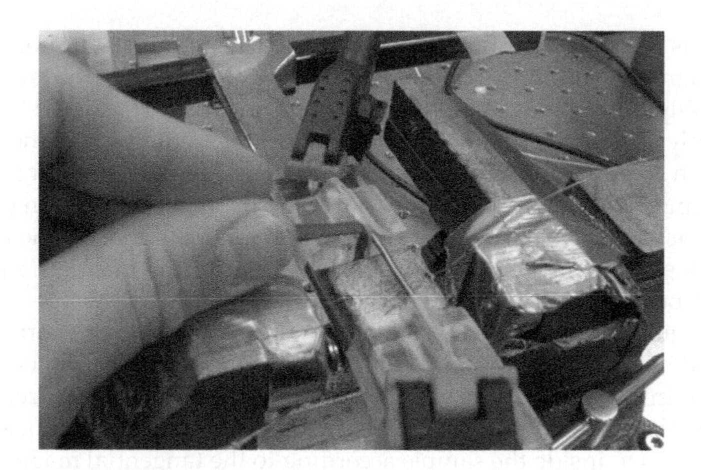

FIGURE 2.11
Measurement point 3 for field B_3.

magnitude. Before each measurement, we first changed the separation of the magnets and then centered the sample between the magnets. After all components were secured and fixed, we performed magnetic field measurement at three locations (see Figure 2.12). Then we recorded the reflected power spectrum from the wavelength meter and noticed the peak power wavelength of the spectrum. Magnetic field and power spectrum data were collected for eight magnet separation distances (d's) varying from 66 to 37 mm.

From the three-point magnetic field measurement, we estimated the average relative permeability, μ_r, of the Terfenol-D composite to be 4.8 ± 0.5. The

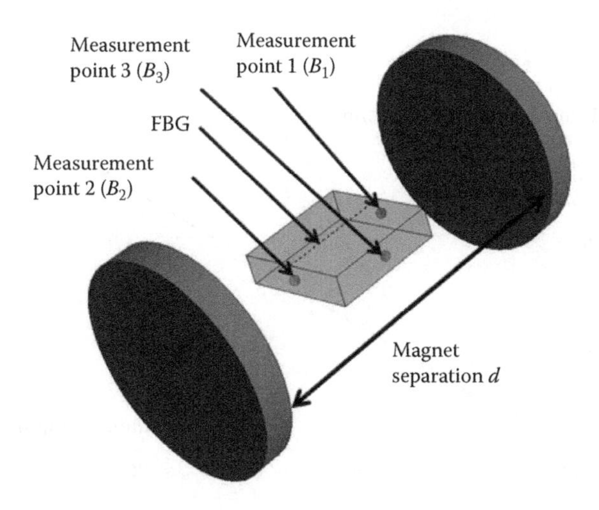

FIGURE 2.12
Schematic of experimental setup and measurement points.

estimate is based on assumption that magnetic field is almost uniform and μ_r is a constant. The uniformity of magnetic is demonstrated by the fairly linear relationship between average magnetic field intensity H_{avg} and $1/d$. Figure 2.13 clearly shows that magnetic field becomes more uniform as the magnet separation is reduced. Although the H_{avg} range is much larger than our usual testing range with a DC coil, it does not reach the nonlinear region (Or et al. 2003, Hudson et al. 2000) and our assumption for μ_r is valid. As the uniform field inside the sample should be perpendicular to the magnets, measurements at points 1 and 2 should give the magnetic flux densities B_1 and B_2 that are equivalent to those inside the sample according to the normal magnetic field boundary condition. These magnetic flux densities should be very close, and in fact, experiment data confirm this prediction. Measurements of magnetic flux density at point 3, B_3, provide values of magnetic field intensity $H_3 = B_3 = \mu_o$ inside the sample according to the tangential magnetic field boundary condition, where μ_o is the free space permeability. With these data, we can use the following procedure to estimate μ_r and average magnetic field intensity H_{avg} (d_i) at magnet separation distance d_i where $i = 1,\ldots,n$ and n is number of magnet separation:

1. Estimate average magnetic flux density at each magnet separation distance, d_i, with $B_{avg}(d_i) = 0.5(B_1(d_i) + B_2(d_i))$.
2. Calculate $\mu_r = \frac{1}{n}\sum_{i=1}^{n} B_{avg}(d_i)/B_3(d_i)$.
3. $H_{avg}(d_i) = \frac{1}{2\mu_o}\left(\frac{B_{avg}(d_i)}{\mu_r} + B_3(d_i)\right)$.

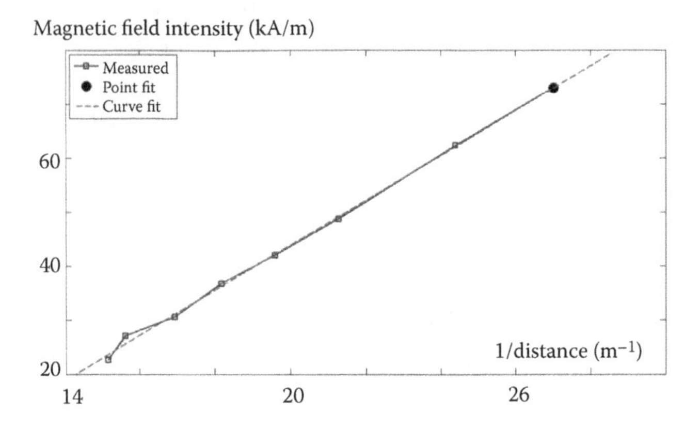

FIGURE 2.13
Magnetic field versus reciprocal of magnet separation.

We performed similar measurements with a monolithic Terfenol-D sample. Using the same procedure to estimate the permeability of Terfenol-D $(\mu_r)_T$, we obtained a value of 5.99 ± 0.25, which is consistent with reported value of 4.5–10 for commercial Terfenol-D. As we had expected, this estimate for monolithic Terfenol-D is higher than that for the composite sample. However, the estimate for the composite is quite a bit higher than that of mixture approximation:

$$\mu_r = (\mu_r)_T V_f + (\mu_r)_e(1 - V_f), \tag{2.18}$$

where $(\mu_r)_e$ is the permeability of epoxy and is assumed (2.18) to be 1. For V_f = 0.4, $\mu_r \approx 3$. On the other hand, if $\mu_r = 4.8$, then V_f should be 0.76 according to Equation 2.18.

We compare the magnetostriction of monolithic Terfenol-D with that of its composite counterpart in Figure 2.14. Since the composite has a V_f of <1, its sensitivity should be diminished. Figure 2.14 confirms the sensitivity dropping to approximately 50% of the monolithic material's values. The reduction in magnetostriction can be attributed to the somewhat random distribution of Terfenol-D particles. As a result, positive strains are localized around regions with clusters of Terfenol-D particles, while negative strains are formed in other sites. On average, the strain of the composite will be lower than that of monolithic Terfenol-D. Recently, researchers found that Terfenol-D composite can provide magnetostriction close to that of monolithic Terfenol-D if particles assume a regular shape with transverse to longitudinal dimensions being 1:4. Such enhancement in magnetostriction is owing to the orderly alignment of particles provided by shape anisotropy (Ching Yin et al. 2006, Altin et al. 2007).

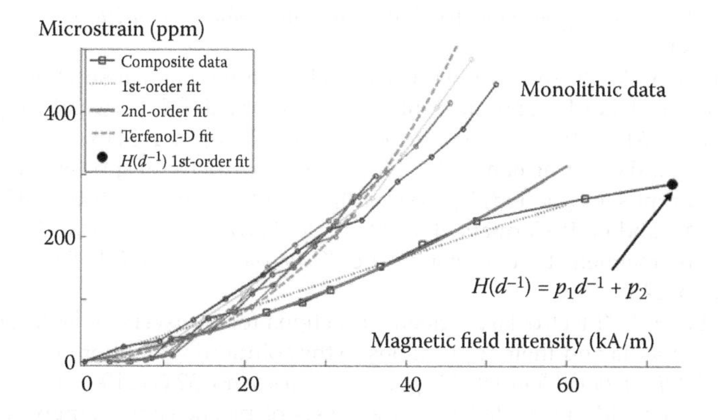

FIGURE 2.14
Magnetostrictive strain versus magnetic field.

2.7 Future Trends and Sources of Further Information

Characterizing the nonlinear behavior of magnetostrictive materials can yield to extending the usable range of magnetostrictive devices. Simulations using the finite difference method have been reported (Engdahl and Bergqvist 1996). Interesting possibilities exist in examining the magnetostrictive behavior of other materials, such as carbon fibers and nanotubes, and their potential application to composites (Guo and Guo 2003, Nai-Xiu et al. 2005, Nai-Xiu and Mao-Sheng 2004). Nonlinearity and hysteresis are characterized using a magneto-mechanical hysteresis model as the constitutive law. Manufacturing of hollow Terfenol-D particles can yield significant weight savings if as preliminary data on Nickel suggests, similar magnetostriction behavior between solid and hollow particles was observed (Nersessian et al. 2004, Guo and Guo 2003).

References

Aboudi, J. 2001. "Micromechanical analysis of fully coupled electro-magneto-thermo-electro-elastic multiphase composites." *Smart Materials and Structures* no. 10 (5):867–877.

Aboudi, J. 2005. "Hysteresis behavior of ferroelectric fiber composites." *Smart Materials and Structures* no. 14:715–726.

Altin, G., K. K. Ho, C. P. Henry, and G. P. Carman. 2007. "Static properties of crystallographically aligned Terfenol-D/polymer composites." *Journal of Applied Physics* no. 101 (3):033537-6.

Chang, C.-M., and G. P. Carman. 2007a. "Experimental evidence of end effects in magneto-electric laminate composites." *Journal of Applied Physics* no. 102 (12):12490-6.

Chang, C.-M., and G. P. Carman. 2007b. "Modeling shear lag and demagnetization effects in magneto-electric laminate composites." *Physical Review B* no. 76 (13):134116.

Ching Yin, L., S. W. Or, and H. L. W. Chan. 2006. "Large magnetostriction in epoxy-bonded Terfenol-D continuous-fiber composite with [112] crystallographic orientation." *Magnetics, IEEE Transactions on* no. 42 (10):3111–3113.

Duenas, T. A., and G. P. Carman. 2000. "Large magnetostrictive response of Terfenol-D resin composites (invited)." *Journal of Applied Physics* no. 87 (9):4696–4701.

Duenas, T. A., and G. P. Carman. 2001. "Particle distribution study for low-volume fraction magnetostrictive composites." *Journal of Applied Physics* no. 90 (5):2433–2439.

Dunn, M. L. 1994. "Electroelastic green's functions for transversely isotropic piezoelectric media and their applications to the solutions and inhomogeneity problems." *International Journal of Engineering Science* no. 32 (1):119–131.

Dunn, M. L., and M. Taya. 1993a. "An analysis of piezoelectric composite materials containing ellipsoidal inhomogeneities." *Proceedings of the Royal Society of London Series A* no. 443:265–287.

Dunn, M. L., and M. Taya. 1993b. "Micromechanics predictions of the effective electroelastic moduli of piezoelectric composites." *International Journal of Solids and Structures* no. 30 (2):161–175.

Elhajjar, R., and D. R. Petersen. 2011. "Adhesive polyvinyl chloride coatings for quantitative strain measurement in composite materials." *Composites Part B-Engineering* no. 42 (7):1929–1936.

Engdahl, G., and A. Bergqvist. 1996. "Loss simulations in magnetostrictive actuators." *Journal of Applied Physics* no. 79 (8):4689–4691.

Gaudenzi, P. 2009. *Smart Structures, Physical Behavior, Mathematical Modeling and Applications*. West Sussex, UK: John Wiley & Sons, Ltd.

Guan, X., X. Dong, and J. Ou. 2009. "Predicting performance of polymer-bonded Terfenol-D composites under different magnetic fields." *Journal of Magnetism and Magnetic Materials* no. 321 (18):2742–2748. doi: 10.1016/j.jmmm.2009.03.084.

Guo, W., and Y. Guo. 2003. "Giant axial electrostrictive deformation in carbon nanotubes." *Physical Review Letters* no. 91 (11):115501.

Hartmut, J. 2007. *Adaptronics and Smart Structures. Basics, Materials, Design, and Applications*. Vol. 2nd, rev. ed. Springer-Verlag, Berlin, Germany.

Ho, K. K., C. P. Henry, G. Altin, and G. P. Carman. 2006. "Crystallographically aligned Terfenol-D/polymer composites for a hybrid sonar device." *Integrated Ferroelectrics* no. 83 (1):121–138. doi: 10.1080/10584580600949642.

Hudson, J., S. C. Busbridge, and A. R. Piercy. 2000. "Dynamic magneto-mechanical properties of epoxy-bonded Terfenol-D composites." *Sensors and Actuators A: Physical* no. 81 (1–3):294–296. doi: 10.1016/s0924-4247(99)00178-8.

Hudson, J., S. C. Busbridge, and A. R. Piercy. 1999. "Magnetomechanical properties of epoxy-bonded Terfenol-D composites." *Ferroelectrics* no. 228 (1):283–295. doi: 10.1080/00150199908226142.

Kim, J.-Y. 2011. "Micromechanical analysis of effective properties of magneto-electro-thermo-elastic multilayer composites." *International Journal of Engineering Science* no. 49 (9):1001–1018.

Kim, Y., and Y. Y. Kim. 2007. "A novel Terfenol-D transducer for guided-wave inspection of a rotating shaft." *Sensors and Actuators A: Physical* no. 133 (2):447–456. doi: 10.1016/j.sna.2006.05.006.

Kondo, K. 1997. "Dynamic behaviour of Terfenol-D." *Journal of Alloys and Compounds* no. 258 (1):56–60.

Lacheisserie, E. du Tremolet de. 1993. *Magnetostriction, Theory and Applications of Magnetoelasticity*. Boca Raton, FL: CRC Press.

Lim, S. H., S. R. Kim, S. Y. Kang, J. K. Park, J. T. Nam, and D. Son. 1999. "Magnetostrictive properties of polymer-bonded Terfenol-D composites." *Journal of Magnetism and Magnetic Materials* no. 191 (1–2):113–121. doi: 10.1016/s0304-8853(98)00315-1.

Linnemann, K., S. Klinkel, and W. Wagner. 2009. "A constitutive model for magnetostrictive and piezoelectric materials." *International Journal of Solids and Structures* no. 46:1149–1166.

Lovisolo, A., P. E. Roccato, and M. Zucca. 2008. "Analysis of a magnetostrictive actuator equipped for the electromagnetic and mechanical dynamic characterization." *Journal of Magnetism and Magnetic Materials* no. 320 (20):e915–e919.

Lu, X., and H. Li. 2010. "Magnetic properties of Terfenol-D film on a compliant substrate." *Journal of Magnetism and Magnetic Materials* no. 322 (15):2113–2116. doi: 10.1016/j.jmmm.2010.01.043.

McKnight, G. P., and G. P. Carman. 2001. Large magnetostriction in Terfenol-D particulate composites with preferred [112] orientation, Newport Beach, CA, USA. Smart Structures and Materials 2001: Active Materials: Behavior and Mechanics, Newport Beach, CA, March 04, 2001.

Miehe, C., B. Kiefer, and D. Rosato. 2011. "An incremental variational formulation of dissipative magnetostriction at the macroscopic continuum level." *International Journal of Solids and Structures* no. 48:1846–1866.

Moffett, M. B., A. E. Clark, M. Wun-Fogle, J. Linberg, J. P. Teter, and E. A. McLaughlin. 1991. "Characterization of Terfenol-D for magnetostrictive transducers." *The Journal of the Acoustical Society of America* no. 89 (3):1448–1455.

Muliana, A. H. 2010. "A micromechanical formulation for piezoelectric fiber composites with nonlinear and viscoelastic constituents." *Acta Materialia* no. 58:3332–3344.

Muliana, A. H. 2011. "Time-temperature dependent behavior of ferroelectric materials undergoing cyclic electric field." *International Journal of Solids and Structures* no. 48 (19):2718–2731.

Nai-Xiu, D., and Z. Mao-Sheng. 2004. "Magnetostrictive properties of carbon black filled polypropylene composites." *Polymer Testing* no. 23 (5):523–526.

Nai-Xiu, D., Z. Mao-Sheng, and W. Peng-Xiang. 2005. "Magnetostrictive properties of carbon fiber filled polypropylene composites." *Polymer Testing* no. 24 (5):635–640.

Nersessian, N., S. W. Or, and G. P. Carman. 2003. "Magneto-thermo-mechanical characterization of 1-3 type polymer-bonded Terfenol-D composites." *Journal of Magnetism and Magnetic Materials* no. 263 (1–2):101–112. doi: 10.1016/s0304-8853(02)01542-1.

Nersessian, N., S. W. Or, G. P. Carman, W. Choe, and H. B. Radousky. 2004. "Hollow and solid spherical magnetostrictive particulate composites." *Journal of Applied Physics* no. 96 (6):3362–3365.

Or, S. W., N. Nersessian, G. P. McKnight, and G. P. Carman. 2003. "Dynamic magnetomechanical properties of [112]-oriented Terfenol-D/epoxy 1—3 magnetostrictive particulate composites." *Journal of Applied Physics* no. 93 (10):8510–8512.

Rodriguez, C., M. Rodriguez, I. Orue, J. L. Vilas, J. M. Barandiarin, M. L. F. Gubieda, and L. M. Leon. 2009. "New elastomer Terfenol-D magnetostrictive composites." *Sensors and Actuators A: Physical* no. 149 (2):251–254. doi: 10.1016/j.sna.2008.11.026.

Rodriguez, C., A. Barrio, I. Orue, J. L. Vilas, L. M. Lein, J. M. Barandiarn, and M. L. Fdez-Gubieda Ruiz. 2008. "High magnetostriction polymer-bonded Terfenol-D composites." *Sensors and Actuators A: Physical* no. 142 (2):538–541. doi: 10.1016/j.sna.2007.05.021.

Sun, L., and X. Zheng. 2006. "Numerical simulation on coupling behavior of Terfenol-D rods." *International Journal of Solids and Structures* no. 43:1613–1623.

Sun, L., and X. Zheng. 2006. "Numerical simulation on coupling behavior of Terfenol-D rods." *International Journal of Solids and Structures* no. 43 (6):1613–1623. doi: 10.1016/j.ijsolstr.2005.06.085.

Verhoeven, J. D., J. E. Ostenson, E. D. Gibson, and O. D. McMasters. 1989. "The effect of composition and magnetic heat treatment on the magnetostriction of TbxDy-xFey twinned single crystals." *Journal of Applied Physics* no. 66 (2):772–779.

3

Graphitic Carbon Nanomaterials for Multifunctional Nanocomposites

Mohammad Naraghi

CONTENTS

3.1 Introduction

"There is plenty of room at the bottom," said the well-known physicist and Nobel laureate, Richard Feynman, when he was referring to the manipulation of individual atoms to achieve arrangements perfectly suited for specific applications. While such perfect atomic arrangement will no doubt depend on the desired application of the materials, for a wide range of applications, graphene and, in general, allotropes of graphitic carbon, carbon nanofibers (CNFs) and carbon nanotubes (CNTs), graphitic particles, etc., have a close resemblance to an optimized arrangement of atoms. Among these allotropes, graphene is a single atomic layer of carbon atoms, in a hexagonal structure, each connected to three neighboring carbon atoms by σ bonds with sp^2 hybridized orbitals, and delocalized electrons of the Pi orbitals above and below the atomic layer. Many of these single atomic layers, stacked on top of each other and connected to each other via van der Waals (vdW) interactions, will form graphite particles. The delocalization of electrons in the Pi orbitals provides the graphene with remarkable electrical properties, and the low-energy σ bonds between carbon atoms make this hexagonal arrangement of atoms a thermally stable, rather chemically inactive structure, with exceptional mechanical performance as measured experimentally at the nanoscale and demonstrated by atomistic simulations [1–10]. These remarkable physical properties of graphenes have paved the way for their use in nanocomposites to enhance mechanical properties and achieve multifunctional capabilities. Another allotrope of graphitic carbon are CNTs, which may be thought of as concentric rolls of graphene sheets, with diameters ranging from <1 nm to several tens of nanometers, and aspect ratios exceeding 10^5. In comparison with graphene, on the one hand, the nearly edge-free structure of CNTs improves its structural stability, which is why graphene degradation, for instance during oxidation at high temperatures, starts from defect sites such as edges. On the other hand, the bending energy stored in the rolled structures of CNTs, especially at sufficiently thin CNFs, such as single-walled carbon nanotubes (SWCNTs), reduces the structural stability of the tubes. Another nearly one-dimensional (1D) allotrope of graphene is CNF, which may be composed of amorphous carbon and graphitic or turbostratic domains, inclined with respect to the nanofiber axis. The misalignment of the basal planes of the graphitic and turbostratic domains with the fiber axis has a profound effect on the mechanical performance of CNFs, especially in contrast to CNTs, as will be discussed in the subsequent sections. Similar to graphene, CNTs and CNFs are also known for their remarkable mechanical and electrical performance, and their thermal stability, and they have been incorporated as building blocks in multifunctional nanocomposites [11–13]. In addition, the high aspect ratio of CNTs and CNFs also facilitates the application of these materials in the development of advanced yarns, as a means to realize the superior mechanical performance of these nanomaterials in macroscale [1,8,14–17].

In this chapter, a summary of the recent progress on the incorporation of graphitic carbon–based nanomaterials, such as graphene particles and CNTs, in composites to develop multifunctional materials with remarkable mechanical performance is presented. Among the functionalities discussed in this chapter are self-sensing (damage detection), self-healing, and actuation. Most of the research proposed in this chapter is yet in the research and development stage, and the utilization of those materials in industrial scales requires further material property optimization and development of more economic production techniques.

3.2 Types of Carbon Nanomaterials and Their Properties

The presence of four valence electrons in the atomic carbon and the different types of hybridizations of valence orbitals allows for the formation of a variety

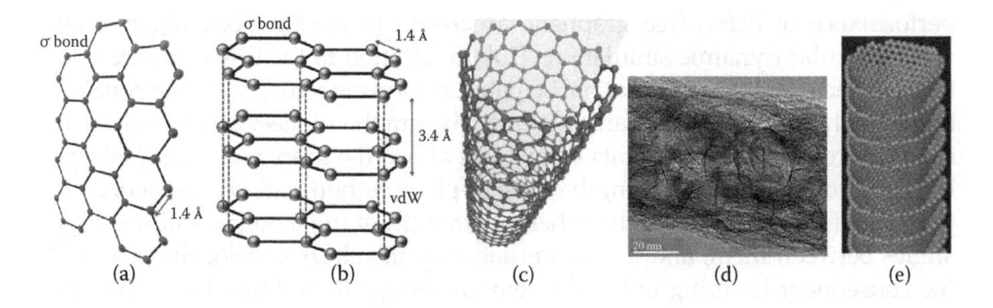

FIGURE 3.1
Schematics of the atomic structure of some allotropes of carbon: (a) graphene; (b) graphite; (c) carbon nanotube; (d) electrospun (reproduced from Arshad, S.N., M. Naraghi, and I. Chasiotis, *Carbon*, 49, 1710, 2011. With permission of Pergamon); and (e) vapor-grown carbon nanofibers (Reprinted with permission from Endo, M. et al., *Applied Physics Letters*, 80, 1267. Copyright 2002, American Institute of Physics). In all these allotropes, carbon atoms with sp^2 hybridization are covalently bonded to three carbon atoms in their neighbors.

of carbon-based structures with distinctly different atomic arrangements, crystallinity, and, as a consequence, very different physical properties. Among the allotropes of carbon that have received considerable attention in developing multifunctional hybrid materials are graphite and graphene particles, CNTs, and electrospun and vapor-grown CNFs (ESCNFs and VGCNFs, respectively), as shown in Figure 3.1. To realize the extent of the variations of the physical properties among allotropes of carbon, one may consider their electrical properties. For instance, graphite and diamond in general lay on the two sides of the spectrum of electrical conductivity of materials, i.e., the electrical conductivity of the former being several orders of magnitude higher than the latter [18,19]. The wide range of physical properties, such as electrical conductivity, and the remarkable thermal stability and mechanical properties of allotropes of carbon make them suitable as building blocks for variety of multifunctional hierarchical nanocomposites. In this section, we will discuss the atomic structure and properties of some of the allotropes of carbon, such as graphene, CNTs, and nanofibers, which are most widely used in the fabrication of nanocomposites.

3.2.1 Graphene and Graphite

As pointed out in Section 3.1, graphene is a single-atom-thick sheet of carbon atoms, in a hexagonal lattice structure, which are connected to each other via strong σ bonds, with valence electrons in sp^2 hybridized orbitals (see Figure 3.1a). The minimum distance between carbon atoms in graphene is ~1.42 Å. Because of the strong in-plane bonds between carbon atoms, graphene is the strongest material found on Earth, as confirmed both in experiments and computationally. Nanoindentation experiments on monolayers of graphene *in situ* atomic force microscope revealed modulus and strength of ~1 TPa and ~130 GPa, respectively [4], which is consistent with the predictions of the mechanical performance of defect-free graphene according to the tight-binding method and molecular dynamic simulations [20]. In addition to the three valence electrons of each carbon atom that participates in forming strong in-plane covalent bonds with the three other carbon atoms, the graphene sheet contains delocalized electrons in the pi orbitals below and above the graphene sheet (delocalized electrons). The high strength of the in-plane carbon–carbon interactions in graphene is owed partly to the inherent strength of the σ bonds and the large angles between them, and it is accentuated by the electron delocalization and the consequent lowering of bond potential energy. In addition to mechanical strength, the stability of the bonds also enhances the thermal stability of pristine graphene and enhances its resistance to oxidation. The delocalized electrons also contribute to the electrical conductivity of graphene, making graphene, in its defect-free condition, one of the best electrical conductors on Earth [21].

Just like any other crystalline structure, the mechanical, electrical, and thermal properties of graphene is highly controlled by the presence of defects, such as grain boundaries, vacancies, dislocations, and topological defects, including five- and seven-membered rings [22]. In general, defects will lead to a substantial loss in mechanical strength and electrical conductivity. For instance, via finite element simulation of graphene, Tserpes [23] has estimated that a 50% loss in strength of graphene may occur when only 4.4% of the carbon atoms are missing. Similarly, coupled quantum, molecular, and continuum mechanics simulations of graphene pointed to close to 60% loss in graphene strength with a slit size of ~40 Å in a 400-Å-wide graphene sheet [24]. Moreover, the additional bond energy and dangling bonds at the location of defects increase the likelihood of chemical reactions between graphene and functionalizing and oxidizing agents.

The planar structure of graphene allows for their natural stacking in the out-of-plane direction. In this arrangement of carbon atoms, known as graphite (see Figure 3.1b), the graphene layers maintain their in-plane integrity via σ bonds, while the out-of-plane interactions between them are mainly weak vdW forces, at an equilibrium distance of ~3.4 Å. Therefore, graphite demonstrates anisotropic physical properties. From a mechanics point of view, each layer is generally expected to be as strong and stiff as an isolated monolayer of graphene in the in-plane direction. On the other hand, weak vdW interactions between layers impose limited hindrance to mutual sliding between layers, which facilitates the use of graphite as solid lubricants [25]. Similar to graphene, graphite can conduct electricity within each layer via the delocalized electrons [26]. However, significantly poorer electrical conductivity along the *c*-axis (perpendicular to the graphene plane) has been reported [27].

3.2.2 Carbon Nanotubes

One may think of CNTs as rolled sheets of graphene (Figure 3.1c) [28]. The direction of rolling and diameter of each shell is typically defined by a roll-up vector, which is defined as a linear sum of the two base vectors of graphene lattice, each multiplied by an integer. The set of the two integers, m and n, defines the chirality of the shell. On the basis of chirality, CNTs are generally divided into three groups: zigzag, armchair, and chiral. In the first and second category, a third of the carbon–carbon bonds are perpendicular and parallel to the CNT axis, respectively, while in chiral CNTs no covalent bond is either parallel or perpendicular to the CNT axis [28,29]. The physical properties of CNTs depend on their chirality, as will be discussed later in this section.

Another criterion to categorize CNTs is based on the number of shells, according to which CNTs are typically divided into two groups: single-wall carbon nanotubes (SWNTs) and multiwall carbon nanotubes (MWNTs). In SWNTs, each carbon atom is connected to its three neighboring carbon atoms via strong covalent bonds, similar to graphene. The thinnest SWNT observed experimentally is ~3 Å, while SWNTs with diameters of larger than ~4 nm tend to collapse, forming dog-bone cross sections [30–33]. On the other hand, it is expected that smaller tube diameter will increase the energy stored in carbon–carbon bonds, decreasing CNT stability [17]. In MWNTs, the shells are separated by a distance of ~3.4 Å, which is about the separation distance between graphene layers in graphite [34]. While a chemical bond structure similar to SWNTs exists within each shell of MWNTs, the interactions between shells are mainly via vdW forces, which are significantly weaker than the in-plane covalent bonds. Therefore, tensile load on MWNTs is generally carried out almost entirely by the outmost shell by stretching the in-plane covalent bonds, and the failure of this shell is typically followed by the pull out of the inner shells, referred to as "sword-in-sheath failure," during which the load is transferred via vdW interactions between shells [5,35]. Hence, the true strength of each CNT shell, in both SWNTs and MWNTs, is comparable to the strength of the graphene sheet with a similar defect density. One of the earliest studies that addressed the strength of CNTs is the one by Yu et al. [5], in which they measured modulus and strength of CNTs to be in the range of 0.32–1.5 TPa and 10–60 GPa, respectively. In addition, Peng et al. [2] reported higher mechanical strength of individual shells of CNTs (reaching 100 GPa), by employing a microelectromechanical-based *in situ* transmission electron microscopy tension test, potentially due to lower defect density. However, similar to graphene, the presence of different types of defects, such as vacancies, dislocations, and topological defects, can substantially lower the strength of CNT shells [36–38].

Another factor that affects the mechanical performance of CNTs is their chirality. However, the effect of chirality on mechanical properties of CNTs is typically very marginal, compared with other factors such as defect density. As shown by Zhao et al. [20], on the basis of molecular dynamics simulations, the elastic modulus of the graphene sheet is equal for loadings both in the zigzag and armchair directions, and it is equal to 0.91 TPa. On the other hand, the strength of graphene loaded in the zigzag direction is ~107

GPa, which is ~19% higher than the strength of graphene in the armchair direction. Considering CNTs as rolled-up graphene sheets with a chiral vector that is normal to tube axis, one should note that in a zigzag CNT that is loaded axially, the equivalent graphene sheet is loaded in the armchair direction, and vice versa [20]. Therefore, pristine armchair CNTs are stronger than zigzag ones [20,39].

Despite its marginal effect on the mechanical performance of CNTs, chirality can significantly influence the charge transport in CNTs. For instance, armchair CNTs are metallic, while zigzag CNTs are generally semiconductor. The band gap in semiconductor CNTs is in the range of ~0.1–1.6 eV, and it decreases with the tube diameter [19,29,40,41]. Moreover, similar to mechanical properties of CNTs, their electrical properties are also highly influenced by defect [42].

3.2.3 Carbon Nanofibers

Another class of carbon-based nanomaterials is CNF. The microstructure of CNFs is composed of both nanoscale graphene sheets in turbostratic/graphitic arrangements. Moreover, they may be partly amorphous. Different methods of fabrication of CNFs will result in markedly different microstructures (Figure 3.1d and e). Two main scalable manufacturing techniques implemented in the past to fabricate CNFs are chemical vapor deposition with carbon-carrying feedstock such as methane (with CNFs commonly known as vapor grown carbon nanofibers—VGCNF), and thermal stabilization and carbonization/graphitization (pyrolysis) of electrospun polymer nanofibers (referred to herein as electrospun carbon nanofibers—ESCNF) [6,7,43–47]. The turbostratic particles, composed of stacked graphene sheets, are randomly oriented in ESCNFs (see Figure 3.1d). In contrast, graphene sheets in VGCNFs typically form stacked cones (Figure 3.1e) that are inclined with respect to fiber axis, at an angle ranging from a few degrees to a few tens of degrees and have hollow cores. In some cases, graphene sheets are folded and connected to each other at the inner and outer edges of the nanofiber [45,48,49].

Owing to the inclination of the graphene sheets with respect to the nanofiber axis in both types of CNFs, the load transfer path in CNFs during axial loading includes weak vdW interactions between graphene sheets in series with stronger in-plane covalent bonds within each graphene sheet [48,49]. Moreover, defects and stress concentration sites, such as vacancies and loops, in graphene sheets, and amorphous carbon in between graphene sheets can further reduce the strength of CNFs [23,50]. As a result, the strength and modulus of CNFs are only a small fraction of the in-plane properties of defect-free graphene sheets [48,49]. In this regard, one may compare the highest measured strength and modulus of CNFs, ~5 and 300 GPa, respectively, to the corresponding values of pristine graphene sheets, reaching values as high as 100 GPa and 1 TPa, respectively [4,6,7,48]. It is, however, to be emphasized that despite the aforementioned flaws in CNFs, their specific strength and modulus are among the highest values in engineering materials and advanced fibers, such as Kevlar©, glass, and carbon fibers. In addition, strong sp^2 bonds between carbon atoms and their delocalized electrons result in high thermal stability and electrical conductivity [6,7,10,48,51–53]. Studies on electrical conductivity of individual ESCNFs point to conductivities of as high as 10^4–10^5 S/m [52,53], which are controllable by the degree of crystallinity and carbonization temperature [53,54]. More specifically, the study of Wang and Santiago-Aviles [53,54] revealed that charge transport in ESCNFs has a semiconducting nature with very small band gaps of the order of 10^{-2} eV or less. Similarly, Zhang et al. [55] reported a conductivity of ~2×10^4 S/m for VGCNFs, lower than the conductivity of graphite within the basal plane and higher than the graphite conductivity along the *c*-axis. Therefore, the main charge conductivity path is proposed to be the electron transport within the basal plane of graphitic layers, suppressed by the electron transport between graphitic planes. These remarkable physical properties of CNFs have boosted the research in the field of nanocomposites to utilize CNFs as a means to enhance the mechanical, electrical, and thermal properties of nanocomposites and yarns [11,46,56,57].

3.3 Carbon Nanomaterials–Based Yarns and Nanocomposites

Given their desirable physical properties, such as their remarkable strength, electrical conductivity, and thermal stability, carbon-based nanomaterials have been extensively utilized as building blocks of nanocomposites to enhance different functionalities in macroscale. Depending on the volume concentration of carbon-based nanomaterials, the research in this field has been carried out on two types of material systems [17]. The first type of carbon-based hierarchical structures discussed here is their twisted yarns, composed of an intertwined network of carbon-based 1D nanomaterials such as CNTs or CNFs. The second type, referred to herein as carbon-based nanocomposites or simply nanocomposites, typically includes low volume fractions of nanomaterials, not exceeding the percolation threshold by far, in a base material such as a polymer matrix. The added nanomaterial may then have a reinforcing effect, or lead to improved electrical and thermal properties, etc., with respect to the properties of the matrix. In nanocomposites, due to the low volume fraction, the reinforcements are surrounded by the matrix through which the load, electricity, and heat are transferred. In contrast, in yarns, the nanoscale constituents (e.g., CNTs) are in direct contact with each other. This fundamental difference in the load, charge, and heat transfer mechanism results in very different physical properties in yarns and nanocomposites, and will require different processing methods to enhance their physical properties [48,58,59].

3.3.1 CNT- and CNF-Based Multifunctional Yarns

Mats and forests of high-aspect-ratio carbon-based nanomaterials such as CNTs and ESCNFs can be twisted into yarns [1,8,16,60–62] (Figure 3.2a through c). Owing to the high volume fraction of nanomaterials in yarns, CNTs and CNFs are considered platforms to bridge the gap between the remarkable mechanical properties of carbon-based materials at the nanoscale (e.g., the strength of individual CNFs and CNTs, measured to be as high as 4 GPa and more [5,6]) and the mechanical properties measured in the macroscale, the scale of yarns [1,8,16,60]. However, a major challenge in developing CNT and CNF yarns as structural elements is to induce sufficient shear interactions between neighboring nanomaterials. Insufficient load transfer will lead to premature failure of yarns, due to excessive sliding between nanomaterials, at stresses that are substantially below the strength of individual CNTs and CNFs [58,63]. As a result, several approaches have been implemented to enhance the shear interactions in CNT and CNF such as e-beam-induced cross-linking of CNT shells, surface functionalization of CNTs, and infiltration of yarns with polymer matrices to induce short- and long-range interactions between CNTs and CNFs [2,3,57,64]. In addition to their remarkable mechanical properties, CNT yarns demonstrate outstanding thermal stability. As shown by Zhang et al. [8], CNT yarns retained a substantial amount of their mechanical strength and ductility after being heated in air to temperatures of as high as 450°C, which reflect the stability of C–C covalent bonds in CNTs.

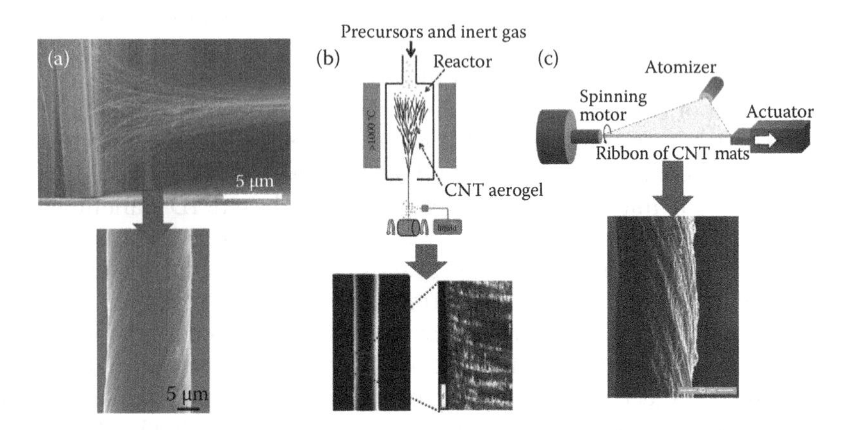

FIGURE 3.2
Several techniques have been used to fabricate CNT yarns, such as (a) spinning yarns from CNT forests (Reproduced from Lepro, X., M.D. Lima, and R.H. Baughman, *Carbon*, 48, 3621, 2010. With permission of Pergamon); (b) *in situ* chemical vapor deposition spinning of yarns (From Vilatela, J.J. and A.H. Windle, *Journal of Engineered Fibers and Fabrics*, 7, pp. 23–28, 2012. Special Issue—Fibers. www.jeffjournal.org); and (c) spinning CNT mats into yarns. (From Naraghi, M. et al., *ACS Nano*, 4, 6463, 2010.)

Electrical properties of CNTs and yarns have also been the target of many studies aimed at developing multifunctional CNT yarns. Of special considerations in this regard is the utilization of CNTs to fabricate supercapacitor yarns, which is motivated not just by the remarkable electrical conductivity of CNT yarns but also by their high surface-to-volume ratio. As shown by Zhong et al. [65], CNT yarns in the electrolyte NaCl solution demonstrate the remarkable capacitance of ~80 F/g, potentially due to the formation of double layers on CNTs, which makes them suitable for energy storage applications. The formation of double layers and the injection of charges to CNT yarns, when used as electrodes in an electrolyte, will change the length of C–C bonds. The bond length variation induced by charge injection was the basis of the CNT yarn actuators developed by Mirfakhrai et al. [66,67]. As shown by Viry et al. [68], the generated stress of the CNT yarns, as a quantitative measure of the electrochemical actuation, was enhanced by inducing CNT alignment through mechanical stretching. It is speculated that the "unbundling" of the CNTs due to mechanical stretching and the consequent increase in CNT free surfaces is the origin of the enhancement in the electrochemical capacity [68].

In addition to structural elements, CNT yarns can be used as strain sensors [8,69]. For example, Zhao et al. [69] have demonstrated the piezoresistive behavior of CNT yarns with a gage factor of ~0.5, significantly weaker than the gage factor of individual CNTs, but with no hysteresis up to strains of ~1%. The changes in the electrical conductivity of CNT yarns as a function of strain are attributed to the piezoresistivitiy of individual CNTs, i.e., the changes in their bandgap of the CNTs due to applied strain [69,70]. As pointed out by Minot et al. [71], mechanical loads on individual CNTs result in the modification of the electronic structure of CNTs. According to their studies, strain can generate a band gap in metallic CNTs, and alter the band gap structure of semiconducting CNTs. Moreover, the changes in the band gap in semiconducting CNTs can be positive or negative for CNTs under tension, depending on their chirality [71,72]. In addition, Monte Carlo simulations predict that the strain-induced changes in the bandgap structure of bundles of CNTs with random chiralities under uniform strain will result in a relative increase in the resistivity by an equivalent of 78 times for every unit of strain [72].

Owing to their structural stability, CNT yarns have also been used as the host for other functional materials. A great example related to this capability of CNT yarns is the study by Baughman and his coworkers [60], who developed a method to bi-scroll forests of CNTs into yarns while trapping nanoparticles of interest within the spun yarn. By choosing the nanomaterials with desired physical properties, they managed to fabricate CNT-hosted superconductors, catalytic nanofibers for fuel cells, etc. For instance, the superconductivity was achieved by embedding magnesium and boron particles inside CNT yarns, and exposing it to Mg vapor, which results in the formation of MgB_2 with superconducting capabilities. Owing to the remarkable strength of CNTs, the yarns were almost entirely composed of MgB_2, while their integrity was maintained by CNTs, which composed only 1 wt.% of the yarn.

A recent trend in the research related to carbon nanomaterial yarns has been focused on developing graphene/graphene oxide yarns. The main motivation of this aspect of research is to replace CNTs with the more cost-efficient graphene, given the fact that in both cases, the hexagonal lattice structure of carbon atoms provides remarkable mechanical, electrical, and thermal properties at the nanoscale [4,5,9,73]. One of the successful efforts to this end was reported by Dong et al. [73], who fabricated graphene yarns by molding graphene oxide into tubular templates, followed by chemical reduction of the graphene oxide fibers. They managed to develop graphene yarns with diameters of ~33–35 μm, with remarkable specific strength reaching ~800 MPa/(g/cm^3). Moreover, graphene yarns can be used as platforms to incorporate functional materials in developing multifunctional yarns. For instance, inclusion of magnetic particles (Fe_3O_4) in the yarns transforms them into magnetic yarns.

3.3.2 Carbon-Based Nanocomposites with Sensing Capability

The piezoresistive behavior of CNTs, rooted in the changes in their bandgaps in response to mechanical loads, can be used to develop composite films with strain-sensing capability. As shown by Dharap et al. [74], the piezoresistive behavior of CNT films at sufficiently low strains (less than ~1%) is linear, with the same slope under tension and compression. The changes in electrical resistance of CNT films as a function of strain, in addition to the piezoresistive properties of individual CNTs, are due to changes in the contact resistance between CNTs [74,75]. In addition, the reversible formation of local defects such as kinks may contribute to the phenomenon [75]. The gage factor of typical CNT films can be as high as 75 [75]. Of special interest for practical use of such strain sensors, the electrodes used to measure the change in resistance may not need to be permanently connected to the film; rather, a mobile four-point probe station for electrical measurements can be used to extract the strain at various locations [76].

Owing to their remarkable electrical properties, carbon-based nanomaterials such as CNTs and CNFs have been extensively used in developing conductive nanocomposites (with a nonconductive polymer matrix) with strain-sensing capabilities, in which the nanocomposites' electrical resistance will change with strain (piezoresistive nanocomposites). This type of piezoresistivity, which is typically nonlinear with respect to strain, is often observed when the concentration of CNTs/CNFs is close to the percolation threshold. In such nanocomposites, the electrons are transferred in between the reinforcements through a tunneling effect (see Figure 3.3a) [77,78]. Although individual CNTs are piezoresistive, the basis for the sensing capability of CNT nanocomposites is attributed to the relative changes

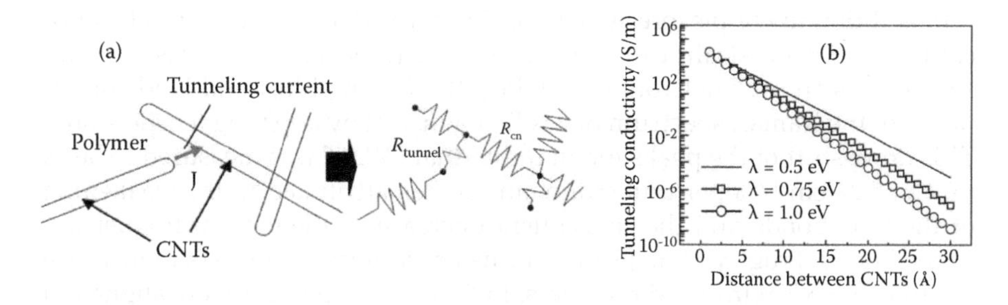

FIGURE 3.3
(a) Modeling of the conductive network of CNT-/CNF-reinforced nonconductive matrix composites that are composed of conductive CNT/CNFs and the electron tunneling (hopping) between them. (b) The effective conductivity of the tunneling effect is inversely related to the distance between reinforcements (λ is the height energy barrier for tunneling effect). (Reprinted from *Acta Materialia*, 56, Hu, N. et al., Tunneling effect in a polymer/carbon nanotube nanocomposite strain sensor, 2929–2936, Copyright 2008, with permission from Elsevier.)

in the distances between CNTs in a deforming nanocomposite as a result of mechanical loads and the consequent increase in the effective tunneling resistance [77]. The nonlinear piezoresistive behavior is attributed to the high sensitivity of the tunneling resistance with respect to the average distance between two nanoscale reinforcements, as shown in Figure 3.3b [77,79].

Hu et al. [77] utilized the tunneling effect in CNT-reinforced epoxy nanocomposites to generate strain-sensing capability. According to their study, the relative change in the electrical resistance of CNT nanocomposites per unit strain can be as high as ~3, and the ratio increases with reducing the CNT concentration from ~5 to 1 wt.%. It is, however, to be emphasized that the electrical resistivity was measured by utilizing two probes. Therefore, the initial resistivity was overestimated, leading to underestimations of the relative changes in the resistivity. Moreover, it was concluded that the contribution of the piezoresistivity of CNTs to the overall resistance changes are insignificant owing to poor load transfer between the matrix and the CNTs.

Similar to CNT nanocomposites, CNF and graphene-reinforced nanocomposites in a nonconductive matrix will demonstrate a nonlinear piezoresistive effect [78,80]. Adding a few percentages of graphene particles or VGCNFs to the nonconductive polymer matrix leads to orders of magnitude reduction in volume resistivity [78,80]. Moreover, such nanocomposites demonstrate a significant piezoresistive effect with gage factors of ~10–100 (relative change of the electrical resistivity to the applied strain) [78,80]. Furthermore, in case of VGCNFs, incorporating different types of epoxies, brittle and ductile, the tangential gage factors, measured at strains of <1%, were unchanged, pointing to the significance of electron tunneling in establishing the piezoresistive effect in nanocomposites. Moreover, in ductile epoxies and at larger strains, the piezoresistivity became nonlinear with the slope of the tangential piezoresistive effect increasing, a further indication of the electron tunneling in nanocomposites and its dependence on strain [78].

In addition to the piezoresistivity of CNTs, the dependence of CNTs' optical properties on strain can also be utilized to develop CNT-based strain sensors. As shown by Cronin et al. [81], the G (graphitic) and D (disorder) peaks of the Raman spectrum of CNTs downshift by applying tensile strain. This downshift of the peaks for an individual SWNT was measured to be as much as 24 cm^{-1}/% per unit strain, and it was attributed to the weakening of the C–C bonds and the consequent decrease in the natural frequencies of the bonds. Frogley et al. [82] used this characteristic of CNTs to measure strain in SWNT-reinforced polymers. In their study, SWNTs were aligned in polyurethane acrylate matrix using an *in situ* polymerization and curing in a shear flow. It was demonstrated that by using polarized light to excite CNTs in specific orientations, the axial and transvers normal strains in a uniaxially drawn sample could be measured. This method of strain measurement does not require direct contact with the sample. However, it could lead to underestimation of strains in a polymer matrix especially after ~1% of strain, when the matrix yields at the interface between the CNT and the matrix. Subsequent to interface yielding, no more net load is transferred from the matrix to the CNTs. This effect will appear as a plateau in the CNT Raman peak location as a function of nanocomposite strain [82,83]. This effect is particularly due to elastic mismatch between CNTs and the matrix.

3.3.3 Carbon Nanomaterials for Structural Health Monitoring and Self-Healing

The remarkable electrical and thermal properties of CNTs and CNFs have also fostered substantial research efforts to develop nanocomposites with structural damage detection and self-healing capabilities. The magnitude of damage detected in these studies may be subcritical or catastrophic. An example of the former is the local debonding and matrix yielding at the matrix–CNT interface, detected by the changes in the sensitivity of the Raman G peak location with respect to overall strain in CNT nanocomposites, as discussed in Section 3.3.2 [82,83]. On the other hand, more critical damages, such as coalesced microcracks and delamination, can be detected by monitoring the changes in material electrical resistivity.

For instance, Zhang et al. [84] demonstrated that the magnitude of the fatigue-induced crack growth and delamination in nanocomposites reinforced with CNTs directly correlates with the effective volume resistivity and through the thickness resistance of composites. This correlation is primarily due to the disruption of the electrical conductive path in the material as a result of the crack growth. Thonstenson and Chou [85] observed significant hysteresis in the electrical resistance–strain curves of glass fiber–reinforced composites that contained CNTs, attributed to the opening and closing of the cracks, such as the debonding between the matrix, the fibers, and CNTs, under cyclic loads (see Figure 3.4a and b). The formation of the crack manifests itself as steep changes in resistance as a function of strain.

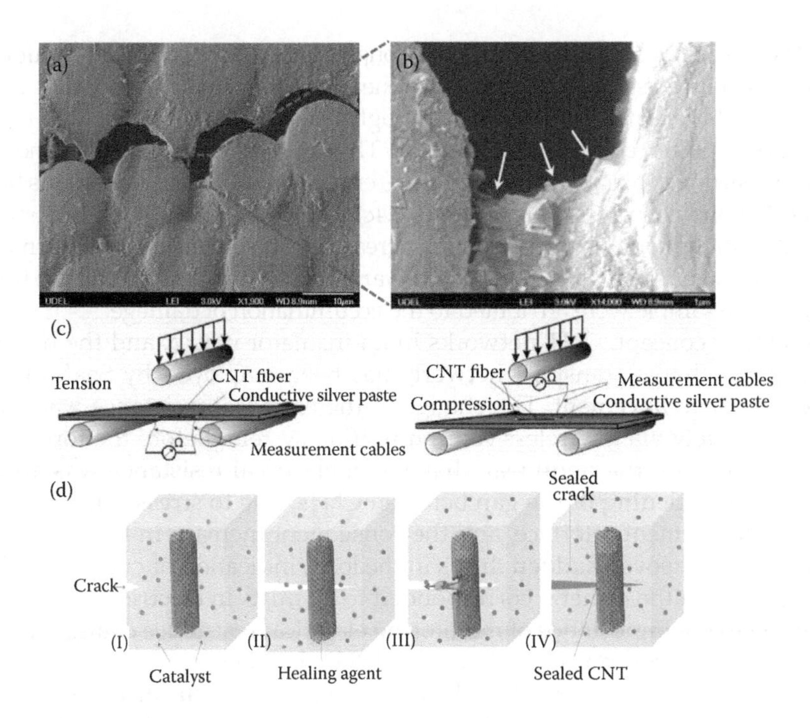

FIGURE 3.4
(a) SEM images of the fracture site in CNT/epoxy/glass fiber composites. (Reproduced from Thostenson, E.T. and T.W. Chou, *Nanotechnology*, 19, 2008. With permission.) (b) Arrows point to the pulled-out CNTs and their broken network, leading to substantial reduction in electrical conductivity. (Reproduced from Thostenson, E.T. and T.W. Chou, *Nanotechnology*, 19, 2008. With permission.) (c) CNT yarns are used as sensors to monitor the degree of damage in the structure in tension and compression. (Reproduced from Alexopoulos, N.D. et al., *Composites Science and Technology*, 70, 260, 2010. With permission of Pergamon.) (d) The concept of utilizing CNTs as nanoreservoirs containing self-healing agents in a composite. The propagation of cracks (I and II) will lead to the rupture of the CNT and the release of curing agents (III and IV). (Reproduced from Lanzara, G. et al., *Nanotechnology*, 20, 2009. With permission.)

Moreover, it was demonstrated that under cyclic loads with sufficiently low amplitudes, the resistance is completely recovered on unloading. However, increasing the load amplitude beyond a threshold will result in permanent changes in electrical resistance (not recoverable upon unloading), an indication of the accumulation of defects in the sample.

In addition, Thonstenson and Chou [85] used CNT networks embedded in glass fiber–reinforced epoxy joint to monitor the evolution of damage. CNTs were mixed with epoxy using a calendaring technique. Electrodes, across which electrical resistance was measured, were placed on opposite sides of the joint. In this case, the sudden increase (change in slope) of the resistance as a function of displacement is a likely indication of formation of cracks in the sample.

CNT yarns have also been used in conjunction with microscale reinforcements such as glass fibers for structural health monitoring. Alexopoulos et al. [86] used CNT yarns manufactured through coagulation and embedded into a glass fiber–reinforced polymer composite. The sample was tested in 3P bending, and the yarn was placed near the surface, experiencing peak axial strains both in tension and compression (see Figure 3.4c). After a nominal maximum strain of ~1.25% in the composites, a residual increase in the resistance of the yarn was observed. Given the high ductility of the yarns (>200%), the changes in the resistance of the sample were attributed to the accumulation of damage.

A similar concept, CNT networks in an insulator matrix and the disruption of the charge transport network, has been employed by Saafi [87] to detect damage in cement. The change in the electrical resistance was measured remotely via a wireless communication system. Before the formation of visible cracks, the nonlinear changes in electrical resistance was attributed to the widening of the gap between CNTs, due to stress concentration at the CNT–cement interface, and the consequent increase in the tunneling resistance. Moreover, sudden drops in the load, indication of crack initiation and growth in the cement, corresponded to upshifts in electrical resistance, pointing to the importance of monitoring electrical resistance changes of the cement–CNT composites as a means to detect damage.

In addition to structural health monitoring, CNTs can also be used to improve the healing process of a composite. For example, Zhang et al. [84] used the network of CNTs to accelerate the heat transfer in polymers and induce faster healing in the cracked composite. This aspect of the application of CNTs benefits from their remarkable thermal conductivity. It was demonstrated that the presence of CNTs results in an order of magnitude faster healing, as a result of accelerated heat transfer in CNT–composite samples.

A rather ambitious idea to utilize CNTs for self-healing was proposed by Lanzara et al. [88], who suggested to include CNTs not just as reinforcements of a matrix but also as "nanoreservoirs" containing a potential healing agent stored in their otherwise hollow core. The catalytic trigger molecules could be dispersed in the matrix or coated on the exterior of the CNTs. Subsequent to the crack growth from the matrix into the CNT, the healing agent will be released into the matrix and initiate a curing chemical reaction with the catalyst particles, as shown schematically in Figure 3.4d. While the practicality of this idea requires further investigations and experimental implementation, molecular dynamics simulations revealed that the percentage of the total methane molecules encapsulated inside CNTs, which will be released upon crack formation, is very small (below 1%), and it scales with the size of the crack.

3.3.4 Carbon Nanomaterials–Based Solid State Actuators

A major limitation in utilizing CNT yarn actuators, which work based on the formation of double layers and charge injection as presented by Mirfakhrai et al. [66,67] (described in Section 3.3), is the presence of liquid electrolytes to facilitate the charge transport. To overcome this limitation, Cottinet et al. [89] utilized CNT bucky papers (BPs) as electrodes onto sides of Nafion, which serves as a solid electrolyte, as shown in Figure 3.5a. Upon applying electric fields onto the papers, cations such as H^+ will be absorbed to the negatively charged electrode, leading to swelling of the Nafion and bond expansion in CNTs. The net effect is the change in length of the electrodes and the bending of the BP–Nafion–BP structure (see Figure 3.5b).

At sufficiently low electric fields, ~20 kV/m, authors observed a linear relation between the electric field density and the strain. However, at a field intensity of ~45 kV/m, the strain reached a plateau, mostly due to the saturation of the induced electric charges. A major challenge in the design of such actuators is the small range of strains achieved (in the order of 0.01%) and the low electrical to mechanical energy conversion rate (of ~0.05%) mostly due to the high compliance of the BPs and relatively low ionic transport in Nafion.

CNTs and CNFs have also been used to enhance actuation speed, shape fixity, and shape recovery in shape-memory polymers. One of the most informing studies of this kind was carried out by Koerner et al. [90], who developed thermoplastic shape-memory polymer nanocomposites reinforced with CNTs. It was demonstrated that 1–5 vol.% CNTs can increase the rubbery modulus by 1–5 times, resulting in higher blocking force. Moreover, the presence of CNTs enhanced strain-induced crystallinity in the polymer, resulting in higher shape fixity. The presence of CNTs also enhanced infrared absorption, thus allowing for remote actuation via infrared radiation. Moreover, the electrically conductive network of CNTs inside the matrix also allows for accelerated actuation through Joule heating. Viry et al. [68]

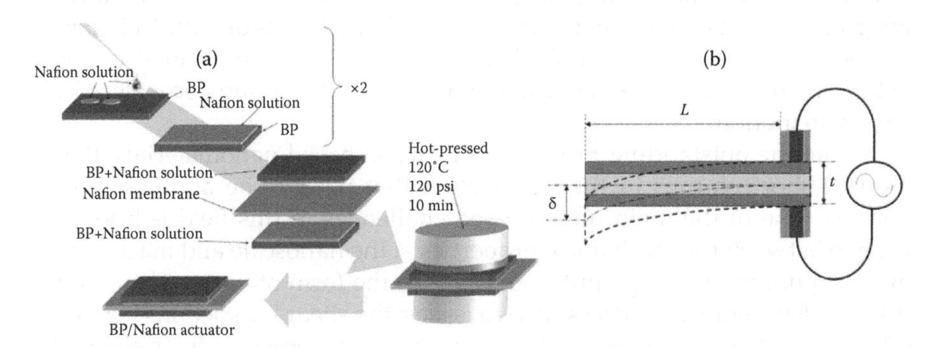

FIGURE 3.5
(a) Schematic of the fabrication process of the CNT bucky paper/Nafion actuator. Nafion is sandwiched between two bucky papers and works as a solid electrolyte. (b) Connecting the two CNT papers to opposite electric poles will induce nonuniform bond stretching in the two CNT papers due to the formation of the double layers, leading to lateral deformation. (Reprinted from *Sensors and Actuators A-Physical*, 170, Cottinet, P.J. et al., Nonlinear strain–electric field relationship of carbon nanotube bucky paper/Nafion actuators, 164–171, Copyright 2011, with permission from Elsevier.)

used the latter concept to develop CNT-reinforced polymer composite fibers. The fibers in their study were fabricated through the coagulation of polymer solutions containing up to 10 wt.% CNFs. The polymer used was polyvinyl alcohol, and the drawing of the fibers enhanced CNT alignment. The CNT composite yarns demonstrate a shape memory behavior, being able to induce recovery stresses of 30–150 MPa, depending on the actuation temperature (see Figure 3.5c).

Similar to CNT bucky paper, solutions of CNFs can also be vacuum filtered into CNF papers and incorporated into polymers as reinforcements and electrically conductive networks to enhance the shape-memory behavior. As pointed out by Lu et al. [91], the conductive network of the CNF papers can be used to generate heat inside the polymer via Joule heating to induce shape recovery.

3.4 Concluding Remarks

Arguably, nanotechnology all started when Richard Feynman said, "there is plenty of room at the bottom." A significant boost to this field came with the discovery of Iijima [34] of CNTs, and later studies pointing to their remarkable, but expected mechanical, electrical, and thermal properties, all rooted in the hexagonal crystalline structure of carbon atoms [5,71,92]. As mentioned in this chapter, such crystalline arrangement is not a unique feature of CNTs but can also be found in other allotropes of carbon, such as graphene and CNFs. All of these factors have inspired many researchers to focus on utilizing graphitic carbon nanomaterials as the building blocks of multifunctional composites, to develop materials that are not only strong but can also sense deformation and damage, can heal themselves, and can transform electrical energy to motion.

Despite the outstanding properties of carbon-based nanomaterials, there are still shortcomings in their utilization that need to be addressed. For instance, one of the main unaddressed challenges in this field is bridging the gap between the mechanical properties at the nanoscale and macroscale, by incorporating large quantities of CNTs in the form of yarns. The cure to this problem requires addressing two issues: the shear interactions between CNTs toward strong yet tough junctions, and developing industrially scalable (cost-efficient) CNT production techniques that can develop large quantities of CNTs with low defect density. The latter will lead to higher strength, electrical conductivity, and thermal stability in the nanoscale, while the former will help in developing high-strength yarns. Similar challenges exist on the path of developing multifunctional CNF yarns. However, the production cost of CNFs is not as prohibitive as it is in the case CNTs. Obviously, the downside to this argument is that the strengths of CNFs is lower than the strength of CNTs (at least for low-defect-density CNTs), due to the inclination of the graphitic layers with respect to nanofiber axis and other imperfections in the structures of CNTs, such as amorphous carbon. Thus, theoretically, the highest strength of CNF yarns is lower than the corresponding values for CNT yarns, although both can be considered as the building blocks for the next generations of advanced fibers with exceptional mechanical performance.

Other aspects of the research proposed in this chapter also require further studies. While many of the presented researches have passed the feasibility test, future research can now be focused on system optimization and parametric studies related to maximizing desirable device properties. As an example, one needs to address the question of how much deformation can be extracted from a CNT bucky paper actuator that works on the basis of ionic transport in electrodes, by controlling the fabrication parameters. Another example would be related to finding the parameters that would maximize device sensitivity in the case of CNT-/CNF-based strain gages.

References

1. Naraghi, M. et al. A multiscale study of high performance double-walled nanotube-polymer fibers. *ACS Nano*, 2010. **4**(11): pp. 6463–6476.
2. Peng, B. et al. Measurements of near-ultimate strength for multiwalled carbon nanotubes and irradiation-induced crosslinking improvements. *Nature Nanotechnology*, 2008. **3**(10): pp. 626–631.
3. Filleter, T. et al. Ultrahigh strength and stiffness in cross-linked hierarchical carbon nanotube bundles. *Advanced Materials*, 2011. **23**(25): pp. 2855–2860.
4. Lee, C. et al. Measurement of the elastic properties and intrinsic strength of monolayer graphene. *Science*, 2008. **321**(5887): pp. 385–388.
5. Yu, M.F. et al. Strength and breaking mechanism of multiwalled carbon nanotubes under tensile load. *Science*, 2000. **287**(5453): pp. 637–640.
6. Arshad, S.N., M. Naraghi, and I. Chasiotis. Strong carbon nanofibers from electrospun polyacrylonitrile. *Carbon*, 2011. **49**(5): pp. 1710–1719.
7. Zussman, E. et al. Mechanical and structural characterization of electrospun PAN-derived carbon nanofibers. *Carbon*, 2005. **43**(10): pp. 2175–2185.
8. Zhang, M., K.R. Atkinson, and R.H. Baughman. Multifunctional carbon nanotube yarns by downsizing an ancient technology. *Science*, 2004. **306**(5700): pp. 1358–1361.
9. Fedorov, G. et al. Tuning the band gap of semiconducting carbon nanotube by an axial magnetic field. *Applied Physics Letters*, 2010. **96**(13): pp. 132101 (3 pages).
10. Zhang, X., S. Fujiwara, and M. Fujii. Measurements of thermal conductivity and electrical conductivity of a single carbon fiber. *International Journal of Thermophysics*, 2000. **21**(4): pp. 965–980.
11. Al-Saleh, M.H. and U. Sundararaj. A review of vapor grown carbon nanofiber/polymer conductive composites. *Carbon*, 2009. **47**(1): pp. 2–22.

12. Huang, Z.M. et al. A review on polymer nanofibers by electrospinning and their applications in nanocomposites. *Composites Science and Technology*, 2003. **63**(15): pp. 2223–2253.
13. Spitalsky, Z. et al. Carbon nanotube-polymer composites: Chemistry, processing, mechanical and electrical properties. *Progress in Polymer Science*, 2010. **35**(3): pp. 357–401.
14. Zhang, M. et al. Strong, transparent, multifunctional, carbon nanotube sheets. *Science*, 2005. **309**(5738): pp. 1215–1219.
15. Koziol, K. et al. High-performance carbon nanotube fiber. *Science*, 2007. **318**(5858): pp. 1892–1895.
16. Li, Y.-L., I. Kinloch, and A. Windle. Direct spinning of carbon nanotube fibers from chemical vapor deposition synthesis. *Science*, 2004. **304**: pp. 276–278.
17. Espinosa, H.D., T. Filleter, and M. Naraghi. Multiscale experimental mechanics of hierarchical carbon-based materials. *Advanced Materials*, 2012. **24**(21): pp. 2805–2823.
18. Chen, C.F., S.H. Chen, and K.M. Lin. Electrical properties of diamond films grown at low temperature. *Thin Solid Films*, 1995. **270**(1–2): pp. 205–209.
19. Ebbesen, T.W. et al. Electrical conductivity of individual carbon nanotubes. *Nature*, 1996. **382**(6586): pp. 54–56.
20. Zhao, H., K. Min, and N.R. Aluru. Size and chirality dependent elastic properties of graphene nanoribbons under uniaxial tension. *Nano Letters*, 2009. **9**(8): pp. 3012–3015.
21. Morozov, S.V. et al. Giant intrinsic carrier mobilities in graphene and its bilayer. *Physical Review Letters*, 2008. **100**(1): pp. 016602 (4 pages).
22. Fan, B.B., X.B. Yang, and R. Zhang. Anisotropic mechanical properties and Stone–Wales defects in graphene monolayer: A theoretical study. *Physics Letters A*, 2010. **374**(27): pp. 2781–2784.
23. Tserpes, K.I. Strength of graphenes containing randomly dispersed vacancies. *Acta Mechanica*, 2012. **223**(4): pp. 669–678.
24. Khare, R. et al. Coupled quantum mechanical/molecular mechanical modeling of the fracture of defective carbon nanotubes and graphene sheets. *Physical Review B*, 2007. **75**(7): pp. 075412-(1-12).
25. Lin, J.S., L.W. Wang, and G.H. Chen. Modification of graphene platelets and their tribological properties as a lubricant additive. *Tribology Letters*, 2011. **41**(1): pp. 209–215.
26. Soule, D.E. Magnetic field dependence of the Hall effect and magnetoresistance in graphite single crystals. *Physical Review Letters*, 1958. **1**(9): pp. 347–347.
27. Tsuzuku, T. Anisotropic electrical-conduction in relation to the stacking disorder in graphite. *Carbon*, 1979. **17**(3): pp. 293–299.
28. Qian, D. et al. Mechanics of carbon nanotubes. *Applied Mechanics Reviews*, 2002. **55**: pp. 495–533.
29. Saito, R. et al. Electronic-structure of chiral graphene tubules. *Applied Physics Letters*, 1992. **60**(18): pp. 2204–2206.
30. Zhao, X. et al. Smallest carbon nanotube is 3 angstrom in diameter. *Physical Review Letters*, 2004. **92**(12): pp. 125502 (3 pages).
31. Xiao, J. et al. Collapse and stability of single- and multi-wall carbon nanotubes. *Nanotechnology*, 2007. **18**(39).
32. Elliott, J.A. et al. Collapse of single-wall carbon nanotubes is diameter dependent. *Physical Review Letters*, 2004. **92**(9).

33. Motta, M. et al. High performance fibers from 'dog bone' carbon nanotubes. *Advanced Materials*, 2007. **19**(21): pp. 3721.
34. Iijima, S. Helical microtubules of graphitic carbon. *Nature*, 1991. **354**(6348): pp. 56–58.
35. Yu, M.F., B.I. Yakobson, and R.S. Ruoff. Controlled sliding and pullout of nested shells in individual multiwalled carbon nanotubes. *Journal of Physical Chemistry B*, 2000. **104**(37): pp. 8764–8767.
36. Zhang, S.L. et al. Mechanics of defects in carbon nanotubes: Atomistic and multiscale simulations. *Physical Review B*, 2005. **71**(11): pp. 115403-(1-12).
37. Belytschko, T. et al. Effect of defects on the mechanical properties of carbon nanotubes. *Abstracts of Papers of the American Chemical Society*, 2007. **233**.
38. Mielke, S.L. et al. The effects of extensive pitting on the mechanical properties of carbon nanotubes. *Chemical Physics Letters*, 2007. **446**(1–3): pp. 128–132.
39. Duan, W.H. et al. Molecular mechanics modeling of carbon nanotube fracture. *Carbon*, 2007. **45**(9): pp. 1769–1776.
40. Tanaka, K. et al. Electronic-properties of bucky-tube model. *Chemical Physics Letters*, 1992. **191**(5): pp. 469–472.
41. White, C.T., D.H. Robertson, and J.W. Mintmire. Helical and rotational symmetries of nanoscale graphitic tubules. *Physical Review B*, 1993. **47**(9): pp. 5485–5488.
42. Dai, H.J., E.W. Wong, and C.M. Lieber. Probing electrical transport in nanomaterials: Conductivity of individual carbon nanotubes. *Science*, 1996. **272**(5261): pp. 523–526.
43. Wang, Y., S. Serrano, and J.J. Santiago-Aviles. Raman characterization of carbon nanofibers prepared using electrospinning. *Synthetic Metals*, 2003. **138**(3): pp. 423–427.
44. Chun, I. et al. Carbon nanofibers from polyacrylonitrile and mesophase pitch. *Journal of Advanced Materials*, 1999. **31**(1): pp. 36–41.
45. Endo, M. et al. Structural characterization of cup-stacked-type nanofibers with an entirely hollow core. *Applied Physics Letters*, 2002. **80**(7): pp. 1267–1269.
46. Tibbetts, G.G. et al. A review of the fabrication and properties of vapor-grown carbon nanofiber/polymer composites. *Composites Science and Technology*, 2007. **67**(7–8): pp. 1709–1718.
47. Chun, I.S. et al. Carbon nanofibers from polyacrylonitrile and mesophase pitch. *43rd International SAMPE Symposium and Exhibition on Materials and Process Affordability—Keys to the Future*, Vol. 43, 1998: pp. 718–729.
48. Ozkan, T., M. Naraghi, and I. Chasiotis. Mechanical properties of vapor grown carbon nanofibers. *Carbon*, 2010. **48**(1): pp. 239–244.
49. Uchida, T. et al. Morphology and modulus of vapor grown carbon nano fibers. *Journal of Materials Science*, 2006. **41**(18): pp. 5851–5856.
50. Suk, J.W. et al. Mechanical measurements of ultra-thin amorphous carbon membranes using scanning atomic force microscopy. *Carbon*, 2012. **50**(6): pp. 2220–2225.
51. Mostovoi, G.E., L.P. Kobets, and V.I. Frolov. Study of the thermal stability of the mechanical properties of carbon fibers. *Mechanics of Composite Materials*, 1979. **15**(1): pp. 20–25.
52. Sharma, C.S. et al. Fabrication and electrical conductivity of suspended carbon nanofiber arrays. *Carbon*, 2011. **49**(5): pp. 1727–1732.
53. Wang, Y. et al. Pyrolysis temperature and time dependence of electrical conductivity evolution for electrostatically generated carbon nanofibers. *IEEE Transactions on Nanotechnology*, 2003. **2**(1): pp. 39–43.

54. Wang, Y. and J.J. Santiago-Aviles. Low-temperature electronic properties of electrospun PAN-derived carbon nanofiber. *IEEE Transactions on Nanotechnology*, 2004. **3**(2): pp. 221–224.
55. Zhang, L. et al. Four-probe charge transport measurements on individual vertically aligned carbon nanofibers. *Applied Physics Letters*, 2004. **84**(20): pp. 3972–3974.
56. Zhou, Z.P. et al. Graphitic carbon nanofibers developed from bundles of aligned electrospun polyacrylonitrile nanofibers containing phosphoric acid. *Polymer*, 2010. **51**(11): pp. 2360–2367.
57. Moon, S. and R.J. Farris. Strong electrospun nanometer-diameter polyacrylonitrile carbon fiber yarns. *Carbon*, 2009. **47**(12): pp. 2829–2839.
58. Wei, X., M. Naraghi, and H.D. Espinosa. Optimal length scales emerging from shear load transfer in natural materials—Application to carbon-based nanocomposite design. Accepted, *ACS Nano*, 2012. **6**: pp. 2333–2344.
59. Ozkan, T., Q. Chen, and I. Chasiotis. Interfacial strength and fracture energy of individual carbon nanofibers in epoxy matrix as a function of surface conditions. *Composites Science and Technology*, 2012. **72**(9): pp. 965–975.
60. Lima, M.D. et al. Biscrolling nanotube sheets and functional guests into yarns. *Science*, 2011. **331**(6013): pp. 51–55.
61. Lepro, X., M.D. Lima, and R.H. Baughman. Spinnable carbon nanotube forests grown on thin, flexible metallic substrates. *Carbon*, 2010. **48**(12): pp. 3621–3627.
62. Vilatela, J.J. and A.H. Windle. A multifunctional yarn made of carbon nanotubes. *Journal of Engineered Fibers and Fabrics*, 2012. Special Issue—Fibers. **7**: pp. 23–28.
63. Vilatela, J.J., J.A. Elliott, and A.H. Windle. A model for the strength of yarn-like carbon nanotube fibers. *ACS Nano*, 2011. **5**(3): pp. 1921–1927.
64. Munoz, E. et al. Multifunctional carbon nanotube composite fibers. *Advanced Engineering Materials*, 2004. **6**(10): pp. 801–804.
65. Zhong, X.H. et al. Continuous multilayered carbon nanotube yarns. *Advanced Materials*, 2010. **22**(6): pp. 692–696.
66. Mirfakhrai, T. et al. Electrochemical actuation of carbon nanotube yarns. *Smart Materials & Structures*, 2007. **16**(2): pp. S243–S249.
67. Mirfakhrai, T. et al. Carbon nanotube yarn actuators: An electrochemical impedance model. *Journal of the Electrochemical Society*, 2009. **156**(6): pp. K97–K103.
68. Viry, L. et al. Nanotube fibers for electromechanical and shape memory actuators. *Journal of Materials Chemistry*, 2010. **20**(17): pp. 3487–3495.
69. Zhao, H.B. et al. Carbon nanotube yarn strain sensors. *Nanotechnology*, 2010. **21**(30): pp. 305502 (5 pages).
70. Yang, X. et al. Measurement and simulation of carbon nanotube's piezoresistance property by a micro/nano combined structure. *Indian Journal of Pure & Applied Physics*, 2007. **45**(4): pp. 282–286.
71. Minot, E.D. et al. Tuning carbon nanotube band gaps with strain. *Physical Review Letters*, 2003. **90**(15): pp. 156401-(1-4).
72. Cullinan, M.A. and M.L. Culpepper. Carbon nanotubes as piezoresistive microelectromechanical sensors: Theory and experiment. *Physical Review B*, 2010. **82**(11): pp. 115428-(1-6).
73. Dong, Z.L. et al. Facile fabrication of light, flexible and multifunctional graphene fibers. *Advanced Materials*, 2012. **24**(14): pp. 1856–1861.
74. Dharap, P. et al. Nanotube film based on single-wall carbon nanotubes for strain sensing. *Nanotechnology*, 2004. **15**(3): pp. 379–382.

75. Cao, C.L. et al. Temperature dependent piezoresistive effect of multi-walled carbon nanotube films. *Diamond and Related Materials*, 2007. **16**(2): pp. 388–392.
76. Li, Z.L. et al. Carbon nanotube film sensors. *Advanced Materials*, 2004. **16**(7): pp. 640–643.
77. Hu, N. et al. Tunneling effect in a polymer/carbon nanotube nanocomposite strain sensor. *Acta Materialia*, 2008. **56**(13): pp. 2929–2936.
78. Yasuoka, T., Y. Shimamura, and A. Todoroki. Electrical resistance change under strain of CNF/flexible-epoxy composite. *Advanced Composite Materials*, 2010. **19**(2): pp. 123–138.
79. Simmons, J.G. Generalized formula for electric tunnel effect between similar electrodes separated by a thin insulating film. *Journal of Applied Physics*, 1963. **34**(6): pp. 1793.
80. Eswaraiah, V., K. Balasubramaniam, and S. Ramaprabhu. One-pot synthesis of conducting graphene-polymer composites and their strain sensing application. *Nanoscale*, 2012. **4**(4): pp. 1258–1262.
81. Cronin, S.B. et al. Measuring the uniaxial strain of individual single-wall carbon nanotubes: Resonance Raman spectra of atomic-force-microscope modified single-wall nanotubes. *Physical Review Letters*, 2004. **93**(16): pp. 167401-(1-4).
82. Frogley, M.D., Q. Zhao, and H.D. Wagner. Polarized resonance Raman spectroscopy of single-wall carbon nanotubes within a polymer under strain. *Physical Review B*, 2002. **65**(11): pp. 113413 (4 pages).
83. Ma, W.J. et al. Monitoring a micromechanical process in macroscale carbon nanotube films and fibers. *Advanced Materials*, 2009. **21**(5): pp. 603–608.
84. Zhang, W., V. Sakalkar, and N. Koratkar. *In situ* health monitoring and repair in composites using carbon nanotube additives. *Applied Physics Letters*, 2007. **91**(13): pp. 133102 (3 pages).
85. Thostenson, E.T. and T.W. Chou. Real-time *in situ* sensing of damage evolution in advanced fiber composites using carbon nanotube networks. *Nanotechnology*, 2008. **19**(21): pp. 215713 (4 pages).
86. Alexopoulos, N.D. et al. Structural health monitoring of glass fiber reinforced composites using embedded carbon nanotube (CNT) fibers. *Composites Science and Technology*, 2010. **70**(2): pp. 260–271.
87. Saafi, M. Wireless and embedded carbon nanotube networks for damage detection in concrete structures. *Nanotechnology*, 2009. **20**(39): pp. 395502 (4 pages).
88. Lanzara, G. et al. Carbon nanotube reservoirs for self-healing materials. *Nanotechnology*, 2009. **20**(33): pp. 335704 (4 pages).
89. Cottinet, P.J. et al. Nonlinear strain–electric field relationship of carbon nanotube bucky paper/Nafion actuators. *Sensors and Actuators A-Physical*, 2011. **170**(1–2): pp. 164–171.
90. Koerner, H. et al. Remotely actuated polymer nanocomposites—Stress-recovery of carbon-nanotube-filled thermoplastic elastomers. *Nature Materials*, 2004. **3**(2): pp. 115–120.
91. Lu, H. et al. Electroactive shape-memory polymer nanocomposites incorporating carbon nanofiber paper. *International Journal of Smart and Nano Materials*, **1**(1): pp. 2–12.
92. Lima, A.M.F. et al. Purity evaluation and influence of carbon nanotube on carbon nanotube/graphite thermal stability. *Journal of Thermal Analysis and Calorimetry*, 2009. **97**(1): pp. 257–263.

4

Active Fiber Composites: Modeling, Fabrication, and Characterization

Yirong Lin and Henry A. Sodano

CONTENTS

4.1 Introduction

The past few decades have seen significant growth in the development and application of active materials to a wide range of host structures owing to their superior sensing and actuation properties (Williams et al. 2002). While there exist many types of useful active materials, such as shape-memory alloys, electrostrictives, and magnetorheological fluids, piezoelectric materials remain the most widely used "smart" material for a number of reasons. First, piezoceramics have a high stiffness, providing them with strong, voltage-dependent actuation authority. Additionally, piezoceramics are capable of interacting with dynamic systems at frequencies spanning six orders of magnitude, from about 1 Hz to 1 MHz. In the past, a good deal of success was achieved in the field of intelligent structures using monolithic wafers of piezoceramic material. However, there are several practical limitations to implementing this delicate type of material; namely, the brittle nature of ceramics makes them vulnerable to accidental breakage during handling and bonding procedures, as well as their extremely limited ability to conform to curved surfaces and the large mass associated with using a typically lead-based ceramic.

To resolve the inadequacies of monolithic piezoceramic materials, researchers have developed composite piezoelectric materials consisting of an active piezoceramic fiber embedded in a polymer matrix. Configuration of material in this way is advantageous because typical crystalline materials have much higher strengths in the fiber form due to reduced volume fractions of flaws during fabrication (Williams et al. 2002). Also, in addition to providing improved robustness by protecting the fragile fibers, the flexible nature of the polymer matrix allows the material to be able to more easily conform to the curved surfaces found in more realistic industrial applications. These advantages have been capitalized on by the development of a new group of devices called piezoelectric fiber composites (PFCs). Research in this area led to the development of a broad range of active PFC actuators, including active fiber composites (AFCs) (Bent and Hagood 1993a,b, Bent et al. 1995), macrofiber composites (MFCs) (Wilkie et al. 2000), 1–3 composites, and the hollow fiber composite (Cannon and Brei 2000). These PFCs are designed to fulfill the specific purpose of structural sensing and actuation and are typically constructed in the form of a patch of material that can be bonded to the surface of a structure or laid up as "active layers" along with conventional fiber-reinforced lamina. While the PFC provides significant advantages over monolithic piezoceramics, they are separated from the structural components and are not intended to provide any load-bearing functionality.

This chapter presents a new piezoceramic fiber that is fabricated by coating a conductive structural fiber with a piezoceramic film, as illustrated in Figure 4.1. The structural fiber serves two purposes, first to provide strength

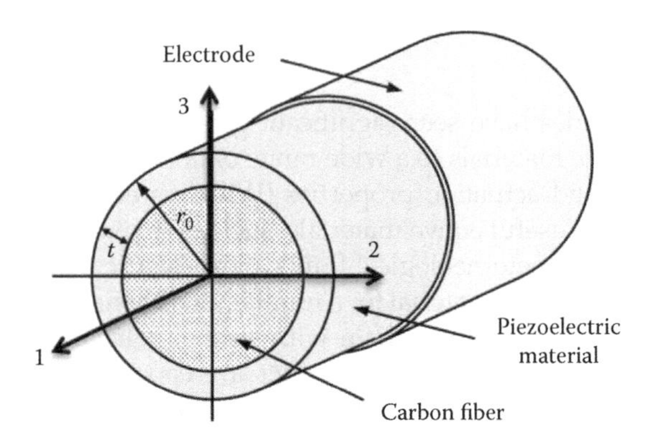

FIGURE 4.1
Schematic showing the cross-section of the novel ASF.

to the active fiber and second to act as an embedded electrode to electrically interact with the piezoceramic. The result of this body of work offers a novel active piezoelectric structural fiber that can be laid up in a composite structural material to perform sensing and actuation, in addition to providing critical load-bearing functionality. The sensing and actuation aspects of the novel multifunctional material will allow composites to be designed with embedded structural health monitoring, power generation, vibration sensing and control, damping, and shape control through anisotropic actuation. Each of these potential applications resulting from the fundamental development of the proposed active structural fiber (ASF) will have broad impacts on the performance and safety of modern structures. Furthermore, the advances made through the development of multifunctional material systems such as those here will advance the way in which adaptive structures are designed and the modeling of composite materials with an active interphase layer.

4.2 Micromechanics Modeling of Effective Electroelastic Properties

The first step for the development of the ASF-reinforced multifunctional composites shown in Figure 4.1 is to understand how design parameters such as choice of materials, fiber aspect ratio, and volume fraction of each phase influence the performance of the entire composites system. To accomplish this, analytical models need to be derived to accurately estimate the overall electroelastic properties of multifunctional composites based on these design parameters. A reduced model based on the rule of mixture and an analogy between the thermal and piezoelectric responses of the composite will be developed first to estimate the effective piezoelectric coupling coefficient in the fiber axis of the ASF lamina with different design parameters. This easy and simple model will show its capability to accurately predict the longitudinal piezoelectric response of the composites; however, the one-dimensional model is limited to predict the effective longitudinal piezoelectric strain-coupling coefficient only. Therefore, the second part of the modeling task will develop a three-dimensional micromechanics model to predict the full electroelastic properties of the three-phase multifunctional composites.

4.2.1 One-Dimensional Micromechanics Modeling

For the ASF considered here, the electric field will be applied along the radial axis or through the thickness of the piezoelectric shell. Because the inner electrode will have a smaller surface area than the outer electrode, the electric field will vary nonlinearly through the thickness of the piezoceramic. This nonlinear field variation must be accounted for such that the breakdown voltage at the inner wall of the fiber is not reached. From Gauss' law, the electric field along the radial direction of the active fiber can be expressed as (Halliday and Resnick 1988)

$$E(r) = \frac{-V}{r\ln(1-\alpha)} \tag{4.1}$$

where V is the voltage applied across the fiber thickness, r is the radial position, and α is the aspect ratio of the piezoelectric portion of the fiber, equal to t/r_0, where t is the thickness of the piezoelectric coating and r_0 is the total radius of the fiber. According to Equation 4.1, the local electric field is proportional to $1/r$, leading to a higher electric field at the inner wall than that of the outer wall. This leads to a higher actuation strain at the inner wall of the fiber and limits the magnitude of the electric field applied before depoling occurs.

The longitudinal piezoelectric stress of the piezoelectric shell can be expressed as

$$\sigma(r) = Y^{P}\varepsilon(r) = Y^{P}d_{31}E(r) \tag{4.2}$$

where Y^{P} is the longitudinal modulus of elasticity of the piezoelectric shell, σ is the piezoelectric shell longitudinal stress, ε is the piezoelectric shell longitudinal strain, d_{31} is the piezoelectric coupling, the subscript 31 represents the electric field applied in –3 (transverse of ASF) direction while the generated strain is in –1 (longitudinal of ASF) direction as specified in Figure 4.1. The total piezoelectric force is determined by integrating the stress over the cross-section area of the piezoelectric shell,

$$F = \int_0^{2\pi}\int_{r_0-t}^{r_0} Y^{P}d_{31}E(r)r\,dr\,d\theta = \frac{-2\pi d_{31}Y^{P}Vt}{\ln(1-t/r_0)} \tag{4.3}$$

therefore, the free strain resulting from the total piezoelectric force can be derived by Hook's law:

$$\varepsilon = \frac{\sigma}{Y^{P}} = \frac{F}{AY^{P}} = \frac{-2\pi d_{31}Y^{P}Vt}{\pi[r_0^2-(r_0-t)^2]Y^{P}\ln(1-t/r_0)} = \frac{-E_{tw}d_{31}}{(r_0/t-0.5)\ln(1-t/r_0)} \tag{4.4}$$

where $E_{tw} = V/t$ is the electric field derived by thin wall approximation, A is the piezoelectric shell cross-section area, and F is the piezoelectric force.

Express the electric field–induced free strain as the product of thin wall electric field E_{tw} and the effective ASF piezoelectric coupling $d_{31,eff}^{f}$

$$\varepsilon = \left(\frac{-d_{31}}{\ln(1-\alpha)(1/\alpha - 0.5)} \right) E_{tw} = d_{31,eff}^{f} E_{tw} \tag{4.5}$$

where d_{31} is the piezoelectric coupling coefficient and $d_{31,eff}^{f}$ is the effective coupling of the piezoelectric shell incorporating the thin wall electric field approximation, $E_{tw} = V/t$. From Equation 4.5, it can be seen that at a certain electric field, the aspect ratio is the only parameter that will influence the effective $d_{31,eff}^{f}$ of the piezoelectric shell.

The coupling for the piezoelectric shell must be then combined with the core fiber to determine the effective piezoelectric coupling of the piezoelectric structural fiber. Assume the ASF is perfectly axisymmetric, and there is perfect bonding between the core fiber and piezoelectric shell, the longitudinal elastic modulus of the active fiber containing a core fiber can be defined using the rule of mixtures and written as (Hyer 1998)

$$Y^{multi} = Y^{p} v^{p} + Y^{f}(1 - v^{p}) \tag{4.6}$$

where Y is the longitudinal modulus of elasticity; v is the volume fraction; and the superscripts f, p, and multi represent the core fiber, piezoelectric, and complete multifunction piezoelectric structural fiber, respectively. The relationship between the fiber aspect ratio and fiber volume fraction is shown in Figure 4.2. According to Equations 4.4 and 4.5, the piezoelectric force generated from the piezoelectric shell can be expressed as

$$F = A\varepsilon Y^{p} = \frac{-E_{tw} d_{31}}{(r_0/t - 0.5)\ln(1 - t/r_0)} A Y^{p} = E_{tw} d_{31,eff}^{f} A Y^{p} \tag{4.7}$$

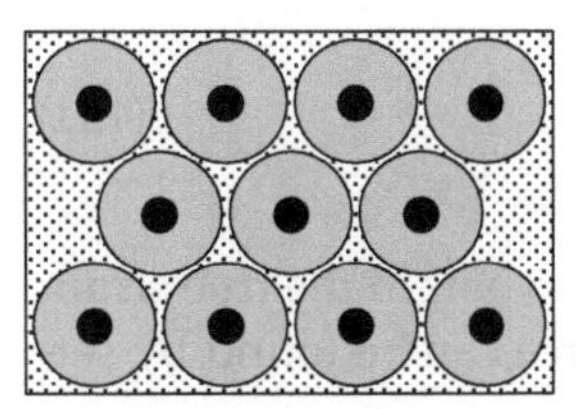
High aspect ratio
High volume fraction

Low aspect ratio
High volume fraction

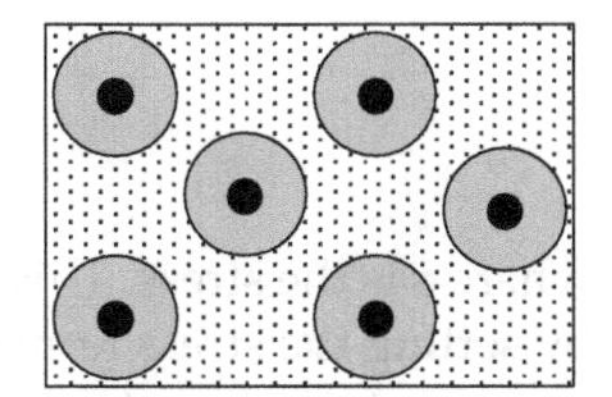
High aspect ratio
Low volume fraction

FIGURE 4.2
Schematic demonstrating the relationship between the fiber aspect ratio and fiber volume fraction. (From Cannon, B.J., Brei, D., *Journal of Intelligent Material Systems and Structures*, 11, 659, 2000.)

the total strain of the ASF caused by the piezoelectric force is then defined as

$$\varepsilon^{\text{multi}} = \frac{\sigma^{\text{multi}}}{Y^{\text{multi}}} = \frac{\frac{F}{A/v^{p}}}{Y^{\text{multi}}} = \frac{d_{31,\text{eff}}^{\text{f}} Y^{\text{p}} v^{\text{p}} E_{\text{tw}}}{Y^{\text{multi}}} = d_{31}^{\text{multi}} E_{\text{tw}} \tag{4.8}$$

the electromechanical coupling of a piezoelectric structural fiber with a piezoelectric coating can then be defined as

$$d_{31}^{\text{multi}} = \frac{d_{31,\text{eff}}^{\text{f}} Y^{\text{p}} v^{\text{p}}}{(Y^{\text{p}} - Y^{\text{f}}) v^{\text{p}} + Y^{\text{f}}} \tag{4.9}$$

where Y^{f} and Y^{p} are the elastic modulus of the fiber and piezoelectric material, respectively; v^{p} is the volume fraction of piezoelectric material; d_{31} is the piezoelectric coupling coefficient; and $d_{31,\text{eff}}^{\text{f}}$ is the effective coupling of the piezoelectric shell as defined from Equation 4.5.

The piezoelectric coupling term in Equation 4.9 predicts the response of a single active fiber; however, to determine the coupling when multiple active fibers are embedded in a polymer matrix, the rule of mixtures can be applied a second time by taking the piezoelectric shell to be an interphase layer. Using the same derivation as Equation 4.9, the resulting coupling of the ASF lamina can be written as

$$d_{31}^{\text{Lam}} = \frac{d_{31,\text{eff}}^{\text{f}} Y^{\text{p}} v^{\text{p}}}{(Y^{\text{p}} - Y^{\text{m}}) v^{\text{p}} + (Y^{\text{f}} - Y^{\text{m}}) v^{\text{f}} + Y^{\text{m}}} \tag{4.10}$$

where Y^{m} is the modulus of elasticity of the matrix material and v^{f} is the volume fraction of core fiber. Considering the piezoelectric constitutive equations, the lamina stress–strain relationship in the 31-direction can be identified as

$$\varepsilon^{\text{Lam}} = \frac{\sigma}{Y^{\text{Lam}}} + d_{31}^{\text{Lam}} E \tag{4.11}$$

where σ is the stress. This equation can then be used to obtain the free strain by setting the stress term to zero and the blocked force can be found by setting the strain term to zero written as

$$\text{Free strain: } \varepsilon^{\text{Lam}} = d_{31}^{\text{Lam}} E \tag{4.12}$$

$$\text{Block force: } F_{\text{bl}} = -A Y^{\text{Lam}} d_{31}^{\text{Lam}} E \tag{4.13}$$

The equations defining the electromechanical coupling of the piezoelectric structural fiber can now be applied to study the effect the fiber geometry has on the response of the fiber. The free strain equation can then be used in finite element analysis (FEA) or experiments to validate the theoretically predicted electromechanical coupling along the fiber length.

4.2.2 Three-Dimensional Micromechanics Modeling

The reduced order model presented above will be shown to provide excellent prediction on the composite properties in the fiber axis in the subsequent sections of this chapter; however, it is not capable of predicting the response in the out-of-plane axis due to limitation in the rule of mixtures. Thus, a three-dimensional model has also been developed to extend the double inclusion approach for the prediction of the entire set of electroelastic constitutive properties for the multiphase piezoelectric composite. Considering a transversely isotropic piezoelectric material, the linear constitutive equations used to describe the coupled interaction between the electrical and mechanical variables can be expressed as (Odegard 2004)

$$\sigma_{ij} = C_{ijmn}\varepsilon_{mn} - e_{nij}E_n \tag{4.14}$$

$$D_i = e_{imn}\varepsilon_{mn} + \kappa_{in}E_n \tag{4.15}$$

where σ_{ij}, ε_{mn}, E_n, and D_i are the stress, strain, electric field, and electric displacement tensors, respectively, and C_{ijmn}, e_{nij}, and κ_{in} are elastic (at a constant electric field), piezoelectric field-stress (in a constant strain or electric field), and dielectric (at a constant strain) tensors, respectively.

For the modeling of inhomogeneous composites with piezoelectric inclusions, it is convenient to combine the mechanical and electrical variables such that the two equations can be expressed in a single constitutive equation (Dunn and Taya 1993). This notation is identical to conventional indicial notation with the exception that lowercase subscripts are in the range of 1–3, while the capitalized subscripts are in the range of 1–4 and repeated capitalized subscripts summed over 1–4. With this notation, the elastic strain and electric field can be expressed as

$$Z_{Mn} = \begin{cases} \varepsilon_{mn}, & M = 1,2,3, \\ E_n, & M = 4 \end{cases} \tag{4.16}$$

Similarly, the stress and electric displacement can be represented as

$$\Sigma_{iJ} = \begin{cases} \sigma_{ij}, & J = 1,2,3, \\ D_i, & J = 4 \end{cases} \tag{4.17}$$

the electroelastic moduli can then be presented as

$$E_{iJMn} = \begin{cases} C_{ijmn} & J,M = 1,2,3, \\ e_{nij} & J = 1,2,3; M = 4 \\ e_{imn} & J = 4; M = 1,2,3, \\ \kappa_{in} & J,M = 4, \end{cases} \tag{4.18}$$

Therefore, according to Equations 4.16 through 4.18, the piezoelectric constitutive Equations 4.14 and 4.15 can be combined into a single expression as

$$\Sigma_{iJ} = E_{iJMn} Z_{Mn} \tag{4.19}$$

Considering the orthotropic nature of the piezoceramic material and the core fiber, Equation 4.19 can be expressed in matrix form as

$$\begin{bmatrix} \sigma_{11} \\ \sigma_{22} \\ \sigma_{33} \\ \sigma_{23} \\ \sigma_{13} \\ \sigma_{12} \\ D_1 \\ D_2 \\ D_3 \end{bmatrix} = \begin{bmatrix} C_{11} & C_{12} & C_{13} & 0 & 0 & 0 & 0 & 0 & -e_{31} \\ C_{12} & C_{22} & C_{23} & 0 & 0 & 0 & 0 & 0 & -e_{32} \\ C_{13} & C_{23} & C_{33} & 0 & 0 & 0 & 0 & 0 & -e_{33} \\ 0 & 0 & 0 & C_{44} & 0 & 0 & 0 & -e_{15} & 0 \\ 0 & 0 & 0 & 0 & C_{55} & 0 & -e_{15} & 0 & 0 \\ 0 & 0 & 0 & 0 & 0 & C_{66} & 0 & 0 & 0 \\ 0 & 0 & 0 & 0 & e_{15} & 0 & \kappa_1 & 0 & 0 \\ 0 & 0 & 0 & e_{15} & 0 & 0 & 0 & \kappa_2 & 0 \\ e_{31} & e_{32} & e_{33} & 0 & 0 & 0 & 0 & 0 & \kappa_3 \end{bmatrix} \begin{bmatrix} \varepsilon_{11} \\ \varepsilon_{22} \\ \varepsilon_{33} \\ \gamma_{23} \\ \gamma_{13} \\ \gamma_{12} \\ E_1 \\ E_2 \\ E_3 \end{bmatrix} \tag{4.20}$$

Assuming perfect bonding between all phases in the composite, the general expression of the volume averaged piezoelectric fields in the multiphase active composites can be expressed as (Odegard 2004)

$$\bar{\Sigma} = \sum_{r=1}^{N} c_r \bar{\Sigma}_r \tag{4.21}$$

$$\bar{Z} = \sum_{r=1}^{N} c_r \bar{Z}_r \tag{4.22}$$

where c is the volume fraction, the subscript r represents the r-th phase of the composites, with 1 representing the matrix phase, and the bars denote the volume average of the quantities. Considering a piezoelectric composite subjected to homogeneous elastic strain and electric potential boundary conditions, Z^0, the volume-averaged strain and electric field $\bar{Z}$ equals Z^0 (Dunn and Taya 1993). Therefore, Equation 4.22 can be represented as

$$\bar{\Sigma} = E\bar{Z} \tag{4.23}$$

noting that the volume averaged strain and electric field in the r-th phase is expressed as

$$\bar{Z}_r = A_r\bar{Z} \tag{4.24}$$

where A_r is the concentration tensor of phase r and has the following properties

$$\sum_{r=1}^{N} c_r A_r = I \tag{4.25}$$

where I is the fourth-order identity tensor. Combining Equations 4.21 through 4.25, the overall electroelastic modulus predicted by the double inclusion model can be expressed as

$$E = E_1 + \sum_{r=2}^{N} c_r(E_r - E_1)A_r \tag{4.26}$$

where E is the extended electroelastic matrix defined in Equation 4.18 and A is the concentration tensor, which is a function of Eshelby's tensor and the electroelastic properties of the each phase. For the double-inclusion model of the three-phase composites shown in Figure 4.1, the concentration tensor is defined as (Dunn and Ledbetter 1995)

$$\begin{aligned} A_3^{di} &= I + \Delta S\Phi_2 + S_3\Phi_3 \\ A_2^{di} &= I + \left[S_2 - \frac{c_3}{c_2}\Delta S \right]\Phi_2 + \frac{c_3}{c_2}\Delta S\Phi_3 \end{aligned} \tag{4.27}$$

where the S is Eshelby's tensor, which is a function of the inclusion geometry as well as the electroelastic properties of the matrix; the explicit expression for a fibrous inclusion can be found elsewhere (Dunn and Taya 1993); and

Φ is the fourth-order tensor, which is a function of Eshelby's tensor and the electroelastic properties of each phase. The expression of Φ is given by

$$\Phi_2 = -\left[\Delta S + (S_2 + F_3)\left(S_3 - \frac{c_3}{c_2}\Delta S + F_3\right)^{-1}\left(S_3 - \frac{c_3}{c_2}\Delta S + F_2\right)\right]^{-1}$$
$$\Phi_3 = -\left[(S_2 + F_3) + \Delta S\left(S_3 - \frac{c_3}{c_2}\Delta S + F_2\right)^{-1}\left(S_3 - \frac{c_3}{c_2}\Delta S + F_3\right)\right]^{-1} \tag{4.28}$$

where the F and ΔS are expressed as

$$F_2 = (E_2 - E_1)^{-1} E_1$$
$$F_3 = (E_3 - E_1)^{-1} E_1 \tag{4.29}$$
$$\Delta S = S_3 - S_2$$

The geometry of the ASF for the multifunctional composite is shown in Figure 4.1 with the coordinate system adopted here (Lin and Sodano 2008). Because the ASFs are poled along the transverse direction, the piezoelectric coupling e_{32} and e_{33} are along the radial direction of the piezoelectric shell, while e_{31} is along the longitudinal direction. To maintain consistency with Eshelby's tensor coordinates system (Dunn and Taya 1993, Odegard 2004), the actual electric field that is applied through the thickness of the piezoelectric layer (along radial direction) must be considered along with the global coordinate system. Note that the standard convention in composites defines the fiber direction as the –3 direction; therefore, in accordance with this convention, the piezoelectric axes have been modified from the traditional directions such that the poling axis occurs in both the –1 and –2 directions due to the concentric electrodes.

A nonuniform local electric field occurs due to the concentric nature of the electrodes and has been evaluated by Lin and Sodano (2008). The authors found the relation between the local and the global electric field is defined as

$$E_{\text{local}} = \frac{1}{(1/\alpha - 0.5)\ln(1-\alpha)} E \tag{4.30}$$

where α is the aspect ratio defined as the ratio of piezoelectric shell thickness to the total radius of the ASF, E_{local} is the local electric field added through the thickness of the piezoelectric shell, and E is the electric field in the global coordinate system to be consistent with previous modeling analysis (Dunn and Taya 1993, Odegard 2004). The material properties used in the modeling are shown in Table 4.1 (Odegard 2004); owing to the orthotropic nature of the

TABLE 4.1

Electroelastic Properties of Reinforcement and Matrix Materials

Material	C_{11} (GPa)	C_{12} (GPa)	C_{13} (GPa)	C_{33} (GPa)	C_{44} (GPa)	C_{66} (GPa)	κ_{11}	κ_{33}	e_{15} (C/m^2)	e_{31} (C/m^2)	e_{33} (C/m^2)
Matrix	8.1	5.4	5.4	8.1	1.4	1.4	2.8	2.8	0	0	0
Carbon_ Fiber	24	9.7	6.7	11	27	11	12	12	0	0	0
PZT-7A	148	76.2	74.2	131	25.4	35.9	460	235	9.5	−2.1	9.2

materials, not all the material properties listed in Equation 4.20 are shown. Note that a common piezoelectric material PZT 7A is used here for general consideration.

4.2.3 Three-Dimensional Finite Element Modeling

To validate the micromechanics models, finite element (FE) analysis of both the piezoelectric structural fiber and a composite containing the ASF was performed using ABAQUS (an FE analysis/modeling software by SIMULIA; www.3ds.com). Because the FE model predicts the stress and strain fields inside the inclusion, piezoelectric shell, and matrix, the predicted properties are very accurate (Odegard 2004). Therefore, the FE results can be used to check the accuracy of the three-dimensional micromechanics model developed. An example of the FE model used is shown in Figure 4.3. A series of simulations with different aspect ratio ASFs and volume fraction representative volume elements (RVEs) were performed to determine the eight independent electroelastic material parameters described in the previous section.

Both of the core fiber/piezoelectric shell and piezoelectric shell/epoxy matrix were constrained by the "TIE" command in ABAQUS, which results in zero relative motion between the contacted surfaces. A square RVE has been used because it is more easily modified to account for changing volume fraction, and both square and hexagonal arrays have been shown to return

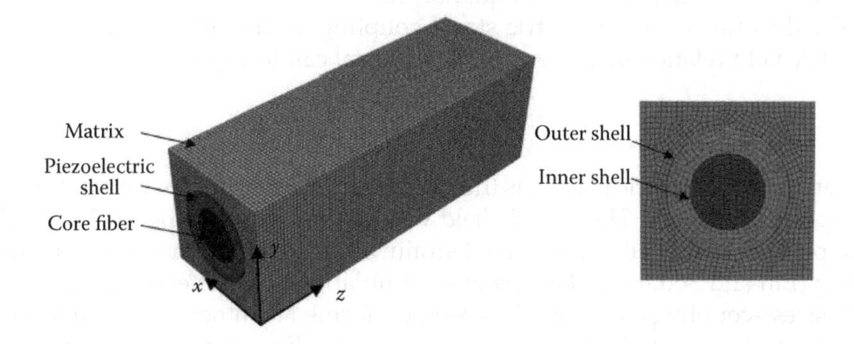

FIGURE 4.3
Finite element model of multifunctional composite with the front view shown on the right.

accurate results under the correct boundary conditions (Lee et al. 2005, Sun and Vaidya 1996, Berger et al. 2006). For each FEA, the volume fraction was obtained by holding the fiber in the same dimensions while adjusting the matrix size. The effective tensile modulus and shear modulus were calculated by the strain energy approach (Odegard 2004). The elastic strain energy of the RVE can be expressed as

$$U_s = \frac{V}{2} C_{ijkl} \varepsilon_{ij} \varepsilon_{kl} \tag{4.31}$$

where V is the volume of the RVE, C_{ijkl} is the effective elastic modulus, and ε_{ij} and ε_{kl} are the strain applied to the RVE. Therefore, the effective tensile modulus ($ij = kl = 11, 22, 33$) and the effective shear modulus ($ij = kl = 44, 55, 66$) can be calculated. Only C_{11}, C_{33}, C_{44}, and C_{66} were simulated due to the transverse isotropic nature of the material. Note that a, c are the dimensions of the FE cell on x and z axes, and the coordinate system adopted for our analysis defines the applied electric field in the radial direction or along the 1–2 plane; thus, the strain-coupling coefficient is defined here as d_{11}.

For the effective dielectric constant, the entire RVE was treated as a capacitor that can store electrostatic energy when exposed to an electric field applied to the surfaces of the RVE. The stored electrostatic energy can be expressed as (Chen et al. 2008)

$$U_e = \frac{1}{2} \kappa_{\text{eff}} \frac{s}{d} (\varphi_2 - \varphi_1)^2 \tag{4.32}$$

where κ_{eff} is the effective dielectric constant of the entire RVE, s is the area of the each surface exposed to the electric field, d is the distance between the two surfaces, φ_1 and φ_2 are the electric potential applied to them. The electrostatic energy was calculated in ABAQUS allowing the effective dielectric constant to be calculated from Equation 4.32.

For the effective piezoelectric strain-coupling coefficient, the free strain–electric field relation of a piezoelectric material can be expressed as

$$\varepsilon_{ij} = d_{nij} E_n \tag{4.33}$$

where E_n is the electric field, ε_{ij} is the strain, and d_{nij} is the piezoelectric strain-coupling coefficient. The electric field was applied through the thickness of the piezoelectric shell and the axial strain was calculated from the FE modeling (Lin and Sodano 2008). Likewise, simulation of the effective piezoelectric stress-coupling coefficient has been performed by other researchers and shown to be in good agreement between modeling and experiment data (Berger et al. 2006, Poizat and Sester 1999).

4.2.4 Modeling Results and Discussion

After the FEA, a validation has been performed through a comparison of the electroelastic properties predicted by the micromechanics model and the FE model. The effective Young's moduli of the multifunctional composites with different aspect ratios are shown in Figure 4.4. The fiber volume fraction is defined here as the ratio of the ASF (core structural fiber and piezoelectric shell) to the entire RVE. For the longitudinal Young's modulus Y_3, the model shows very good agreement with the FEA result for all aspect ratios and volume fractions considered and increased linearly with volume fraction. For transverse Young's modulus $Y_1 = Y_2$, the modulus increases exponentially with the FEA predicting a slightly larger rate of increase than the model. The model overestimates the transverse modulus slightly when the fiber volume fractions are lower than 50%, but underestimates when the fiber volume fractions are 50% and higher. The maximum error (30.6%) occurs when the aspect ratio is 0.8 with the fiber volume fraction of 70%. The three effective shear moduli are shown in Figure 4.5. Figure 4.5a shows the longitudinal shear $G_{13} = G_{12}$ modulus to be highly dependent on the volume fraction and fairly insensitive to the aspect ratio. The transverse shear modulus G_{12} shown in Figure 4.5b has a larger dependence on the aspect ratio and increases with volume fraction. The model shows close agreement with the FEA for both the longitudinal and transverse shear modulus, although the error increases at very high volume fractions. The increased difference at high aspect ratios and high volume fractions might be caused by the invalidity of electric field thin wall approximation. Because in micromechanics models, thin wall approximation was used to determine the electric field applied to ASFs, when the volume fraction of the piezoceramic shell is higher (higher aspect ratio and high volume faction situation), the inaccuracy of this assumption is magnified, leading to inaccurate prediction. Since this is a coupled electrical–mechanical modeling, errors in the electrical modeling part could also lead to inaccurate estimation in mechanical moduli.

In general, piezoelectric materials have high dielectric properties that make the three-phase piezoelectric composite studied here a strong candidate for multifunctional structural capacitor that combine energy storage with load-bearing capability (Chao et al. 2008, South et al. 2004). The longitudinal dielectric constants obtained from both the model and FEA are shown in Figure 4.6. The model has nearly perfect agreement with FEA for all aspect ratios and volume fractions. For the transverse dielectric constant shown in Figure 4.6b, the dielectric constants also increase with the increasing aspect ratio and volume fraction. The transverse dielectric constants acquired from FEA increase dramatically when the volume fraction is above 50%, especially for lower aspect ratio ASF composites. The maximum difference between the two models is 39.3% when the aspect ratio equals to 0.2 and volume fraction is 70%.

To nondimensionalize the results, the effective piezoelectric strain-coupling coefficient of the composite is presented as the ratio of the coupling of the multifunctional composite to that of the active constituent. This ratio thus provides

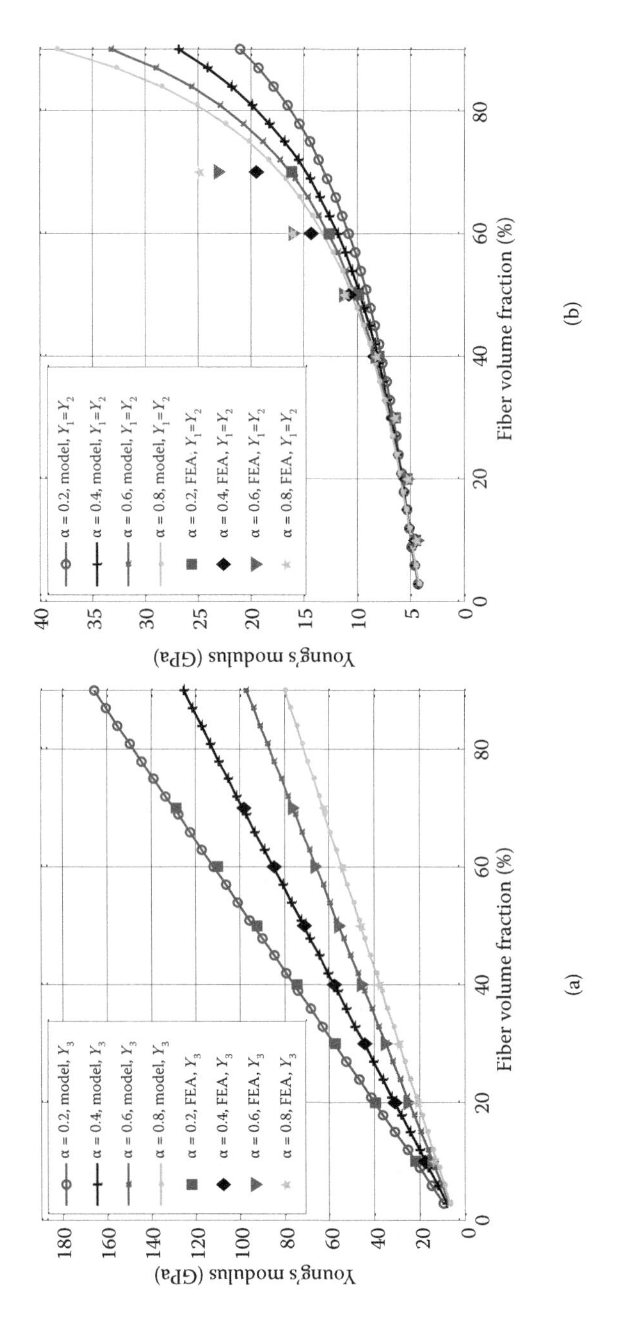

FIGURE 4.4
Effective Young's modulus with respect to ASF volume fraction: (a) longitudinal Young's modulus; (b) transverse Young's modulus.

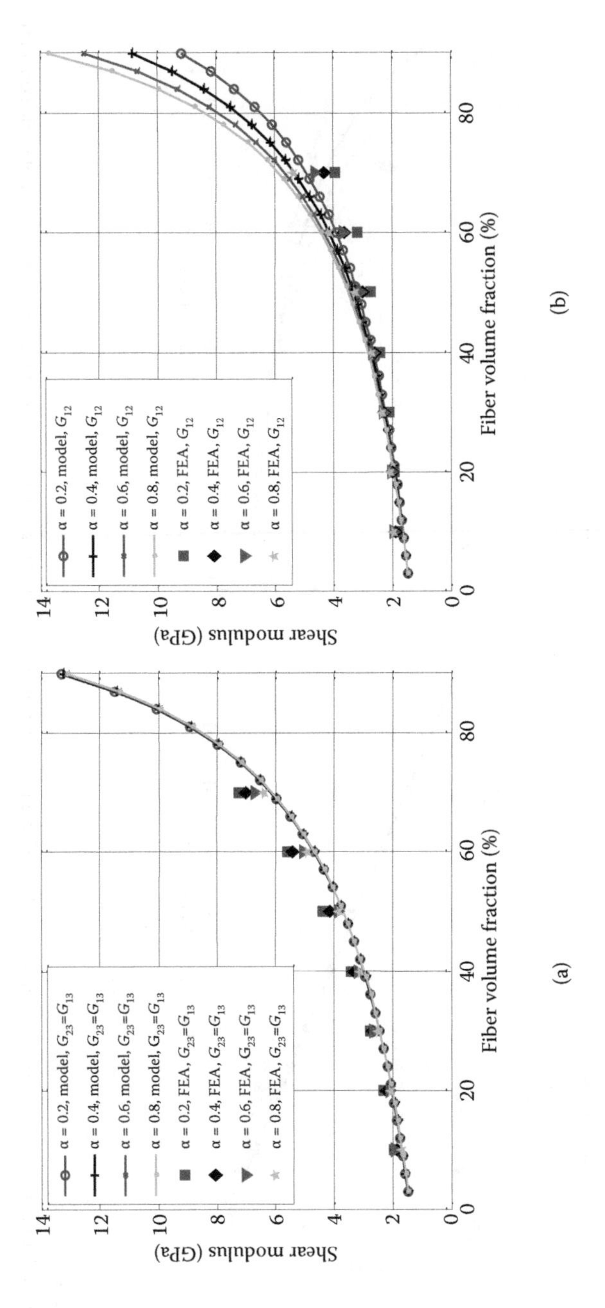

FIGURE 4.5
Effective shear modulus with respect to ASF volume fraction: (a) longitudinal shear modulus; (b) transverse shear modulus.

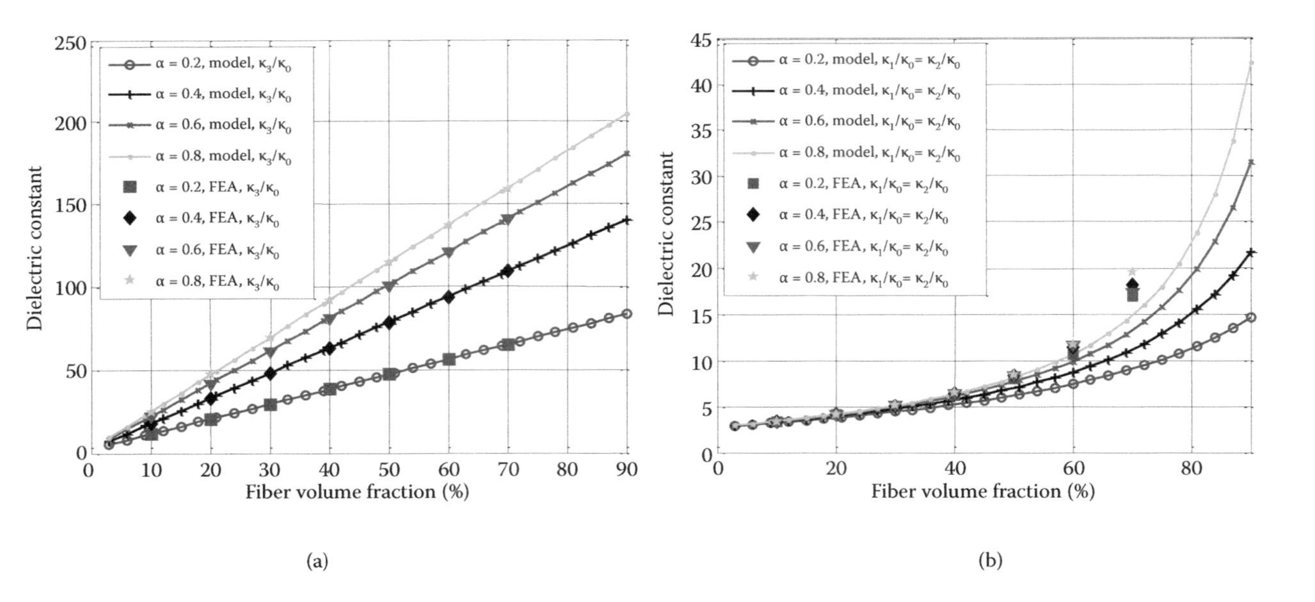

FIGURE 4.6
Effective relative dielectric constant with respect to ASF volume fraction: (a) longitudinal dielectric constant; (b) transverse dielectric constant.

relative performance of the bulk composite to a pure form of the piezoelectric material chosen. A coupling ratio of 70% indicates that a multifunctional composite with ASF reinforcement could achieve 70% of the coupling of the active constituent. The effective longitudinal piezoelectric coupling (along the fiber direction) is shown in Figure 4.7a. The longitudinal piezoelectric coupling ratio increases when the fiber volume fraction is lower than 20% before approaching saturation in the coupling. The maximum longitudinal piezoelectric coupling ratio is higher than 70% of the active continuant, with an aspect ratio of 0.8 and volume fraction of 70%. The high coupling response predicted indicates that structural composite laminates could be fabricated with higher coupling than many pure piezoelectric materials. For instance if PZT-5H4E (d_{13} = –320 pC/N) was used, the structural composite lamina with an aspect ratio of 0.8 and volume fraction of 60% would have a bulk coupling coefficient of greater than –224 pC/N or more than four times that of barium titanate (d_{13} = –49 pC/N). The high longitudinal piezoelectric coupling coefficient indicates the ASF is an excellent candidate for embedded power harvesting, structural sensing, and actuation.

The transverse piezoelectric coupling coefficients are shown in Figure 4.7b and indicate that the coupling increases at a very high rate for low volume fractions then increase at a nearly linear rate above a volume fraction of 20%. It is noted that the coordinate system adopted for our analysis defines the applied electric field in the radial direction or along in the 1–2 plane; thus, the strain-coupling coefficient is defined here as d_{11} (d_{33} in other literature). Similar to the longitudinal piezoelectric coupling ratio, the aspect ratio is a critical factor for the coupling strength. The transverse piezoelectric coupling coefficient increases with both higher aspect ratio and fiber volume fraction. Greater than 65% of the active constituent's coupling can be achieved when the aspect ratio is to 0.8 and the lamina has a fiber volume fraction of 70%. These results shows that the coupling in the longitudinal direction approaches saturation rather quickly allowing high coupling at low volume fractions.

Because both the one-dimensional and three-dimensional models can predict effective longitudinal piezoelectric coupling coefficient, it is interesting to compare the results from the two models. The resulting longitudinal electromechanical coupling of the multifunctional fiber predicted by both the one- and three-dimensional models are plotted in Figure 4.8 with respect to four different aspect ratios and fiber volume fraction ranging from 5% to 95%. The results from the one- and three-dimensional models are having nearly perfect agreement for all the aspect ratio samples along the entire fiber volume fraction range. The maximum difference between the two models is as low as 2.17% when the aspect ratio equals to 0.8 and volume fraction is 5%. Both models show that for all the aspect ratio ASFs, the effective piezoelectric coupling coefficients start to saturate at low fiber volume fraction, which is beneficial for application due to the difficulty to fabricate high fiber volume fraction (>60%) composites. Therefore, the one-dimensional model is an easy, fast, and accurate method to predict the effective piezoelectric coupling coefficients for the design of multifunctional composites.

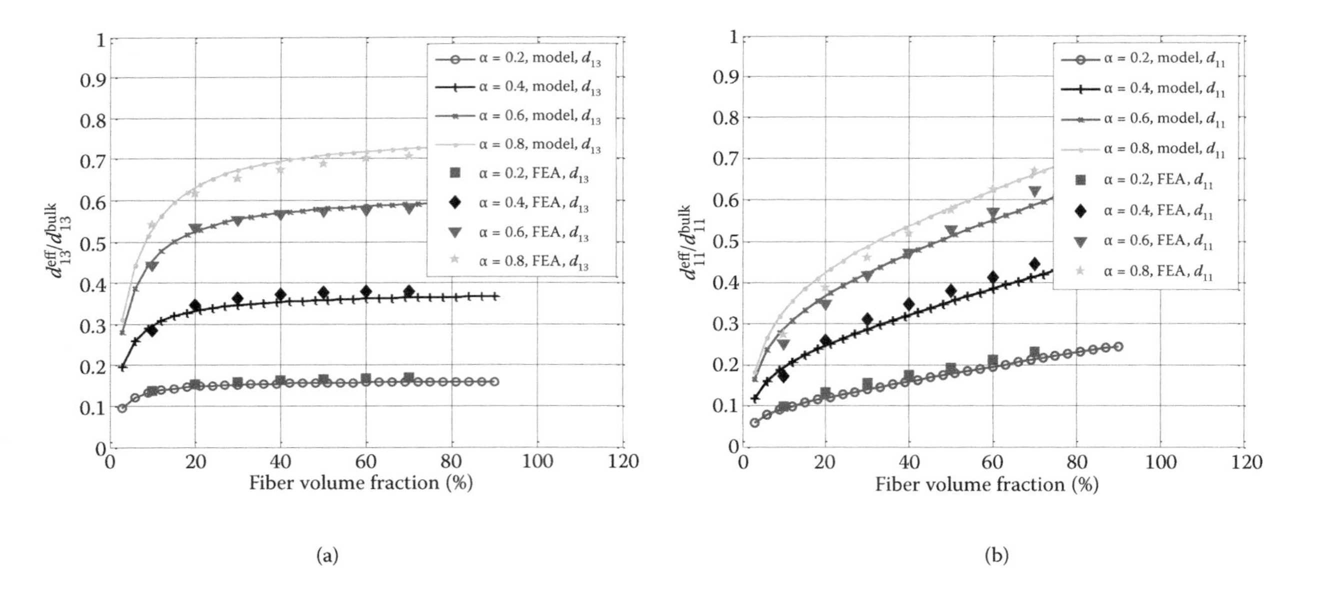

FIGURE 4.7
Effective piezoelectric coupling ratio with respect to ASF volume fraction: (a) longitudinal piezoelectric coupling ratio; (b) transverse piezoelectric coupling ratio.

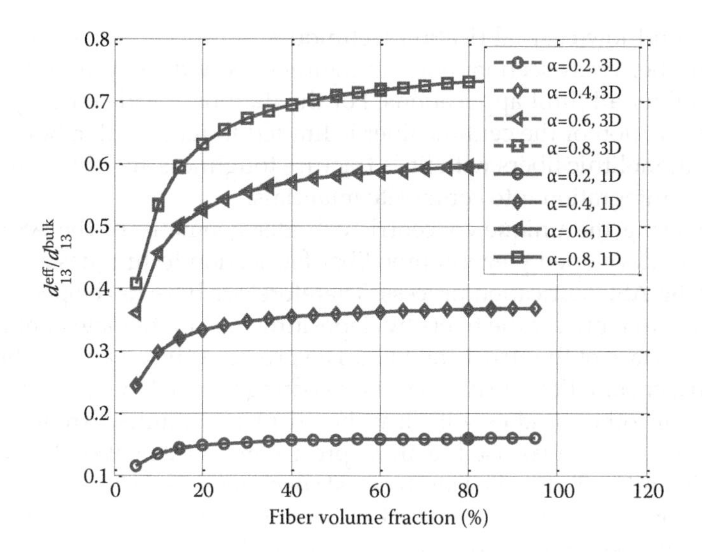

FIGURE 4.8
Comparison of one- and three-dimensional model results for effective piezoelectric coupling coefficients with respect to different fiber volume fraction and aspect ratios.

In this section, the results predicted by both models were validated by a three-dimensional FE model. For all the electroelastic properties, good agreement between the micromechanics models and FE model has been shown. Therefore, the micromechanics models developed in this section are accurate enough to drive the geometry of the fibers for fabrication. For the effective longitudinal piezoelectric coupling coefficients, the two micromechanics models returned almost identical results for composites with any design parameters. This makes the one-dimensional model a simple but accurate tool to predict the effective piezoelectric coupling coefficient of the multifunctional composites with different design parameters. In the following section, laboratory fabrication techniques for the ASF as well as single fiber lamina will be developed. The effective coupling coefficients for single fiber lamina samples will be tested and used to validate those predicted by the micromechanics models. The validation will show the accuracy of the models and demonstrate the high effective coupling coefficients of this multifunctional composite design.

4.3 Fabrication and Electromechanical Characterization

To date, the fabrication techniques for piezoceramic fibers include extrusion (Bent and Hagood 1993a, Gentilman et al. 2003), dicing of monolithic wafers, and soft mold technology (Gebhardt 2000). Extrusion-based methods are easy and low-cost fabrication techniques; however, issues such as non-uniform fiber cross section, poor straightness, low density, and porosity of the final fiber all limit applications. For the diamond blade cutting method, the cross section of the ceramic fiber is limited to rectangular. Soft molding leads to monolithic fibers with little fracture toughness and thus are not well suited for integration into composite materials.

Considering the unique concentric cylinder geometry of the ASF, all the previously developed piezoceramic fiber fabrication techniques are not suitable for the ASF fabrication process. Therefore, we have developed a synthesis process to fabricate the fibers by depositing barium titanate nanoparticles onto the surface of the structural fiber. The process utilized an SiC fiber as the core and a barium titanate piezoceramic coating to avoid the reaction that was observed in other studies with lead based piezoceramics. An atomic force microscope (AFM) was used to allow precise measurement of the piezoelectric strain and such that the length of test specimens could be minimized. The accuracy of the models formulated in Section 4.2 will be validated and the synthesis process will be shown to be valid in the following section.

4.3.1 Active Structural Fiber Synthesis

The ASF used silicon carbide fibers (Type SCS-6, 140 μm diameter; Specialty Materials, Inc. Lowell, MA) as electrodes in the electrophoretic deposition (EPD) process. Barium titanate ($BaTiO_3$) powder was used as the piezoceramic constituent, because it is stable under high temperatures and has a high coupling coefficient. Commercial $BaTiO_3$ nanopowder was used as the starting material ($BaTiO_3$, 99.95%, average particle size: 100 nm; Inframat Advanced Materials, Farmington, CT). The EPD process is schematically shown in Figure 4.9a. The powder was deposited on the fiber using an EPD process in which 3 wt.% of $BaTiO_3$ powder was dispersed in 200 mL organic

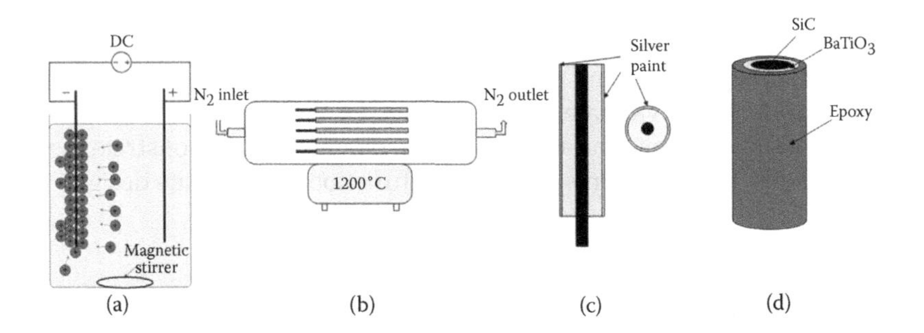

FIGURE 4.9
Schematic of the ASF fabrication processes: (a) piezoceramic layer applied using EPD process; (b) green ceramic layer sintered under inert atmosphere to achieve full density; (c) silver paint applied to form second electrode and subsequently poled at 120°C; (d) epoxy coating applied to each fiber with various thicknesses to obtain desired volume fraction.

solvent mixture composed of acetone and ethanol (1:1 volume ratio, 99.5%; VWR, Radnor, Pennsylvania). To fully positively charge the $BaTiO_3$ particles, the suspension was stirred for 30 min before performing the EPD process. Two 75-mm-long SiC fibers with a 20-mm distance were used as cathode and anode. Since the $BaTiO_3$ particles were positively charged, the deposition occurred on the cathode (Lin and Sodano 2009).

After the application of the green piezoceramic coating, the SiC fibers must be sintered at high temperature to reach full density to maximize the piezoelectric coupling. To protect the SiC core fibers from oxidization, the fibers were sintered in a tube furnace (Thermolyne 79400, Thermo Fisher Scientific, Waltham, Massachusetts) at 1200°C under a nitrogen gas atmosphere. Both ramp up and down rates for the sintering process were set to 6°C/min, which was necessary to ensure crack-free coatings. The crystal structures of the $BaTiO_3$ coating before and after sintering were checked by XRD (Scintag XDS 2000, Scintag Inc., Cupertino, CA). The microstructure and the thickness of the $BaTiO_3$ layer were examined by a field emission scanning electron microscope (Hitachi S-4700, Hitachi High-Tech, Tokyo, Japan).

After sintering the fibers, the outer surface of the $BaTiO_3$ layer was coated with silver paint (SPI Supplies, #5002, West Chester, PA) to form the outer electrode. The silver-coated fibers were heated to 500°C in the tube furnace under a nitrogen atmosphere to anneal the paint onto the ceramic surface. The inner SiC core fiber was used as the other electrode, and the $BaTiO_3$ layer was then transversely poled. For bulk $BaTiO_3$, the poling process can be done under DC electric field (2 kV/cm) at its Curie temperature (120°C) (Li and Shih 1997). Owing to the fiber geometry, the electric field on the inside edge of $BaTiO_3$ is always higher than that of the outside edge; therefore, in order to produce an adequate field on the outside edge of the $BaTiO_3$ coating during poling, it is necessary to apply a higher electric field compared with thin film or plate. For the poling process here, the electric field used was five times higher than that for bulk planar $BaTiO_3$. The coated fibers were poled in a silicon oil bath (Sigma-Aldrich, Milwaukee, WI) at 120°C, and the electric field used was 10 kV/cm with a 60 min holding time. To prevent depoling, the electric field was kept until the fibers were cooled down to room temperature.

Once the ASF was fabricated and poled, the single fiber lamina was fabricated by applying an epoxy layer of a specific thickness to achieve a desired volume fraction. Epon 862 resin and Epikure 9553 harder (100:16.9 by weight; Momentive, Houston, TX) were used as the epoxy coating layer, which is cured under room temperature for 24 h. This epoxy was chosen owing to its room temperature curing property that was critical for piezoceramics that typically have a low Curie temperature above which the piezoelectric will depolarize itself. The ASF lamina was formed by dip coating the fiber with epoxy and carefully smoothing the surface along the entire fiber length, resulting in a concentric cylinder single fiber lamina sample. The fabrication processes for both the ASF and the lamina containing a single ASF and polymer matrix are shown in Figure 4.9.

4.3.2 Characterization of Electrophoretic Deposition and Coating Microstructure

The particle sizes under different sintering conditions are shown in Figures 4.10 and 4.11. For fibers sintered at 1200°C for 1 h, the grain size increased to 500 nm; however, as shown in Figures 4.10b and 4.11a, the $BaTiO_3$ coating is still porous indicating the $BaTiO_3$ coating density is lower than the theoretical value. As the sintering time is increased further, the porosity is reduced until full density is achieved after 3 h. The coating thickness decreases significantly with the increasing sintering holding time, as shown in Figure 4.11. It should be noted that the coating parameter used in Figure 4.11 was 10 V for 1 min and the scale bar is constant for each figure. The reduced cross section is indicative of the high porosity before densification. The diameters of the $BaTiO_3$ coating shown in Figure 4.11 before and after 3 h sintering are 88 and 14 μm, respectively, indicating the density has increased by nearly four times.

The fiber volume fraction of the single fiber lamina sample is defined as the volume ratio between the ASF and the entire single fiber lamina. The fiber volume fraction is controlled by the number of times the epoxy coating is applied and for our testing varied from 11% to 64%. The cross section of typical fibers tested is shown in Figure 4.12. From the cross section of the samples, the ASF aspect ratio and fiber volume fraction can be measured; the samples in Figure 4.12a through 4.12c have aspect ratios of 0.21, 0.42, and 0.60, while the fiber volume fractions are 20.5%, 28.0%, and 31.2%, respectively. Figure 4.12d shows the side view of the sample and the uniformity of the epoxy coating along the length of the ASF.

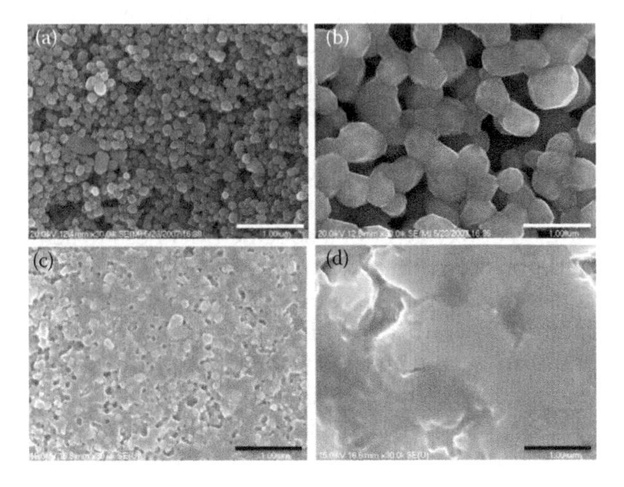

FIGURE 4.10
Microstructure of $BaTiO_3$ coating under different sintering conditions: (a) as deposited; (b) sintered for 1 h; (c) sintered for 2 h; (d) sintered for 3 h. Scale bar: 1 μm.

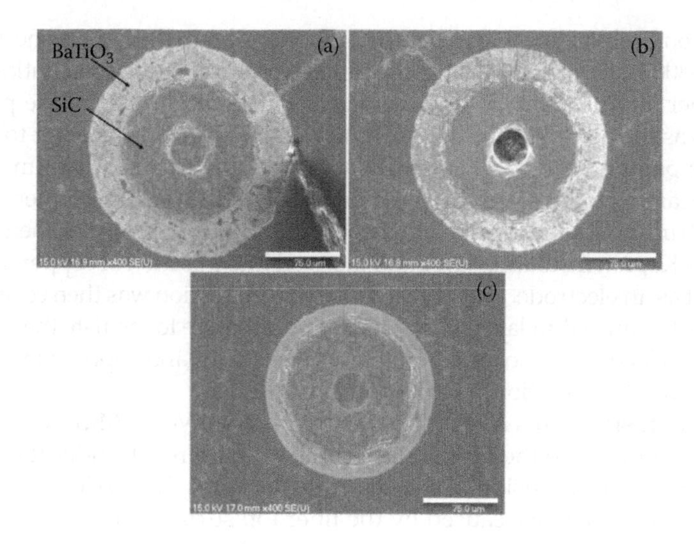

FIGURE 4.11
Cross section of the ASF under different sintering conditions: (a) sintered for 1 h; (b) sintered for 2 h; (c) sintered for 3 h. Scale bar: 75 μm.

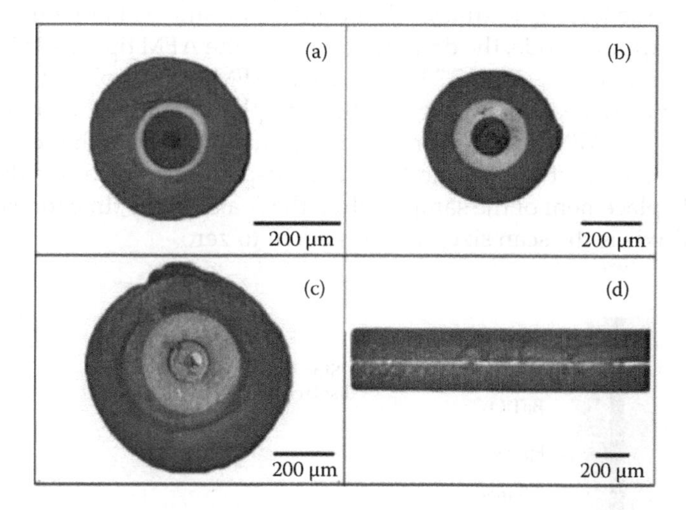

FIGURE 4.12
Cross section and side view of the samples: (a) aspect ratio = 0.21, fiber volume fraction = 20.5%; (b) aspect ratio = 0.42, fiber volume fraction = 28.0%; (c) aspect ratio = 0.60, fiber volume fraction = 31.2%; (d) side view of a sample shows its uniform epoxy coating.

4.3.3 Characterization of Single Fiber Lamina Coupling

For the single fiber lamina, the testing process uses the inverse piezoelectric effect to measure the fiber's electromechanical coupling. In this case, an AC voltage is applied to the active fiber and the resulting sinusoidal displacement is measured as a function of the applied field. The samples were first polished on both ends using a diamond lapping film (Allied, diamond lapping film, #50-30076) to form two flat and parallel surfaces. For the preparation of the single fiber lamina samples, approximately 2 mm of one end of the polished sample was immersed into chloroform to remove the epoxy coating to expose the silver paint for electrical connection. Half of the exposed barium titanate coating near the end was carefully removed, leaving 1 mm of the inner SiC core fiber protruding for electrical connection. The bare SiC fiber was then passed through a Kapton insulating layer and inserted into a section of copper tape that would act as an electrode. The exposed silver paint section was then coated with silver paint using a thin layer of cyanoacrylate adhesive to insulate the two electrodes. Two leads were soldered on to the silver paint and copper tape to form the electrodes for actuation, as shown in Figure 4.13a.

A Digital Instruments AFM (Digital Instruments/Veeco MultiMode AFM) was used to measure the fiber longitudinal displacement under the electric field applied on the inside and outside surfaces of the $BaTiO_3$ coating. To minimize possible noise caused by the fiber top surface roughness, the scan area was set to zero. The scan rate was set to 12.2 lines/s (total 512 lines per scan), and the excitation signal used to actuate the fiber was a sine wave (V_{pp} = 10 V, f = 200 mHz) generated by a function generator (Agilent, 33220A). To increase the testing accuracy, AFM used in the single fiber lamina testing was set in tapping mode, the drive frequency of the AFM tip was 352.68 kHz, and the scan rate was set to 5.29 lines/s. The AFM tip was placed on the top of the sample to measure the longitudinal displacement, as shown in Figure 4.13b. Since the AFM returns three-dimensional data while the scan area in the XY plane was set to zero and the Z axis displacement recorded the longitudinal displacement of the sample while the X axis is the time for each scan and the Y axis is the scan size, which is equal to zero.

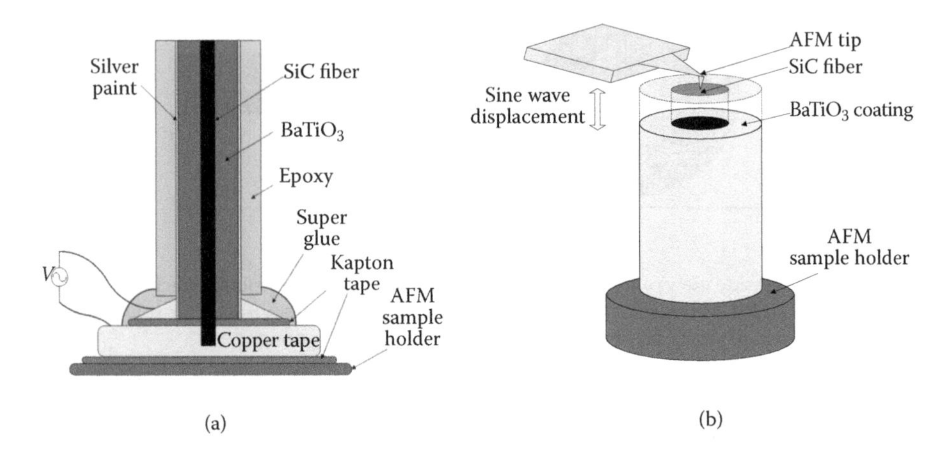

FIGURE 4.13
Schematic of (a) cross section of ASF with electroding for AFM testing and (b) effective piezoelectric coupling d_{31} experiment setup.

Because no external mechanical stress is applied to the sample during the testing, the piezoelectric constitutive equation for the mechanical response in the longitudinal direction can then be reduced and rewritten as

$$\varepsilon_1 = d_{31}^{\text{Lam}} E_3 \tag{4.34}$$

where ε_1 is the fiber longitudinal strain, d_{31}^{Lam} is the effective piezoelectric coupling of the ASF lamina, E_3 is the electric field derived by thin wall approximation, which can be expressed as $E_3 = V/t$, where V is the voltage applied across the coating thickness and t is the thickness of the $BaTiO_3$ coating. The effective piezoelectric constant of the single fiber lamina can be expressed as

$$d_{31}^{\text{Lam}} = \frac{\varepsilon_1}{E_3} = \frac{t \cdot \Delta l}{V \cdot l} \tag{4.35}$$

therefore, once the longitudinal displacement of the samples are measured, the effective coupling for single fiber lamina samples could be calculated from Equation 4.35. Typical displacement data measured by the AFM are shown in Figure 4.14, where Figure 4.14a is an *XZ* plane data (random *Y* coordinate) of Figure 4.14b. The results of the tested values are compared with those calculated through the model with respect to the fiber volume fraction and shown in Figure 4.15.

As can be seen in Figure 4.15, there is excellent agreement between the model and experimental measurements for samples with aspect ratio of 0.21 and 0.42 for the entire fiber volume fraction range. Although the tested d_{31} values for aspect ratio of 0.60 follow the same trend as that of the model, all the testing results fall below the estimated coupling for the entire fiber volume fraction range. The results are consistent with the previous findings of ASF testing. The low electromechanical coupling is attributed to low density

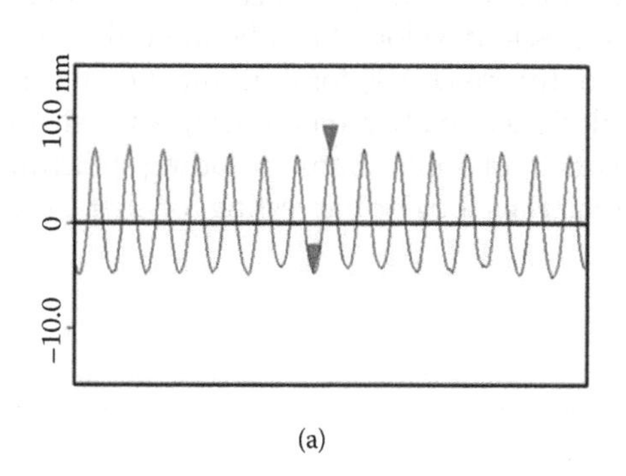

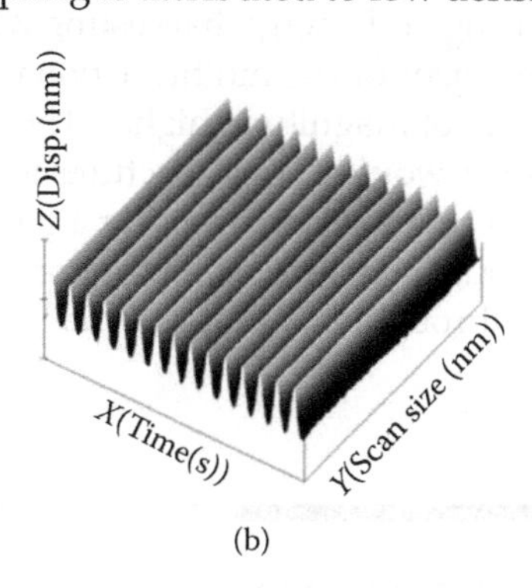

FIGURE 4.14
Single fiber lamina longitudinal displacement measured by tapping mode AFM: (a) longitudinal displacement; (b) three-dimensional data recorded by AFM.

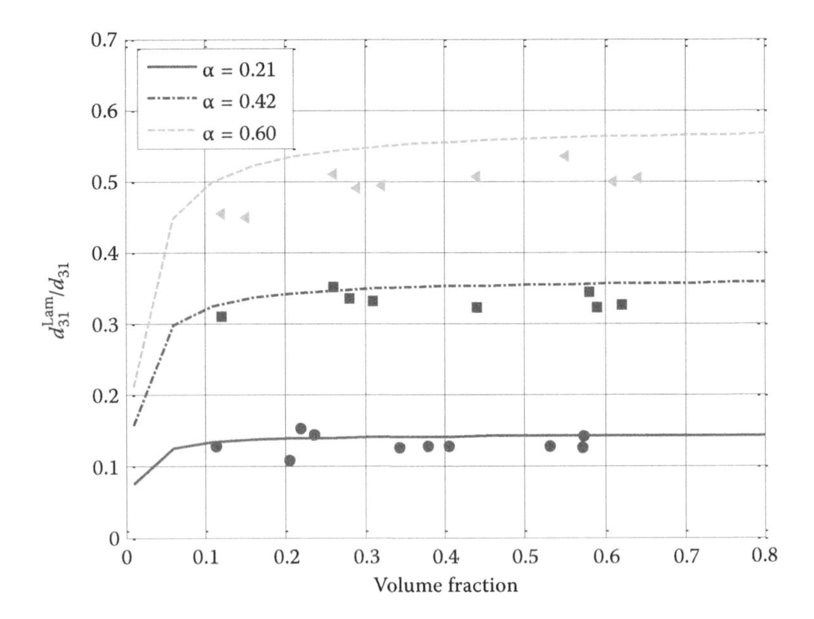

FIGURE 4.15
Comparison of micromechanics model and measured result for active lamina coupling with respect to fiber volume fraction of ASF.

of sintered $BaTiO_3$ layer and the partial poling of $BaTiO_3$ layer due to the electric field difference at the inner and outer walls.

The experimental analysis performed here has demonstrated the validity of the theoretical models developed in Section 4.2. The high coupling response indicates that structural composite laminates could be fabricated with higher coupling than many pure piezoelectric materials. Furthermore, when compared with ZnO, which has received significant attention for microscale and nanoscale sensing and energy harvesting (Qin et al. 2008, Wang et al. 2007), the coupling coefficient of the multifunctional composite developed can be more than two orders of magnitude higher. The use of this new ASF for multifunctional composites would allow structures to be designed to offer load bearing and sensing and actuation properties for a wide variety of applications, including structural sensing, actuation, self-monitoring materials, power harvesting, or shape control through anisotropic actuation.

4.4 Conclusions

The demand for material systems with multiple integrated functionalities is rapidly rising due to the enhanced performance and safety provided by the use of this type of new material. A multifunctional material is typically a composite, in which each phase provides unique performance-related functionality, such as structural reinforcement, sensing, thermal management, self-healing, actuation, energy storage, and power harvesting. However, composites have numerous intrinsic material failure mechanisms that make the detection of damage difficult. Furthermore, composites typically have low damping and poor electrical and thermal conductivity. To increase the performance of the composite with respect to the aforementioned limitations, numerous multifunctional materials have been developed although embedded sensing has draw extensive attention in past years.

Piezoceramic materials exhibit excellent coupling between energy in the mechanical and electric domains, which has led to their widespread use as sensing and actuation materials. However, in the monolithic form, the fragile nature of the material makes it difficult to apply to curved surfaces and easy to damage during handling or when deployed in harsh environments. To overcome the issues related with monolithic piezoceramics, piezoceramic fiber composites have been developed by embedding the fibrous form of the material into an epoxy matrix. The compliance of the epoxy protects the fiber from breaking under mechanic loading, allowing its application to curved surfaces. The piezoceramic fiber composites developed in the past have shown strong structural sensing and actuation capabilities; however, they have poor mechanical properties and provide little strength increase to the host structure. Additionally, these materials typically use surface-bonded interdigitated electrodes that require a patch of material rather than single fibers that greatly limit the piezoceramic fiber composites for embedment into host structures. These two issues limit the ability to embed the material and leave them as surface-bonded patches for sensing and actuation.

To overcome the limitations of previously developed piezoceramic fiber composites, this chapter has discussed the development of a new piezoceramic fiber composite with load-bearing functionality that can be embedded into a composite material. The ASF is fabricated by coating a high-strength carbon-based fiber with a piezoceramic layer, for which the structural fiber acts as the inner electrode for the piezoceramic layer as well as carries mechanical loading to protect the piezoceramic layer from breakage. The excellent mechanical strength and modulus of the structural fiber and superb sensing and actuation properties of the piezoceramic layer make this two-phase fiber an excellent candidate for the next-generation multifunctional material systems.

Acknowledgments

The authors gratefully acknowledge support from the Air Force Office of Scientific Research (award FA9550-08-1-0383) and the National Science

Foundation (grant no. CMMI-0700304). The authors also thank Dr. Gregory Odegard for the insightful discussion in three-dimensional modeling, and Specialty Materials, Inc. for providing the SiC fibers used for this work.

References

Bent, A.A. and Hagood, N.W. 1993a. "Development of piezoelectric fiber composites for structural actuation," In: *34th AIAA/ASME/ASCE/AH SDM Conference*, La Jolla, CA, April, pp. 3625–3638.

Bent, A.A. and Hagood, N.W. 1993b. "Development of piezoelectric fiber composites for structural actuation," In: *Proceeding of the 34th AIAA/ASME/ASCE/AHS Structures, Structural Dynamics and Materials Conference*, La Jolla, CA, pp. 3625–3638.

Bent, A.A., Hagood, N.W. and Rodgers, J.P. 1995. "Anisotropic actuation with piezoelectric fiber composites," *Journal of Intelligent Material Systems and Structures*, 6:338–349.

Berger, H., Kari, S., Gabbert, U., Rodriguez-Ramos, R., Bravo-Castillero, J., Guinovart-Diaz, R., Sabina, F.J. and Maugin, G.A. 2006. "Unit cell models of piezoelectric fiber composites for numerical and analytical calculation of effective properties," *Smart Materials and Structures*, 15:451–458.

Cannon, B.J. and Brei, D. 2000. "Feasibility study of microfabrication by coextrusion (MFCX) hollow fibers for active composites," *Journal of Intelligent Material Systems and Structures*, 11:659–669.

Chao, F., Bowler, N., Tan, X. and Kessler, M.R. 2008. "Three phase composites for multifunctional structural capacitors," In: *Proceedings of SAMPE Fall Technical Conference*, Memphis, TN, Sept. 8–11.

Chen, X., Cheng, Y., Wu, K., Meng, Y. and Wu, S. 2008. "Calculation of dielectric constant of two phase disordered composites by using FEM," In: *Conference Record of the 2008 IEEE International Symposium on Electrical Insulation*, pp. 215–218.

Dunn, M.L. and Ledbetter, H. 1995. "Elastic moduli of composites reinforced by multiphase particles," *Journal of Applied Mechanics*, 62:1023–1028.

Dunn, M.L. and Taya, M. 1993. "Micromechanics predictions of the effective electroelastic moduli of piezoelectric composites," *International Journal of Solids Structures*, 30:161–175.

Gebhardt, S., Schonecker, A., Steinhausen, R., Hauke, T., Seifert, W. and Beige, H. 2000. "Fine scale 1–3 composites fabricated by the soft model process: Preparation and modeling," *Ferroelectrics*, 241:67–73.

Gentilman, R., McNeal, K. and Schmidt, G. 2003. "Enhanced performance active fiber composites," In: *Proceeding of SPIE in Smart Structures and Materials*, San Diego, CA, pp. 350–359.

Halliday, D. and Resnick, R. 1988. *Fundamentals of Physics*. John Wiley & Sons, New York.

Hyer, M.W. 1998. *Stress Analysis of Fiber-Reinforced Composite Materials*. The McGraw-Hill Companies, New York.

Lee, J., Boyd, J.G. and Lagoudas, D.C. 2005. "Effective properties of three-phase electro-magneto-elastic composites," *International Journal of Engineering Science*, 43:790–825.

Li, X.P. and Shih, W.H. 1997. "Size effect in barium titanate particles and clusters," *Journal of the American Ceramic Society*, 80:2844–2852.

Lin, Y. and Sodano, H.A. 2008. "Concept and model of a piezoelectric structural fiber for multifunctional composites," *Composites Science and Technology*, 68:1911–1918.

Lin, Y. and Sodano, H.A. 2009. "Fabrication and electromechanical characterization of a piezoelectric structural fiber for multifunctional composites," *Advanced Functional Materials*, 19:592–598.

Odegard, G.M. 2004. "Constitutive modeling of piezoelectric polymer composites," *Acta Materialia*, 52:5315–5330.

Poizat, C. and Sester, M. 1999. "Effective properties of composites with embedded piezoelectric fibers," *Computational Materials Science*, 16:89–97.

Qin, Y., Wang, X. and Wang, Z.L. 2008. "Microfiber-nanowire hybrid structure for energy scavenging," *Nature*, 451:809–814.

South, J.T., Carter, R.H., Snyder, J.F., Hilton, C.D., O'Brien, D.J. and Wetzel, E.D. 2004. "Multifunctional power-generating and energy-storing structural composites for U.S. army applications," In: *Proceedings of the 2004 MRS Fall Conference*, Boston.

Sun, C.T. and Vaidya, R.S. 1996. "Prediction of composite properties from a representative volume element," *Composites Science and Technology*, 56:171–179.

Wang, X., Song, J., Liu, J. and Wang, Z.L. 2007. "Direct-current nanogenerator driven by ultrasonic waves," *Science*, 316:102–105.

Wilkie, W.K., Bryant, R.G., High, J.W., Fox, R.L., Hellbaum, R.F., Jalink, A., Little, B.D. and Mirick, P.H. 2000. "Low-cost piezocomposite actuator for structural control applications," In: *Proceedings of 7th SPIE International Symposium on Smart Structures and Materials*, Newport Beach, CA, March.

Williams, R.B., Park, G., Inman, D.J. and Wilkie, W.K. 2002. "An overview of composite actuators with piezoceramic fibers," *Proceeding of the 20th International Modal Analysis Conference*, Orlando, FL.

Section II

Structures

5

Mechanics and Design of Smart Composites: Modeling and Characterization of Piezoelectrically Actuated Bistable Composites

Chris Bowen and H. Alicia Kim

CONTENTS

5.1 Introduction

Asymmetric composite laminates can have multiple equilibrium states in which actuator materials such as piezoelectrics or shape-memory alloys (SMAs) can be used to induce "snap-through" between each stable shape. One advantage of this approach of inducing a shape change in composite structures is that a continuous energy input is not required to maintain a structural shape. Such structures have recently found interest for a number of applications, in particular shape change, morphing, and energy harvesting applications.

Owing to their asymmetric stacking sequence, asymmetric laminates exhibit an anisotropic response to the elevated cure temperatures experienced during manufacture. This leads to the development of residual thermal stresses on cooling to room temperature, resulting in a curved deformation. We will now consider the thermal distortion of an asymmetric laminate that leads to three static equilibrium states. Figure 5.1 shows a square $[0/90]_T$ laminate with the three possible states, a saddle shape (A) and two approximately cylindrical curvatures (B and C). It is well understood that the number of stable states is critically dependent on the geometry of the laminates, and a plate with a low edge-to-thickness ratio exhibits only one saddle equilibrium state. However, a plate with an edge-to-thickness ratio greater than a threshold value of ~84 [1] develops three equilibrium states where the saddle state is no longer stable and two cylindrical states become stable. This is shown in the bifurcation diagram of Figure 5.1D and 5.1E. The out-of-plane displacement and shape change between these two states can be appreciable and does not continuously induce a high stress in the material, which is favorable in terms of minimizing failure of fatigue. The snap-through between these two stable states is generally achieved by applying an in-plane strain that can be induced by piezoelectric materials or SMAs.

This chapter will concentrate on using piezoelectric actuation in the form of a macro-fiber composite (MFC) [2], which consists of piezoelectric elements with a fine-scale interdigitated electrode to apply the electric field along the length of the piezoelectric elements and induce a strain related to the d_{33} piezoelectric coefficient (strain per unit electric field). The composite configuration of interest can then be defined as Figure 5.2. The laminate is of arbitrary orthogonal layup as this offers the greatest out-of-plane deflection between two states. There is, therefore, an even number of plies of $2n$ with angle θ_i. All plies are made from the same material except for the top and the bottom layers, which are piezoelectric, denoted by subscript "p."

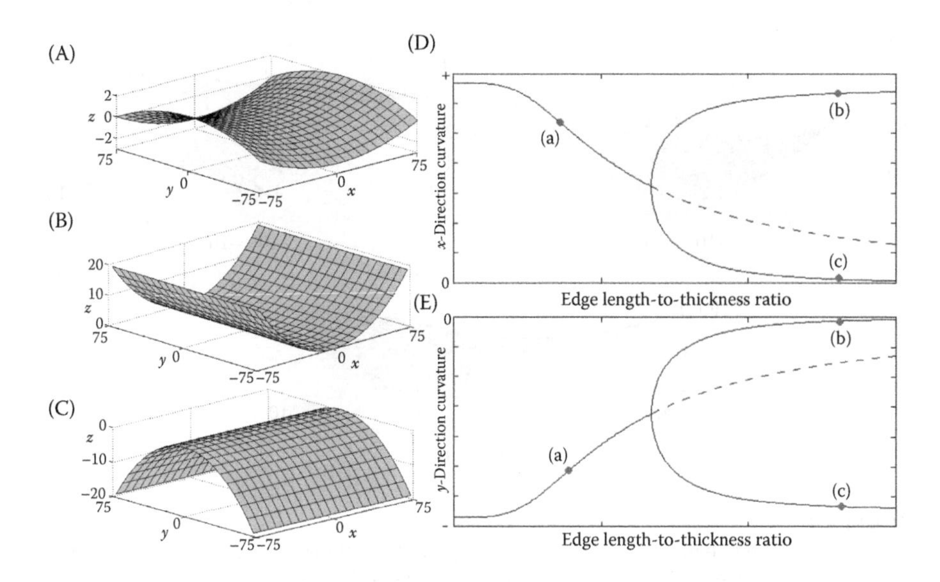

FIGURE 5.1
Equilibrium states of a $[0/90]_T$ laminate: (A) saddle shape; (B) cylindrical shape 1; (C) cylindrical shape 2; (D) bifurcation diagram with curvature in x-direction; (E) bifurcation diagram with curvature in y-direction. (From Betts, D.N. et al., *Applied Physics Letters* 100, 114104, 2012. With permission.)

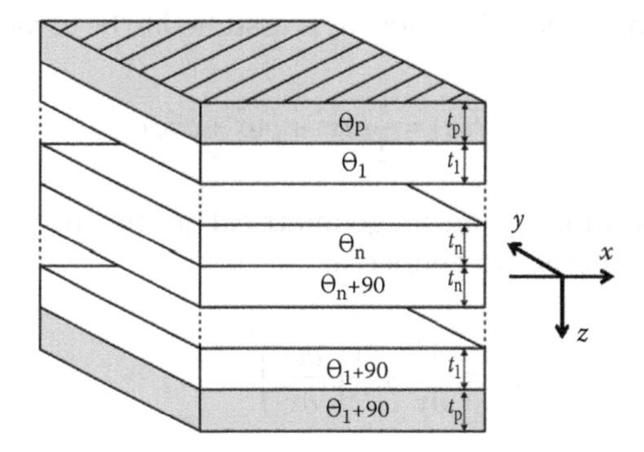

FIGURE 5.2
Orthogonal piezoelectric-composite laminate configuration. (From Betts, D.N. et al., *Applied Physics Letters* 100, 114104, 2012. With permission.)

5.2 Analytical Model

5.2.1 Nonlinear Bistable Composite Model (without Piezoelectric Elements)

The analytical model commonly used for bistable asymmetric laminates was first introduced by Hyer [3]. It has since been modified and developed by a

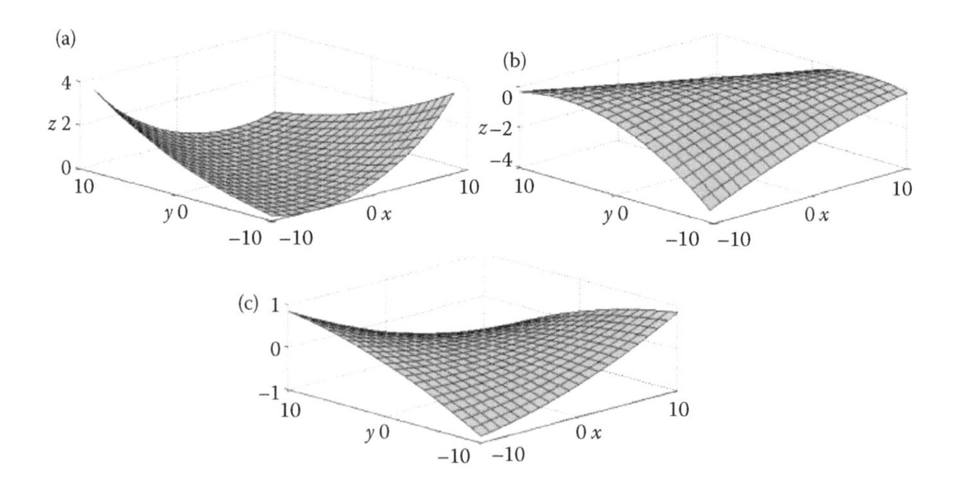

FIGURE 5.3
Shapes of a $[-30/60]_T$ laminate: (a) cylindrical shape 1; (b) cylindrical shape 2; (c) saddle shape. (From Betts, D.N. et al., *Applied Physics Letters* 100, 114104, 2012. With permission.)

number of researchers, and we will follow the model by Dano and Hyer [4] in this chapter. Figure 5.3 shows the three equilibrium states of a generalized asymmetric laminate with a coordinate system in which its origin sits at the geometric center and the ply orientations are measured from the x-axis. The out-of-plane displacement in the z-direction, w, is assumed to be of the quadratic form

$$w(x,y) = \frac{1}{2}(ax^2 + by^2 + cxy) \tag{5.1}$$

The midplane strains, including geometrical nonlinearity according to the von Karman hypothesis, are defined as

$$\begin{aligned}
\varepsilon_x^0 &= \frac{\partial u^0}{\partial x} + \frac{1}{2}\left(\frac{\partial w}{\partial x}\right)^2 \\
\varepsilon_y^0 &= \frac{\partial v^0}{\partial y} + \frac{1}{2}\left(\frac{\partial w}{\partial y}\right)^2 \\
\varepsilon_{xy}^0 &= \frac{\partial u^0}{\partial y} + \frac{\partial v^0}{\partial x} + \frac{1}{2}\frac{\partial w}{\partial x}\frac{\partial w}{\partial y}
\end{aligned} \tag{5.2}$$

where u^0 and v^0 are the in-plane displacements in the x- and y-directions, respectively. The midplane strains are approximated by third-order polynomials. It has been found that terms with powers of x and y that sum to an odd number are always zero. Therefore, the form of the midplane strains can be reduced to the polynomials of Equation 5.3.

$$\begin{aligned} \varepsilon_x^0 &= d_1 + d_2x^2 + d_3xy + d_4y^2 \\ \varepsilon_y^0 &= d_5 + d_6x^2 + d_7xy + d_8y^2 \end{aligned} \tag{5.3}$$

Using Equations 5.1 through 5.3 and introducing the additional shape coefficients d_{9-11} resulting from integration of the midplane strains, expressions for the in-plane displacements u^0 and v^0 can be determined.

$$\begin{aligned} u^0(x,y) &= d_1x + d_9y + \frac{1}{2}\left(d_3 - \frac{1}{2}ac\right)x^2y + \left(d_4 - \frac{c^2}{8}\right)xy^2 \\ &\quad + \frac{1}{3}\left(d_2 - \frac{1}{2}a^2\right)x^3 + \frac{1}{3}d_{11}y^3 \\ v^0(x,y) &= d_9x + d_5y + \frac{1}{2}\left(d_7 - \frac{1}{2}bc\right)xy^2 + \left(d_6 - \frac{c^2}{8}\right)x^2y \\ &\quad + \frac{1}{3}\left(d_8 - \frac{1}{2}b^2\right)y^3 + \frac{1}{3}d_{10}x^3 \end{aligned} \tag{5.4}$$

The total strain energy of the laminate, W, can then be expressed as the integral of strain energy density over the volume of the laminate.

$$W = \int_{-L_x/2}^{L_x/2}\int_{-L_y/2}^{L_y/2}\int_{-H/2}^{H/2} \frac{1}{2}c_{ijkl}\varepsilon_{ij}\varepsilon_{kl} - \hat{\alpha}_{ij}\varepsilon_{ij}\Delta T \, dx dy dz \tag{5.5}$$

where the subscripts ij and kl refer to all combinations of x, y, and xy directions, c_{ijkl}'s are elastic constants, $\hat{\alpha}_{ij}$'s are constants relating to the thermal expansion coefficients, L_x and L_y are the planform side lengths of the laminate, H is the total laminate thickness, ΔT is the temperature change from cure, and ε_{ij}'s and ε_{kl}'s are the total strains defined as

$$\begin{aligned} \varepsilon_x &= \varepsilon_x^0 - za \\ \varepsilon_y &= \varepsilon_y^0 - zb \\ \varepsilon_{xy} &= \varepsilon_{xy}^0 - zc \end{aligned} \tag{5.6}$$

where z is the distance from the laminate midplane. Expansion of Equation 5.5 results in an expression for the total energy, which is a function of the material and geometric properties, the temperature change from cure, and the set of shape coefficients a, b, c, $d_1 \ldots d_{11}$.

For equilibrium, the minimum energy states require

$$f_i = \frac{\partial W}{\partial e_i} = 0; \quad i = 1 \ldots 14 \tag{5.7}$$

where e_i's are the shape coefficients a, b, c, $d_1 \ldots d_{11}$.

Stable equilibrium states are identified by positive definite Jacobian of the solutions

$$J = \frac{\partial(f_1, f_2, \dots, f_{14})}{\partial e_i} = 0; \quad i = 1\dots 14 \tag{5.8}$$

5.2.2 Solution

The model results in 14 nonlinear equations to be solved for 14 shape coefficients. This system can be further reduced given the orthogonality of the laminate. The computation of strain energy in Equation 5.5 requires a representation of stiffness of the material system. For composite laminates, the stiffness is typically represented in terms of in-plane (A), coupling (B), and flexural (D) stiffness matrices, which are in terms of transformed stiffness terms, $\bar{Q}_{ij}$, to account for ply angle orientations. For example,

$$A_{ij} = \int_{-H/2}^{H/2} \bar{Q}_{ij} \, \mathrm{d}z \tag{5.9}$$

In the equivalent discrete form, A_{11} and A_{22} can be expressed as follows:

$$\begin{aligned} A_{11} &= \sum \bar{Q}_{11,i} t_i \\ A_{22} &= \sum \bar{Q}_{22,i} t_i \end{aligned} \tag{5.10}$$

where i denotes the ply number and

$$\begin{aligned} \bar{Q}_{11} &= Q_{11}m^4 + 2(Q_{12} + 2Q_{66})n^2m^2 + Q_{22}n^4 \\ \bar{Q}_{22} &= Q_{11}n^4 + 2(Q_{12} + 2Q_{66})n^2m^2 + Q_{22}m^4 \end{aligned} \tag{5.11}$$

where the Q's are terms of the stiffness matrix and m and n are $\cos\theta$ and $\sin\theta$, respectively. The orthogonality of the ply stacking sequence, hence, leads to

$$A_{11} = A_{22} \tag{5.12}$$

Similarly, a close examination of all stiffness matrix terms simplifies the **A**, **B**, and **D** matrices to the following simplified form:

$$\mathbf{A} = \begin{bmatrix} A_{11} & A_{12} & A_{16} \\ A_{12} & A_{11} & -A_{16} \\ A_{16} & -A_{16} & A_{66} \end{bmatrix}, \quad \mathbf{B} = \begin{bmatrix} B_{11} & 0 & B_{16} \\ 0 & -B_{11} & B_{16} \\ B_{16} & B_{16} & 0 \end{bmatrix}, \quad \mathbf{D} = \begin{bmatrix} D_{11} & D_{12} & D_{16} \\ D_{12} & D_{11} & -D_{16} \\ D_{16} & -D_{16} & D_{66} \end{bmatrix} \tag{5.13}$$

The similar reduction can also be applied to the thermal strain energy, where N_{xT}, N_{yT}, and N_{xyT} denote thermally induced forces and M_{xT}, M_{yT}, and M_{xyT} denote thermally induced moments.

$$[N_{xT}, M_{xT}] = \Delta T \int_{-H/2}^{H/2} \left(\bar{Q}_{11}\alpha_x + \bar{Q}_{12}\alpha_y + \bar{Q}_{16}\alpha_{xy}\right)[1, z]\,dz$$

$$[N_{yT}, M_{yT}] = \Delta T \int_{-H/2}^{H/2} \left(\bar{Q}_{12}\alpha_x + \bar{Q}_{22}\alpha_y + \bar{Q}_{26}\alpha_{xy}\right)[1, z]\,dz \tag{5.14}$$

$$[N_{xyT}, M_{xyT}] = \Delta T \int_{-H/2}^{H/2} \left(\bar{Q}_{16}\alpha_x + \bar{Q}_{26}\alpha_y + \bar{Q}_{66}\alpha_{xy}\right)[1, z]\,dz$$

where α's are the thermal expansion coefficients.

The following relationships can thus be derived:

$$N_{xT} = N_{yT}, \quad N_{xyT} = 0, \quad M_{xT} = M_{yT} \tag{5.15}$$

In addition, experience shows that the linear shape constants of Equation 5.3 exhibit the following relationships:

$$d_6 = d_4; \quad d_7 = d_3; \quad d_8 = d_2 \tag{5.16}$$

which reduces the system of 14 unknowns to 11 unknowns, where the unknown shape coefficients can be defined as $p_i = a, b, c, d_1, d_2, d_3, d_4, d_5, d_9, d_{10}$, and d_{11}. Thus, the equilibrium states can be defined by Equation 5.17.

$$f_i = \frac{\partial W}{\partial p_i} = 0; \quad i = 1 \ldots 11 \tag{5.17}$$

The system of equations with 11 unknown shape coefficients can be expressed in terms of its constitutive parts, namely the material stiffness properties, thermal properties, ply thicknesses, and ply orientations.

$$[\mathbf{U}][\mathbf{Z}][\mathbf{K}][\mathbf{T}] = [A_{11}\ A_{12}\ A_{16}\ A_{66}\ B_{11}\ B_{16}\ D_{11}\ D_{12}\ D_{16}\ D_{66}\ N_{xT}\ M_{xT}\ M_{xyT}] \tag{5.18}$$

where

[**U**] = material stiffness properties
[**Z**] = ply orientations
[**K**] = material thermal expansion properties
[**T**] = ply thickness terms

The equilibrium states of Equation 5.17 can thus be conveniently expressed as

$$f_i = [\mathbf{U}][\mathbf{Z}][\mathbf{K}][\mathbf{T}][\mathbf{P}]_i[\mathbf{L}]; \quad i = 1 \ldots 11 \tag{5.19}$$

where
- $[\mathbf{P}]_i$ = shape coefficient terms for each equation
- $[\mathbf{L}]$ = laminate edge length terms.

The matrices **U**, **Z**, **K**, **T**, and **L** are identical for all 11 equations and only the matrix **P** relating to the shape coefficients vary. It is also observed that the shape coefficients p_{4-11} are linear and independent of each other and only the remaining coefficients p_{1-3}, i.e., a, b, and c lead to nonlinear equations. The analytical solutions can therefore be derived for all shape coefficients. The solutions for nonlinear coefficients a, b, and c are shown below in terms of a set of constants γ, details of which are given in the Appendix for the reader

$$\begin{aligned} a &= \frac{-\gamma_{12}}{4\gamma_{11}} \pm \frac{\gamma_{20} \mp \sqrt{-(3\gamma_{15} + 2\gamma_{19} \pm 2\gamma_{16}/\gamma_{20})}}{2} \\ b &= \frac{\gamma_4 - \gamma_5 a \pm \sqrt{(\gamma_5^2 - 4\gamma_3^2)a^2 - (4\gamma_3\gamma_4 + 2\gamma_4\gamma_5)a + (\gamma_4^2 - 4\gamma_3\gamma_6)}}{2\gamma_3} \\ c &= \pm 2\sqrt{(ab + \beta_6/\beta_2)} \end{aligned} \tag{5.20}$$

5.2.3 Addition of Piezoelectrics to the Bistable Laminate

A piezoelectric actuator attached to the surface of the laminate is treated as an additional layer. However, since the actuator is attached after the curing process at room temperature, the thermal expansion coefficients are set to zero. The piezoelectric effect is included by adding a voltage-dependent strain energy term to the strain energy expression of Equation 5.5.

$$W = \int_{-L_x/2}^{L_x/2}\int_{-L_y/2}^{L_y/2}\int_{-H/2}^{H/2} \frac{1}{2} c_{ijkl}\varepsilon_{ij}\varepsilon_{kl} - \hat{\alpha}_{ij}\varepsilon_{ij}\Delta T - \hat{\beta}_{ij}\varepsilon_{ij}\Delta V \, dx dy dz \tag{5.21}$$

where $\hat{\beta}_{ij}$'s are constants related to piezoelectric coefficients and ΔV is the change in voltage.

Commercial MFC actuators may be limited in size; therefore, if the piezoelectric actuator is smaller than the laminate, different bounds may be applied to the piezoelectric layer.

$$W = W_{\text{lam}} + W_{\text{p}}$$
$$= \int_{z_{\text{lower}}}^{z_{\text{upper}}} \int_{-\frac{L_y}{2}}^{\frac{L_y}{2}} \int_{-\frac{L_x}{2}}^{\frac{L_x}{2}} \omega_{\text{lam}} \, dx\, dy\, dz + \int_{z_{\text{lower}}}^{z_{\text{upper}}} \int_{-\frac{L_{y,\text{p}}}{2}}^{\frac{L_{y,\text{p}}}{2}} \int_{-\frac{L_{x,\text{p}}}{2}}^{\frac{L_{x,\text{p}}}{2}} \omega_{\text{p}} \, dx\, dy\, dz \tag{5.22}$$

where the subscript "lam" refers to the laminate and "p" refers to the piezoelectric, and the integral bounds denote the relevant laminate and piezoelectric dimensions.

This is solved numerically and a good convergence can be obtained using the Newton–Raphson method with the analytical solution of Equation 5.20 as the initial solution. Two sets of solutions are found where there are two stable equilibria. Only one solution of curvature constants indicates that snap-through has occurred.

5.3 Experimental Investigation

5.3.1 Manufacturing of Piezoelectric–Laminate Combination

An asymmetric laminate such as $[0/90]_T$ carbon fiber/epoxy composite can be manufactured using carbon fiber prepreg sheet. The samples are laid on a nonstick pad and a thermocouple is inserted into the prepreg plies to monitor the laminate temperature during the cure cycle (typically ~180°C). Release film is placed over the sample and a breather layer is laid to assist in forming the vacuum during curing.

After curing the $[0/90]_T$ laminate, a piezoelectric (MFC) is bonded to its surface at room temperature with a two-part araldite. The material used for the MFC is a Navy Type II lead zirconate titanate (PZT), which is a relatively "soft" PZT type and exhibits high strain per unit electric field. On the basis of interdigitated electrodes with a 1-mm spacing, the maximum operating voltage is 1500 V (biased to the poling direction) with a maximum reported free strain per volt of 0.75 ppm/V at low electric field (<1 kV/mm) and 0.9 ppm/V at high field (>1 kV/mm) [2]. The higher strain per volt at higher electrical fields is a result of ferroelectric domain motion. Before attachment, the surfaces of both the MFC and carbon fiber composite are cleaned and the surface of the composite roughened to provide better mechanical adhesion. A small quantity of the adhesive is applied to the MFC. Once attached to the composite, the MFC and composite are placed under a weight to keep the composite and MFC flat and in good contact for 24 h while the epoxy cures. Figure 5.4 shows an example of the composite with an MFC and the two stable states.

FIGURE 5.4
Composite in state 1 with MFC attached. Inset shows the second stable state (state 2). (From Betts, D.N. et al., *Composite Structures* 92, 7, 1694–1700, 2010. With permission. Reprinted from *Elsevier Composite Structures*, 92, Giddings, P.F. et al., Bistable composite laminates: Effects of laminate composition on cured shape and response to thermal load, 2220–2225, Copyright 2010, with permission from Elsevier.)

5.3.2 Characterization Method

The three-dimensional shapes of a series of laminates were analyzed and compared against the analytical model. A standard three-dimensional motion analysis technique was employed to obtain a map of coordinates distributed on the laminate surfaces. Details of the experimental methodology can be found in Betts et al. [5].

Each of the square laminates of 150-mm edge length had a set of 145 round colored labels of 8-mm diameter attached to one surface (see Figure 5.5a). In addition to a $[0/90]_T$, a variety of other layups were also evaluated for comparison with the model predictions. Three digital video camera recorders were then set up in an umbrella configuration around a bistable laminate (Figure 5.5b). The cameras were positioned such that the best possible viewing angle could be achieved [6]. The laminates were positioned within an experimental volume, which was calibrated with a 20 × 20 × 10 mm wire frame. The camera views were restricted to a volume just slightly larger than the calibration frame, and the laminates were videotaped simultaneously by the three video cameras. The video clips were then analyzed and the surface coordinates were mapped using a Peak Motus motion analysis system (v. 8.5; Vicon, Los Angeles, California). The center of each of the round labels and the four corners of each laminate were manually digitized from all three camera views using a round cursor matching the size of the round labels. Digitized pixel information from each camera view was then combined with the calibration information to transform these to Cartesian coordinates of the laminate surfaces.

5.3.3 Characterization Results

Table 5.1 compares the experimental and analytical maximum out-of-plane displacements for five different laminates, without piezoelectrics attached. The

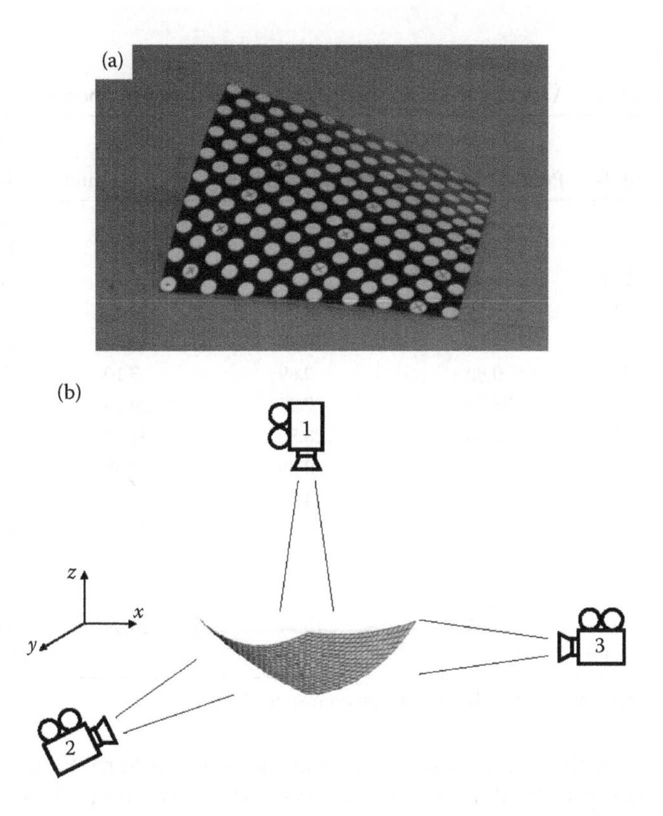

FIGURE 5.5
Three-dimensional shape analysis: (a) round labels distributed on the surface of a square [45/90] laminate; (b) experimental camera setup. (Reprinted from *Elsevier Composite Structures*, 92, Giddings, P.F. et al., Bistable composite laminates: Effects of laminate composition on cured shape and response to thermal load, 2220–2225, Copyright 2010, with permission from Elsevier.)

analytical model assumes a standard room temperature of 21°C. It is noted that the analytical model predicts the same curvature for both states while each state have different curvatures in practice; the difference between two state curvatures (2–1) are shown in Table 5.1. This is primarily due to a layer of resin that seeps through during manufacture (this is discussed further in Section 5.3.3.1). This additional resin layer adds stiffness asymmetrically leading to the difference in curvature between two states. The effect of this resin layer can be added to the analytical model, and the results are shown also in Table 5.1.

The measured Cartesian coordinates and the fitted surface are plotted alongside the laminate shape predicted by the analytical model. The analytical model and experimental shapes generally show good agreements for all laminates studied, and the $[-45/45]_T$ laminate is presented as a demonstrative example in Figure 5.6. The experimental data are offset only for illustrative purposes. This is more closely examined in Figures 5.7 and 5.8 where the curvature profiles in AB and CD are plotted. These profiles are selected as the errors at the corners are at their maximum. As the reference point for the comparison is set

TABLE 5.1

Experimental and Analytical Maximum Out-of-Plane Displacements

		Maximum Out-of-Plane Displacement (mm)			Error (%)	
Laminate	**State**	**Predicted (Ideal)**	**Predicted (Resin)**	**Experimental**	**Ideal**	**Resin**
[–45/45]	1	38.05	37.25	42.69	–10.9	–12.7
	2	38.05	37.99	43.39	–12.2	–12.4
	2–1	0.00	0.74	0.70	1.3	0.3
[–30/60]	1	35.51	34.76	38.14	–6.9	–8.9
	2	35.51	35.45	41.24	–13.9	–14.0
	2–1	0.00	0.69	3.10	7.0	5.1
[–15/75]	1	28.55	27.95	29.29	–2.5	–4.6
	2	28.55	28.50	31.67	–9.9	–10.0
	2–1	0.00	0.55	2.38	7.4	5.4
[45/90]	1	26.61	24.76	27.77	–4.2	–10.8
	2	19.03	19.29	22.85	–16.7	–15.6
	2–1	–7.58	–5.47	–4.92	12.5	4.8
[30/60]	1	14.63	12.56	12.77	14.6	–1.6
	2	14.63	15.94	15.09	–3.0	5.6
	2–1	0.00	3.38	2.32	17.6	7.2

Note: 2–1 corresponds to the difference between states 1 and 2.

at the center of the laminates, the error at the center is zero. Figure 5.7 shows the curvature profile along the minor axis, AB. Here, the deflections for the analytical model show only a slight anticlastic curvature. The experimental data show near-zero deflections in the nonboundary region but pronounced high curvatures near the boundaries. This clearly demonstrates that free edge

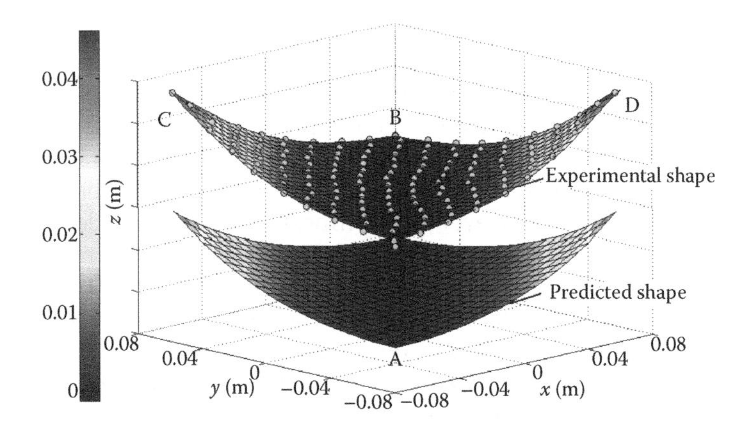

FIGURE 5.6

Predicted room temperature state 1 cylindrical shape and experimental data with spline fitted surface for square [–45/45] laminate. Experimental data are offset. (From Betts, D.N. et al., *Composite Structures* 92, 7, 1694–1700, 2010. With permission.)

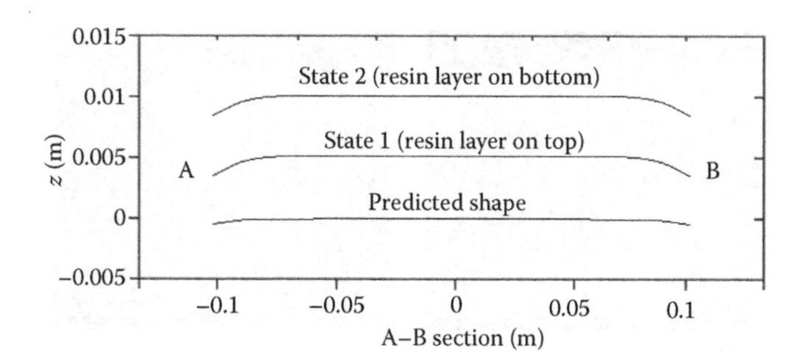

FIGURE 5.7
Cross-section profile of predicted and offset experimental shapes of [−45/45] laminate along line A–B. (From Betts, D.N. et al., *Composite Structures* 92, 7, 1694–1700, 2010. With permission.)

effects can lead to laminate curvatures that are not included in the analytical model. The free edge effects are observed around the boundaries just over 10% of the overall dimension and discussed further in Section 5.3.3.3. The profiles of Figure 5.8 show that the profiles for both states are smooth and the curvature deviation increases smoothly toward the corners of the laminates. The discrepancies in the curvatures between states are also noticeable, which the analytical model does not take into account.

5.3.3.1 Effect of Additional Resin Layer

The effect of the resin layer that is formed during curing is consistently observed, and all laminates exhibit greater out-of-plane displacements in state 2 than those of state 1 (see Table 5.1). The difference in maximum out-of-plane displacement between states for this family of laminates was found to vary by between 0.7 and 3.1 mm, leading to differences in errors between states of 1.3–17.6%. This is in contrast to the analytical model, which does not account for the resin layer, hence the displacements for both states are the same.

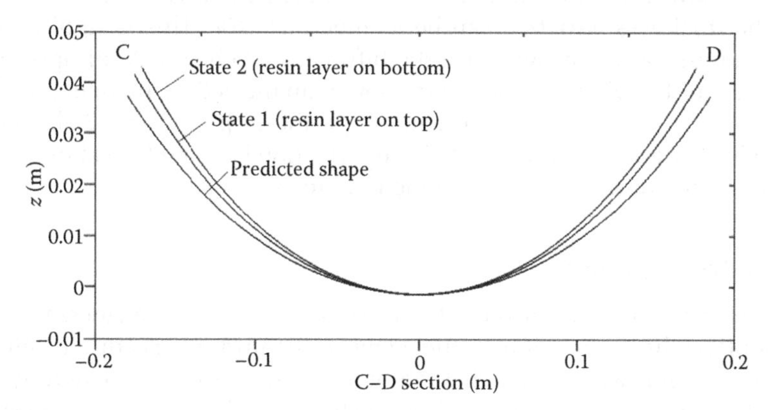

FIGURE 5.8
Cross-section profile of experimental and predicted shapes of [−45/45] laminate along line C–D.

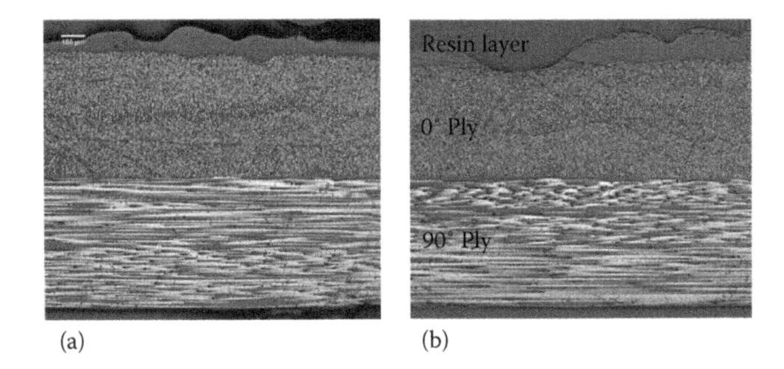

(a) (b)

FIGURE 5.9
Two sample images showing composition for T800/M21 $[0/90]_T$ laminate. White scale bar in (a) is 100 μm, and (b) is at same magnification.

Figure 5.9 shows typical images from optical microscopy of a $[0/90]_T$ laminate, where an uneven layer of resin on the upper surface of the laminate is clearly visible. The observed resin layer varies randomly in thickness between negligible and 0.08 mm. The optical microscopy images of the laminate cross-section such as Figure 5.9 are investigated and the mean thickness of 0.025 mm is added to the analytical model to obtain the results in Table 5.1.

It is also noted that a thin layer of resin exists between the plies. This layer is observed to be of the order of one hundredth of a ply thickness and an inclusion of this inter-ply resin layer (modeled one-twentieth of a ply thickness) resulted in a 1% variation in maximum displacement. The effect of this is, thus, considered minor.

5.3.3.2 Effects of Ply Thickness

The nominal thickness of the initial composite prepreg (before curing) was 0.25 mm; however, the examination of the optical microscopic images reveals that the thickness variation can be as much as ±6%. This is significant as 1% thickness variation can cause the bifurcation behavior to disappear [7]. The effect of the play thickness variation is quantified by modeling each of the laminates with a through-thickness profile of $[t + 2\%/t - 2\%]$, where t is the ideal single ply thickness. This imperfection is found to lead to changes in maximum out-of-plane displacement up to ±4.6%.

5.3.3.3 Free Edge Stress

The effects of free edge stress can be clearly seen in the sudden increase of the out-of-plane displacement near the edges in the minor curvature profile AB of Figure 5.7. It is well known that interlaminar stresses increase rapidly near free edges of laminates, and this leads to the free edge displacement [8,9]. In contrast to the analytical modeling described previously, careful finite element modeling has been shown to be able to capture the local curvature,

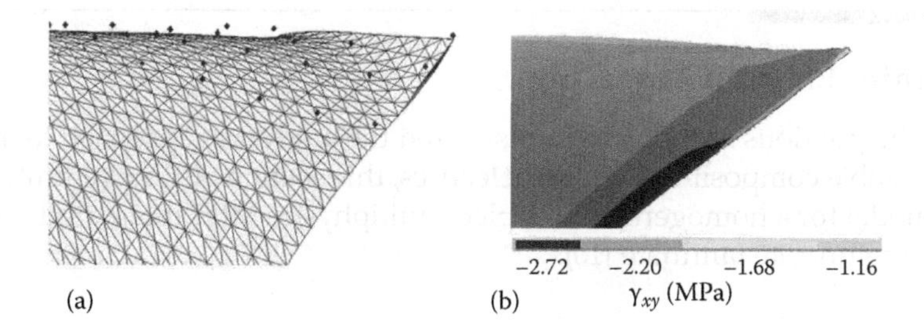

FIGURE 5.10
Close-up view of comer region for $[-30/60]_T$ laminate. (a) Finite element model of displacements with superimposed experimental data; (b) *xy*-shear stress.

which will be discussed in Section 5.4. Details are shown in Figure 5.10, which show good agreement. The local curvatures can be attributed to a rise of through-thickness shear stress, γ_{xz} and γ_{yz}.

5.3.3.4 Temperature

The bistable configuration arises from the anisotropic thermal expansion of the composite materials; thus, the experimental results are dependent on the temperature variation. Away from the bifurcation point, the temperature–curvature relation is approximately linear and a 5°C temperature variation can expect the deflection variation of around 3% of the total deflection. Figure 5.11 shows the profile of a $[0/90]_T$ laminate at a variety of temperatures. As the laminate temperature increases to its original cure temperature, the out-plane-displacement decreases significantly. Clearly, any errors in the temperature range would lead to errors in predicted laminate shape.

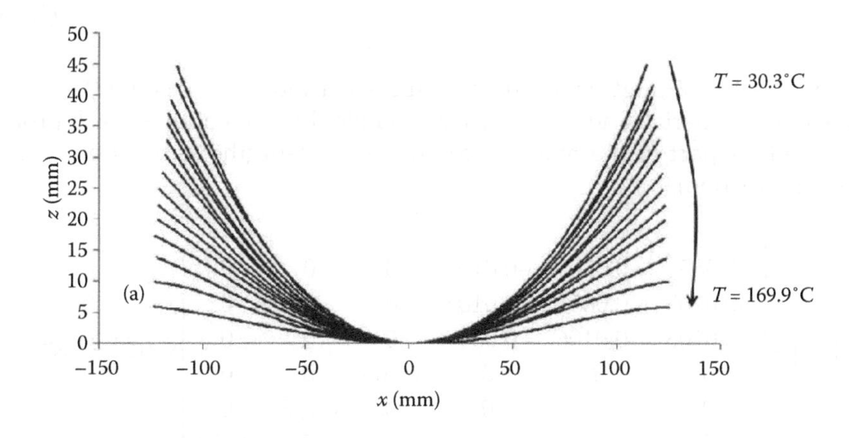

FIGURE 5.11
Variation in laminate profile.

5.4 Finite Element Modeling

While the previous sections have discussed the analytical approach to modeling bistable composites and piezoelectrics, this section outlines a finite element model for a homogenized coupled multiphysics MFC integrated with a bistable composite laminate [10].

5.4.1 MFC Properties

The MFC is considered transversely isotropic with a single axis of rotational symmetry parallel with the poling direction in PZT fibers. The compliance matrix $[s_{ij}^E]$ is thus of the form of Equation 5.23. We used the elastic constants from Williams et al. [11] and this leads to Equation 5.24.

$$[s_{ij}^E] = \begin{bmatrix} 1/E_1 & -\nu_{31}/E_1 & -\nu_{31}/E_3 & 0 & 0 & 0 \\ -\nu_{31}/E_1 & 1/E_1 & -\nu_{31}/E_3 & 0 & 0 & 0 \\ -\nu_{31}/E_3 & -\nu_{31}/E_3 & 1/E_3 & 0 & 0 & 0 \\ 0 & 0 & 0 & 1/G_{31} & 0 & 0 \\ 0 & 0 & 0 & 0 & 2(1+\nu_{31})/E_3 & 0 \\ 0 & 0 & 0 & 0 & 0 & 2(1+\nu_{31})/E_3 \end{bmatrix} \tag{5.23}$$

where E is the Young's modulus, G is the shear modulus, ν is the Poisson's ratio of the material, and the subscripts denote the orientation of each property with respect to the material coordinate system (the 3-direction is the poling direction).

$$[s_{ij}^E] = \begin{bmatrix} 0.065 & -0.0205 & -0.0106 & 0 & 0 & 0 \\ -0.0205 & 0.065 & -0.0106 & 0 & 0 & 0 \\ -0.0106 & -0.0106 & 0.034 & 0 & 0 & 0 \\ 0 & 0 & 0 & 0.165 & 0 & 0 \\ 0 & 0 & 0 & 0 & 0.173 & 0 \\ 0 & 0 & 0 & 0 & 0 & 0.173 \end{bmatrix} \times 10^{-9}\,\mathrm{m^2\,N^{-1}} \tag{5.24}$$

The effective piezoelectric constants were computed using the manufacturer's data [2] and experimentally measured values [11] (Equation 5.25). Shear piezoelectric coefficients (e.g., d_{15}) are considered to be 0, although conventional piezoelectric ceramics d_{15} are non-zero; $|d_{15}| \ll |d_{3j}|$ for many composites. However, since the applied electric field will always be along the poling direction (fiber/rod axis), no contribution for the piezoelectric shear coefficients are expected.

$$[d_{ij}] = \begin{bmatrix} -2.1 & -2.1 & 4.67 & 0 & 0 & 0 \\ 0 & 0 & 0 & 0 & 0 & 0 \\ 0 & 0 & 0 & 0 & 0 & 0 \end{bmatrix} \times 10^{-10}\,\mathrm{m\,V^{-1}} \tag{5.25}$$

The relative permittivity of an active layer is determined in order to fully model its electromechanical coupling, i.e., the magnitude of the induced electric field as a result of mechanical stress. The micromechanical mixing rule for ε_{33}^{T} (Equation 5.26) [12] and the standard mixing rule for dielectric volumes in series representing ε_{22}^{T} (Equation 5.27) are used.

$$\varepsilon_{33}^{T} = \rho\varepsilon_{33}^{T,\mathrm{p}} + (1-\rho)\varepsilon_{33}^{T,\mathrm{m}} \tag{5.26}$$

$$\varepsilon_{22}^{T} = \left(\frac{\varepsilon_{22}^{T,\mathrm{p}}\varepsilon_{22}^{T,\mathrm{m}}}{\rho\varepsilon_{22}^{T,\mathrm{m}} + (1-\rho)\varepsilon_{22}^{T,\mathrm{p}}} \right) \tag{5.27}$$

where ε_{ij}^{T} is the relative permittivity and ρ the volume fraction of PZT within the active layer. The superscripts "m" and "p" denote the piezoelectric and matrix materials, respectively. The relative permittivity of PZT-5A data used to calculate the permittivity of the MFC device was taken from Jaffe and Berlincourt [13], with the value for the epoxy matrix taken from Deraemaeker et al. [12]. The homogenized material properties of the MFC have been validated via an MFC actuated isotropic beam [10].

5.4.2 Finite Element Model of Bistable MFC-Composite Laminates

The $[-30/60/0_{\mathrm{p}}]_{\mathrm{T}}$ laminate (Figure 5.12) was modeled using the commercial finite element software ANSYS [14]. The laminate of Figure 5.13 was modeled by 20-node quadratic SOLID186-layered brick elements, with 1360 elements in region 1 (57 mm × 150 mm) and 816 elements each in regions 2 and 3 (46.5 mm × 150 mm). The MFC patch of 85 mm × 57 mm × 0.3 mm was modeled with 1200 20-node couple field SOLID 226 elements (the purple region in Figure 5.13) with the homogenized MFC material properties from Section 5.4.1. A single element thickness was used, and the laminate and the MFC were rigidly linked. The model was constrained from translation in all three

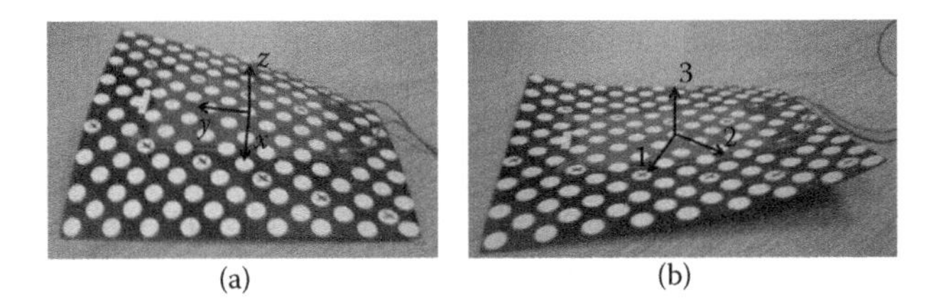

FIGURE 5.12
$[-30/60/0_p]_T$ laminate. (a) State 1 showing global coordinate system. (b) State 2 showing local material coordinate system for uppermost 60° ply.

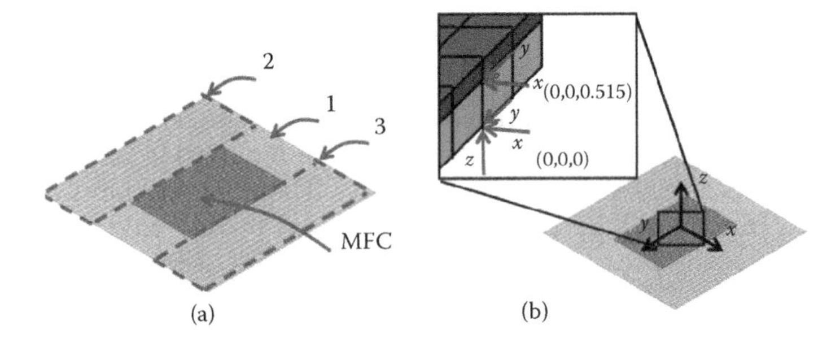

FIGURE 5.13
Finite element model of $[-30/60/0_p]_T$ laminate with a centrally located MFC (in purple): (a) overall mesh; (b) MFC and laminate interface.

orthogonal directions at the origin of the global coordinate system in addition to constraining the (0, 0, 0.515) from in-plane translation (Figure 5.13b). The model is analyzed using geometrically nonlinear finite element analysis with a small out-of-plane displacement as imperfection. A Newton–Raphson solution procedure was adopted as the arc-length procedure was not available for coupled-field SOLID226 elements.

A thermal load of –160 K was applied to the initially flat laminates to represent cooling from the cure temperature to room temperature, and this induced the out-of-plane curvature as observed at room temperature. As the MFC is applied post-cure in practice, an offset voltage, V_0, was applied to compensate for thermal stress in the MFC. The imperfection was then removed. This represented the cured and unloaded shape of the bistable piezocomposites at room temperature. The MFC drive voltage could be applied to obtain electrically loaded curvature shapes. The snap-through event was identified by incrementing the drive voltage at a given state, say state 1 and the model no longer converged to the state 1, i.e., snap-through to state 2 had occurred.

5.4.3 Shapes and Snap-Through

The laminate shapes of $[-30/60/0_p]_T$ and $[0/90/0_p]_T$ are shown in Figure 5.14a and b, respectively, with the wireframe representing the finite element results

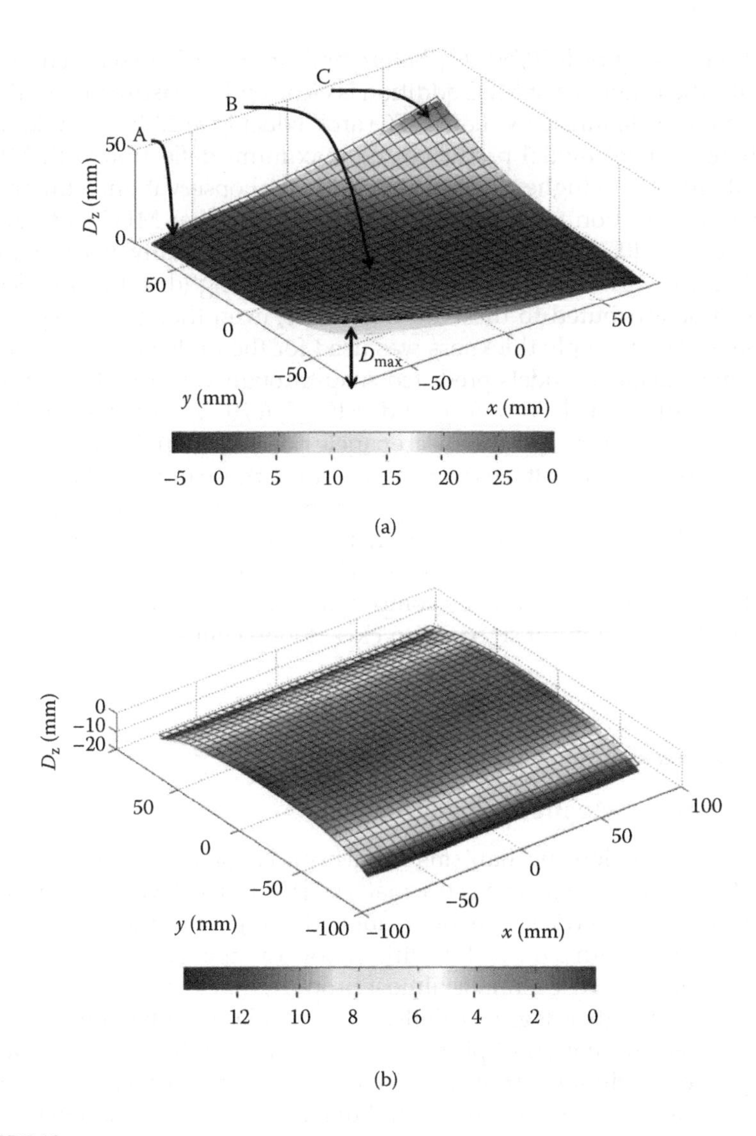

FIGURE 5.14
Laminate shapes from the finite element model (wireframe) and experimental data (contour plot): (a) $[-30/60/0_p]_T$ laminate with interpolated contour plot from 149 measured surface coordinates; (b) $[0/90/0_p]_T$ laminate with interpolated contour plot from 283 measured surface coordinates.

and the contour plot obtained from the experimental data. The two models, overall, show excellent agreements. The attachment of the MFC leads to a significant reduction in curvature in the associated section of the laminate, e.g., point B, in comparison to the cylindrical curvature at point C. The local reversed curvature is also observed at point A. When comparing maximum deflections (D_{max}) of $[-30/60/0_p]_T$ before and after MFC attachment experimentally, the influence of MFC addition is clear with measured deflections of 38.2 mm for the laminate (without MFC) and reducing to 22.48 mm with MFC. The finite element model predicted the maximum deflection with MFC to be 25.20 mm, 12.1% higher than the experimental observation. Similarly, the maximum deflection for the $[0/90/0_{MFC}]_T$ laminate after MFC addition was predicted to be 10.73 mm, 16% lower than the experimentally measured value of 12.77 mm. Variations arising from manufacturing (discussed in Section 5.3.3) can be attributed to these discrepancies, even though an experimentally obtained mean ply thickness was used for the finite element model.

The finite element models predicted snap-through voltages of 677 V (700 V experimentally) for $[-30/60/0_p]_T$ and 645 V (670 V experimentally) for $[0/90/0_p]_T$. They represent the discrepancies of less than 4.5%. The experimental snap-through voltages were greater than those predicted by the finite element models in both cases. Delayed snap-through was observed at voltages immediately below monotonic snap-through voltage when drive voltage was applied for a prolonged period. This can be attributed to the creep of the MFC actuators and could be compensated for in industrial applications using time-varying input signals and closed-loop control.

5.5 Actuation Mechanisms

Two smart actuation mechanisms have been considered to induce an in-plane strain, namely a piezoelectric material [15] and an SMA [16]. An advantage of the piezoelectric strain, often applied using a fiber-based patch such as an MFC, is its high bandwidth, which allows a rapid state change. In addition, since piezoelectric strain is almost proportional to applied electric field (and voltage), a higher degree of deflection control is possible compared with SMA. A major limitation of piezoelectric materials is that they are capable of only a relatively low strain (~0.1%). Owing to the stiffness between the two states, previous research has found that piezoelectric actuation is able to induce a one-way state change but is usually insufficient to reverse the state change. Bowen et al. [17] was only able to achieve a reversible state change by applying an additional compressive mechanical load. Schultz et al. [18] achieved a reversible state change and resetting by attaching two piezoelectric patches on each side of a laminate; however, it was necessary to apply a voltage above the recommended working range of the piezoelectric actuator. An alternative actuation mechanism is SMA, which is able to induce high force and high strain (~8%). However, thermal SMA actuation has received less interest owing to its slow response time and low bandwidth; maximum operating frequencies are lower than 100 Hz for SMA, compared with >10 kHz for piezoelectric ceramics. Dano and Hyer [16] demonstrated the feasibility of the SMA actuation on bistable structures by attaching SMA wires on bridge-like supports above the laminates.

This section discusses a combined actuation mechanism, termed shape-memory alloy–piezoelectric active structures (SMAPAS), that combines the advantages of the piezoelectric and SMA materials to achieve self-resetting bistable composites [19]. The approach uses piezoelectric actuation to provide a rapid snap-through (state 1→2) with a fine degree of control and a relatively slow but high strain SMA actuation to reverse the state change (state 2→1). For fully reversible actuation (1↔2), the snap-through deflection using the piezoelectric actuation must be sufficiently high to deform (twin) the SMA material, thus enabling the shape-memory effect to be utilized.

5.5.1 Configuration of Smart Composites

A thin cantilever beam of $[0/0/90/90]_T$ carbon fiber/epoxy material was used to demonstrate the two-way actuation. The dimension of the cantilever beam was 300 mm × 60 mm × 0.52 mm (with 20 mm × 20 mm corners removed) manufactured using HTA (12k) 913 prepreg sheet, Figure 5.15. An MFC from Smart Material Corp. [2] was used again for piezoelectric actuation, which has an operating range –500 V (contraction) to 1500 V (extension). SMA actuation was achieved by application of a voltage to heat a NiTi SMA wire of 0.175-mm diameter. The SMA wire was electrically and thermally insulated by a series of ceramic thermocouple tubes, and the wire was wrapped around the end tab of the cantilever beam and clamped at the opposite end. A small end mass was added to the cantilever to ensure sufficient deformation of the SMA on snap-through from state 1 to 2. Figure 5.15b shows the two stable states of the bistable cantilever structure (SMA not shown).

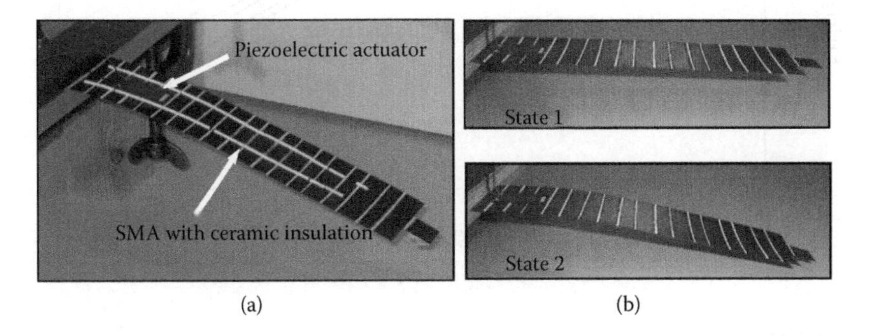

FIGURE 5.15
Cantilever beam: (a) test rig for reversible actuation; (b) two stable states.

5.5.2 Characterization of Actuation

The first part of the experiment was to examine the cantilever deflection due to piezoelectric actuation while no voltage was applied to the SMA. The cantilever beam was initially in its raised state (state 1) at 0 V and the shape of the cantilever beam profile was recorded for the positive voltage range of the MFC piezoelectric actuator (maximum of 1500 V). The results at 0, 300, 600, 900, 1200, and 1300 V are presented in Figure 5.16a. It was observed

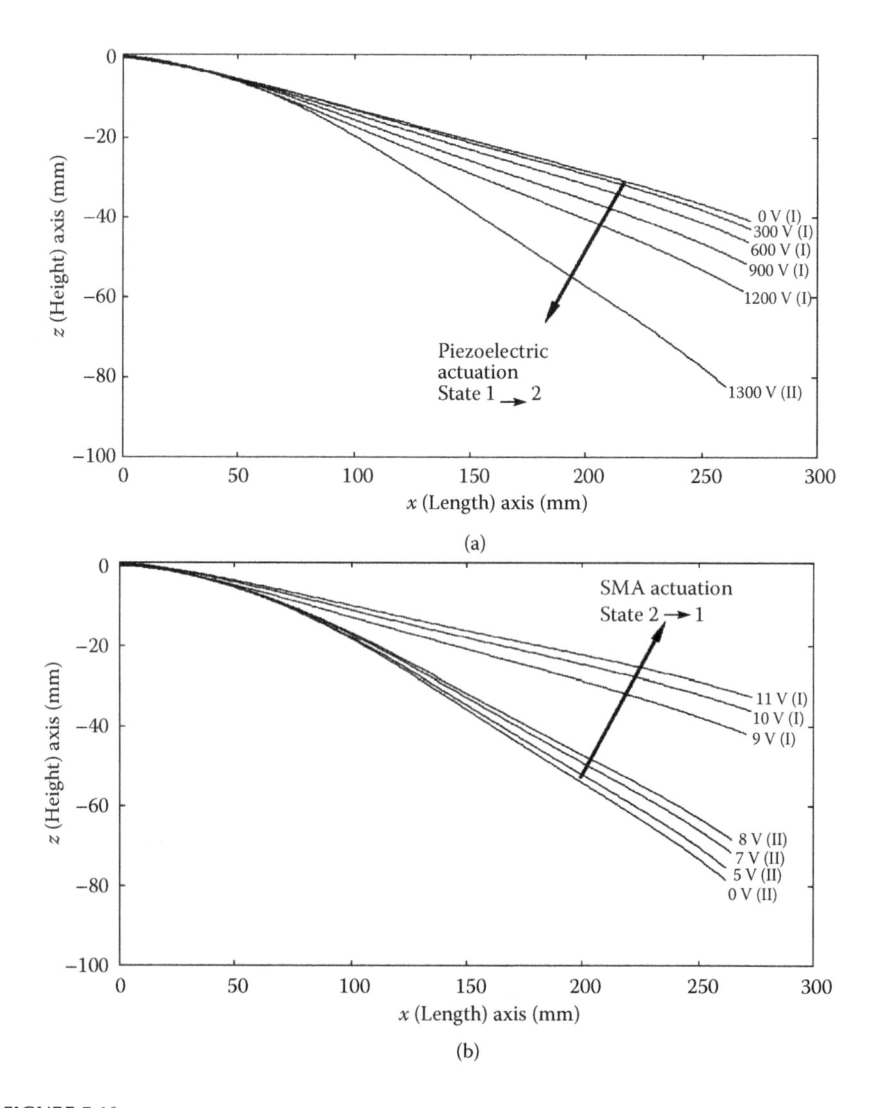

FIGURE 5.16
Shape profile of the cantilever beam: (a) varying applied MFC voltages; (b) varying applied SMA voltages. (I) and (II) indicate states 1 and 2.

that applying voltages between 0 and 1200 V increased the deflection of the cantilever beam but the structure remained in state 1. At 1300 V, there was a significant change in the deflection as a result of snap-through from state 1 to 2. On removal of the applied voltage, a small reduction in the deflection was observed but the structure remained in state 2. The cantilever could not be returned to state 2 using the MFC patch alone, consistent with the previous findings.

The second part of the experiment examined the cantilever beam deflection in response to SMA actuation. After snap-through from state 1 to 2 using the piezoelectric and removing the piezoelectric voltage, the input voltage to the SMA wire was increased from 0 to 11 V at 1-V intervals. The cantilever beam profiles at 0, 5, 6, 7, 8, 9, 10, and 11 V are shown in Figure 5.16b. In this case, the purpose of the applied voltage was to provide sufficient Joule (resistive) heating to achieve a martensite to austenite phase change in the SMA and induce a shape-memory effect. Small deflection changes were seen between 0 and 5 V as the temperature in the wire had not reached the transition temperature. There was a more marked change between 5 and 8 V, although the cantilever beam remained in state 2. Snap-through from state 2 to 1 was observed at 9 V. Upon removal of the voltage to the SMA, the cantilever beam returned to its original (0 V) state 1 profile as in Figure 5.16a. This reversible actuation was completely repeatable, indicating that the piezoelectric snap-through from state 1 to state 2 was sufficient to deform (twin) the SMA wire and enable fully reversible snap-through using the SMA–piezoelectric combination.

A comparison of Figure 5.16 (a and b) reveals the different power requirements of the two actuator materials. The piezoelectric requires a high voltage (>1000 V) with low current, since the piezoelectric is a dielectric. The SMA requires a much lower voltage (<15 V) with high current (up to 1 A), which corresponds to a power of 15 W. The piezoelectric is primarily a reactive (capacitive) load, while the SMA is a resistive load, necessitating different levels of rectification and power. Since piezoelectric strain is proportional to electric field, the voltage requirement of the piezoelectric actuator could be lowered by reducing the interdigitated electrode spacing of the MFC. Interestingly, magnetic SMA materials are becoming commercially available that can achieve strain levels similar to a thermal SMA but higher bandwidth since alternating magnetic fields can be applied at higher frequency.

5.6 Optimization Example

Clearly, the combination of piezoelectric actuators with asymmetric laminates for shape-changing applications will rely on a number of modeling tools to select the appropriate piezoelectric–laminate combination for specific applications. In this section, an optimization approach is introduced.

5.6.1 Morphing Structure

Two stable states in bistable composites enable a structure to achieve a large deflection without continuing to supply energy to maintain the deflection, and a relatively small actuation energy is required to induce the state change. This has been considered favorably in aerospace engineering and many applications have been considered in this context, i.e., morphing wing to reduce drag and to provide load alleviation and aerodynamic control [20,21]. One common criticism, however, is their inherent compliance and poor load-carrying capability. To induce a snap-through state change using a piezoelectric actuator with a low voltage input, the stiffness must be low. This means that normal operating loads can induce an undesired state change. The structural requirement to resist the normal operating loads and a low power requirement to induce and control the state change are, therefore, conflicting. This section develops an optimization study to exploit the anisotropic stiffness of composites to maximize the bending stiffness in one direction (in the direction of normal operating load) while minimizing the bending stiffness in another direction (in the direction of actuation). The numerical study showed that the stiffness in the loading direction can be five times greater than the stiffness in the actuation direction.

5.6.1.1 Optimization Problem

The optimization problem formulation is defined as follows [22]:

Maximize: Bending stiffness in a loading direction in ϕ_1

$$\frac{\delta a_\phi}{\delta M_{x\phi}} \tag{5.28}$$

Subject to: The deflection between states must be greater than a minimum value, representing a significant shape change.

$$w_{\text{def}} = 0.25(a_1 + b_1)L^2 \geq w_{\min} \tag{5.29}$$

Reversible snap-through must be within the working voltage limits of the piezoelectric layers.

$$-500\ \text{V} \leq V \leq 1500\ \text{V} \tag{5.30}$$

Variables: Ply orientations θ_i

Piezoelectric fiber orientation θ_p

Laminate geometry defined by ply thicknesses, t_i, and edge length, L

Loading direction ϕ_1 and snap-through direction ϕ_2

This optimization problem is solved using MATLAB's sequential quadratic programming, *fmincon*, with multiple starting points to capture all optima.

5.6.1.2 Unconstrained Optimization

The optimization study results presented in this section is an unconstrained problem where the constraints of Equations 5.29 and 5.30 are not imposed. The design variables are restricted to two-ply orientations for illustration purposes. The square laminate edge length L is set to 0.15 m, and the uniform single-ply thickness $t_1 = t_2$ is set to 0.1 mm. The piezoelectric layers are assumed to have the same edge length, L, with a thickness of 0.2 mm. The snap-through direction, ϕ_2, is fixed to 0°. The pattern of results is dependent on the relative values of the snap-through direction and loading direction rather than the individual angles. Figure 5.17 shows the design space for this problem for changing values of the loading direction, ϕ_1. The contours show the objective function value; the red dots indicate the local optima and the green dots the global optima.

When the loading direction is initially set to 0°, four distinct local solutions are found for laminates of [0/0/90/90], [90/90/0/0], [0/90/0/90], and [90/0/90/0] (Figure 5.17a). Owing to the periodic nature of the design space, these solutions are repeated at equivalent positions along the boundaries of the design space. Each of these four solutions has a different value of the objective function, with the global optimum of 0.737 found at [0/90/0/90]. As the loading direction, ϕ_1, is changed, the pattern of four local solutions is shifted away from the 0° and 90° ply angles, and the solutions no longer appear orthogonally to one another. When ϕ_1 approaches 20°, one of the four solutions is lost. Of the remaining three solutions, the global optimum of 0.811 is found at [22/–58/32/–68] (Figure 5.17e). When ϕ_1 = 30°, the number of local solutions reduces further to just two (Figure 5.17g). As ϕ_1 tends toward 45°, the two remaining solutions become closer to being orthogonal to one another until settling on [44/–38/52/–46] and [–44/38/–52/46] at ϕ_1 = 45°, with both solutions exhibiting equal objective function value of 0.936 (Figure 5.17j).

In addition to the changing number of optimum solutions, we observe an increase in the global optimum objective function values with increasing ϕ_1. Figure 5.18 shows this relationship for the range of ϕ_1, shown as the blue line. The minimum solution is found when ϕ_1 is 0°. This design represents a laminate layup where the piezoelectric layer orientations are orthogonal to the chosen loading direction. This is expected as the laminate composition makes best use of the directional stiffnesses of composite materials by having the two conflicting stiffness requirements as far apart as possible.

For comparison, the optimum cross-ply solutions, $[\theta_1/\theta_1/\theta_1 + 90/\theta_1 + 90]$ are also shown in Figure 5.18, shown as the red line. Comparing these solutions against the global optima reveals improvement in objective function from 12% to 29%.

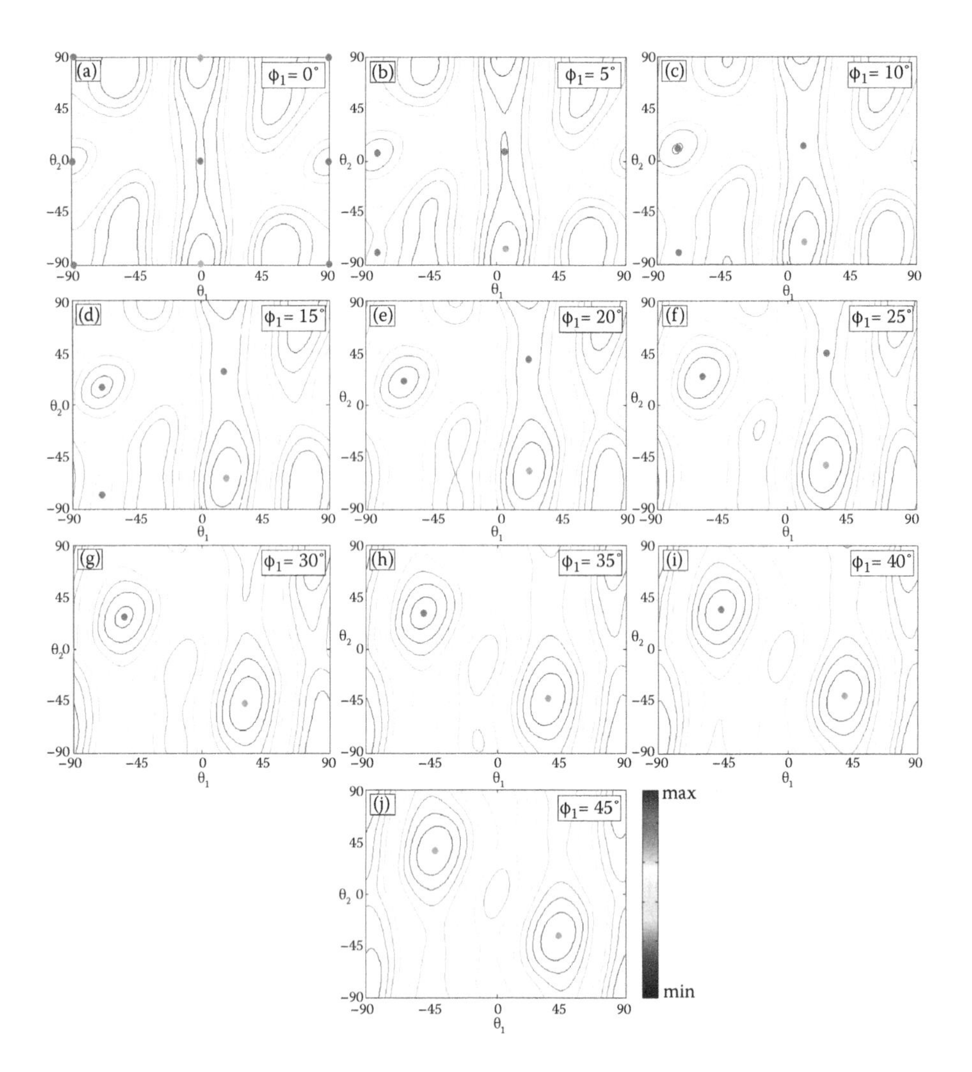

FIGURE 5.17
Variation in design space for a $[0_p/\theta_1/\theta_2/\theta_2 + 90/\theta_1 + 90/90_p]$ laminate; contours represent the objective function, with change in loading direction, ϕ_1.

5.6.1.3 Constrained Optimization

We now impose the constraints of Equations 5.29 and 5.30 for a range of loading directions (0° to 45° at 5° intervals) and a fixed snap-through direction, 0°. The maximum positive voltage of 1500 V represents a limit to avoid dielectric breakdown of the piezoelectric and corresponds to a free strain of 1350 μstrain. The −500 V cannot be exceeded since it can lead to depolarization of the piezoelectric and corresponds to a free strain of −450 μstrain. The design variables include the nonuniform ply thicknesses t_1 and t_2, the

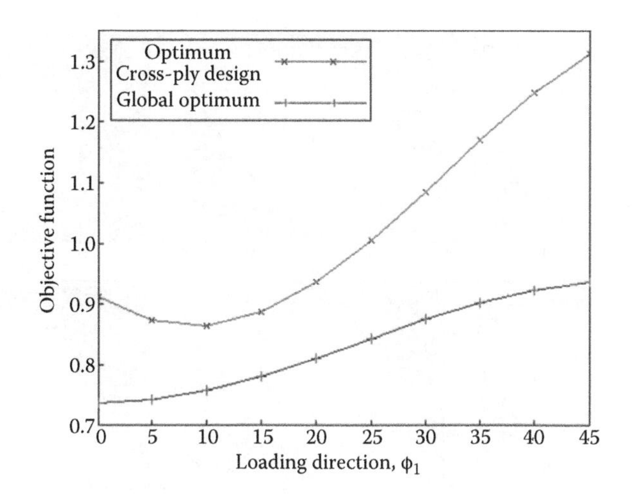

FIGURE 5.18
Variation in objective function with change in loading direction.

laminate edge length *L*, and the ply orientations. However, all locally optimum designs for this problem are found at the upper or lower bounds of the geometric variables. We therefore present results with the square laminate edge length set to 0.15 m, and the uniform single ply thickness set to 0.1 mm, noting that an increase in edge length or a decrease in ply thickness will increase the achievable deflection but not affect the general pattern of results observed.

Figure 5.19 shows the constrained design space for three example loading directions, 0°, 20°, and 45°, chosen to show the behavior across the range shown in Figure 5.17. The infeasible regions are marked gray, with the dark gray representing the deflection constraint of Equation 5.29, 30, 50, and 70 mm shown in Figure 5.19, and the light gray the voltage constraint (Equation 5.30). The local solutions are marked with red dots and the global solutions marked with green dots. The global optimum solutions are summarized in Table 5.2.

Figure 5.19 shows that the design space is highly nonlinear and multimodal, and the interactions between constraints, which drive the optimum solutions, cannot consistently be defined *a priori*. For example, Figure 5.19a shows that the global optimum solution for 0° loading direction and the deflection constraint of 30 mm is not affected by the constraints, with the global optimum of 0.737. As the deflection constraint is increased to 50 mm, this solution becomes infeasible and a new global optimum is found on the deflection constraint boundary at [2/76/–14/–88] with an increased value of 0.754 (Figure 5.19b). Increasing the deflection constraint further to 70 mm continues to move the solution away from the unconstrained optimum. The actuation constraint (light gray region) remains inactive for all deflection values in Figure 5.19a through 5.19c. It is noted that such visualization of the design space is impossible beyond two design variables; however, the

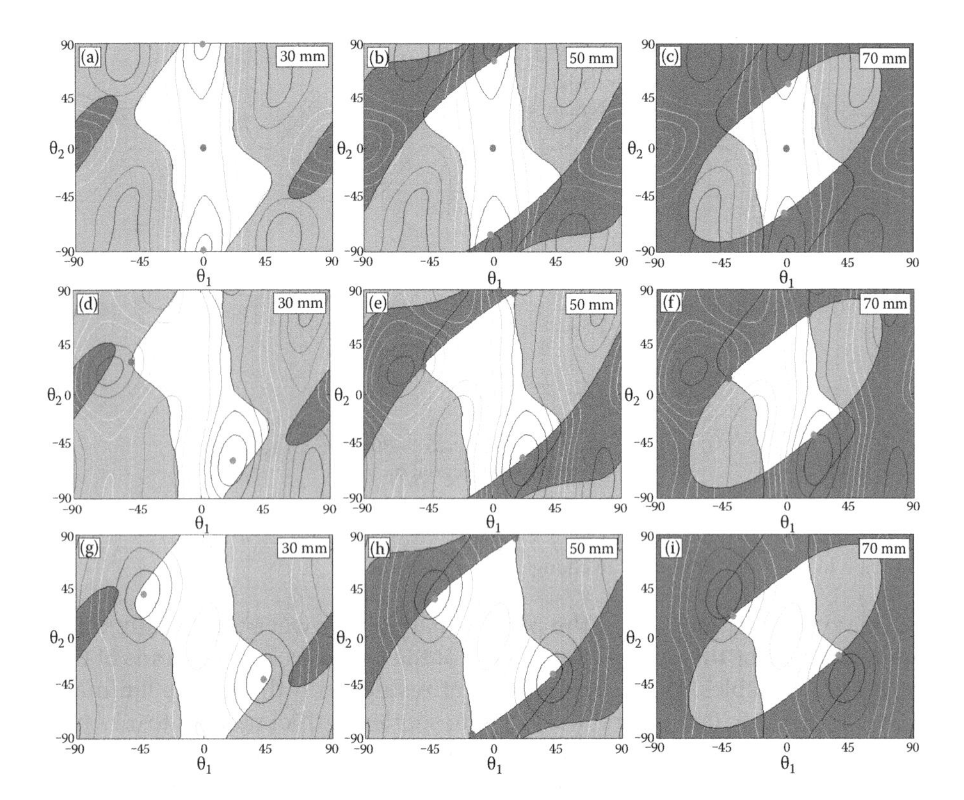

FIGURE 5.19
Constrained design for loading directions, θ_1 (a–c) $\theta_1 = 0°$, (d–f) $\theta_1 = 20°$, and (g–i) $\theta_1 = 45°$, for different deflection constraint values of 30, 50, and 70 mm. Light gray regions are infeasible owing to voltage constraint; dark gray regions are infeasible owing to deflection constraint.

optimization method is able to provide all optimum solutions for a greater number of design variables.

The optimization study thus far makes use of only one of the piezoelectric layers at any one time with the second piezoelectric layer used for the reverse actuation. There is a scope for reducing the total voltage requirement further by a combined use of the top and the bottom piezoelectric layers, and results for all cases are given in Table 5.3. An examination of the optimum solutions reveals that this is beneficial only in some cases. For example, the optimum solutions for the 45° loading direction are found to be 0 V in the bottom layer; thus, no improvement in the total voltage requirement. However, for the 0° loading direction, up to 33.8% reduction of total voltage requirement can be achieved.

5.6.2 Energy Harvesting Device

Most of the examples above have considered the actuation of bistable laminates for shape-control and morphing-type applications. A recent study

TABLE 5.2

Global Optimum Solutions for Morphing Application

Loading Direction (ϕ°)	Deflection Constraint (mm)	First Ply Angle (θ°)	Second Ply Angle (θ°)	Actual Deflection (mm)	Objective Function
0	30	0	90	43.3	0.737
	40	0	90	43.3	0.737
	50	2	76	50.0	0.754
	60	4	68	60.0	0.773
	70	2	56	70.0	0.787
20	30	22	–59	46.7	0.811
	40	22	–59	46.7	0.811
	50	21	–56	50.0	0.813
	60	21	–44	60.0	0.832
	70	20	–36	70.0	0.854
45	30	43	–38	44.2	0.936
	40	43	–38	44.2	0.936
	50	41	–33	50.0	0.943
	60	39	–25	60.0	0.972
	70	36	–19	70.0	1.014

demonstrated that bistable piezocomposites could also harvest electrical energy from mechanical vibration over a wide spectrum of frequencies [23]. This is a significant discovery as the existing vibration energy harvesters are typically tuned to operate near resonance, and its performance falls dramatically outside their resonant frequencies [24]. An alternative bistable

TABLE 5.3

Voltage Requirements Combined Use of Both Piezoelectric Layers

Loading Direction (ϕ°)	Deflection Constraint (mm)	Top Layer (V)	Bottom Layer (V)	Combined Voltage (V)	Reduction (%)
0	40	0	–410	410	33.8
	50	30	–500	530	23.7
	60	215	–500	715	11.7
	70	355	–500	855	1.7
20	40	75	–500	575	2.5
	50	590	0	590	0
	60	675	0	675	0
	70	775	0	775	0
45	40	865	0	865	0
	50	815	0	815	0
	60	940	0	940	0
	70	1085	0	1085	0

energy harvesting mechanism in the literature is a ferromagnetic cantilever with two permanent magnets located symmetrically near the free ends that induce bistability [25]. An advantage of the asymmetric composite configuration considered in this chapter is that the bistability is inherent in the structural mechanics; thus, an obtrusive arrangement of external magnets unwanted electromagnetic fields can be avoided. This section presents the optimal configurations for a bistable piezocomposite energy harvester offered by the static behavior of the device [26].

5.6.2.1 Piezocomposite Energy Harvester Configuration

For converting mechanical to electrical energy, four piezoelectric patches are attached to the top surface, each positioned at the center of one quarter of the laminate surface (Figure 5.20). This pattern is mirrored on the bottom surface by piezoelectrics oriented perpendicularly.

While an increase in piezoelement size increases the area from which electrical energy can be harvested, their additional stiffness reduces laminate curvature and the resulting stress along the piezoelectric polarization direction (Figure 5.21). A high degree of laminate curvature nonlinearly increases the actuation force required to induce snap-through and reduces the overall effectiveness. In a similar manner to the morphing structures, the energy

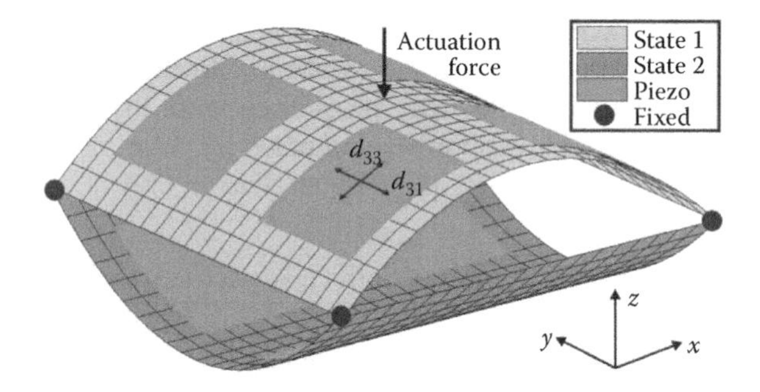

FIGURE 5.20
$[0^P/0_2/90_2/90^P]_T$ laminate with 40% piezoelectric coverage.

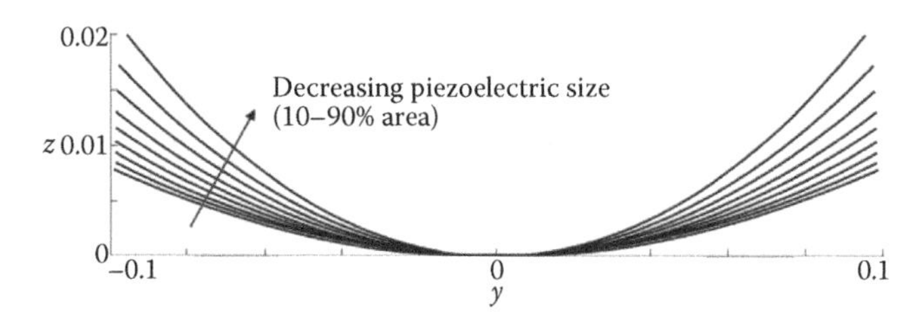

FIGURE 5.21
Major curvature for a square $[0^P/0/90/90^P]_T$ laminate with varying piezoelectric size.

harvesting characteristics are dependent on piezoelectric area (A), laminate ply thickness (t), ply orientations (θ's), and device aspect ratio (AR). To determine the correct combination of these parameters for energy harvesting requires an understanding of the complex interactions of the nonlinear behavior.

5.6.2.2 Optimization for Energy Harvesting

The objective function for energy harvesting is to maximize the electrical energy output. When operating off-resonance, a piezoelectric layer behaves as a parallel plate capacitor. Hence, the electrical energy generated by snap-through is $1/2CV^2$, where C is the piezoelement capacitance and V is the open circuit voltage generated by the direct piezoelectric effect. Under a stress σ, the voltage generated is $\sigma g_{ij} t_p$ where g_{ij} is the piezoelectric voltage constant (electric field per unit stress), t_p is the piezoelectric thickness, and the stresses in x and y are

$$\begin{aligned} \sigma_x &= \bar{Q}_{11}(\varepsilon_x^0 + z\kappa_x) + \bar{Q}_{12}(\varepsilon_y^0 + z\kappa_y) + \bar{Q}_{16}(\varepsilon_{xy}^0 + z\kappa_{xy}) \\ \sigma_y &= \bar{Q}_{12}(\varepsilon_x^0 + z\kappa_x) + \bar{Q}_{22}(\varepsilon_y^0 + z\kappa_y) + \bar{Q}_{26}(\varepsilon_{xy}^0 + z\kappa_{xy}) \end{aligned} \tag{5.31}$$

where κ_x, κ_y, and κ_{xy} are the curvatures. On the basis of the relationship between charge, capacitance, and voltage, the capacitance is equivalent to $d_{ij}\sigma A/V$, where d_{ij} is the piezoelectric strain constant (charge per unit force). The electrical energy based on the static system can thus be expressed as $1/2(d_{ij}g_{ij})\sigma^2(At_p)$. When attached to the laminate surface, the piezoelements are strained in both the poling direction (33 direction, Figure 5.20) and transverse direction (31 direction) due to the anticlastic curvatures. The stress varies across the volume of the piezoelements as a function of the thermally induced strains. The electrical energy for piezoelements positioned at 0° (top surface) and 90° (bottom surface) is therefore

$$U = 4\sum_{m=1}^{2}\left[\frac{1}{2}\int_{v_1}(d_{33}g_{33}\sigma_x^2 + d_{31}g_{31}\sigma_y^2)\,dv_1 + \frac{1}{2}\int_{v_2}(d_{33}g_{33}\sigma_y^2 + d_{31}g_{31}\sigma_x^2)\,dv_2\right] \tag{5.32}$$

where the factor 4 accounts for all piezoelements on one surface, m defines the associated shape, and v_1 and v_2 are the volumes of two layers on opposite laminate surfaces. The material properties d_{ij} and g_{ij} are considered fixed as their optimum values have been identified by Priya [27]. The surface area of the laminate is fixed (0.04 m^2) as in Arrieta et al. [23]. A lower bound on t is set as 0.125 mm, consistent with the typical minimum ply thickness. t_p is fixed for practical reasons. All other variables are unbounded. The optimum solutions are subject to constraints to guarantee bistability, and limiting the piezoelectric strain to below its failure strain (~2000 μstrain). The optimization problem is solved using a sequential quadratic programming method with multiple starting points uniformly distributed throughout the design space to capture all optima.

Maximize: Electrical energy output, U (Equation 5.32).

Subject to: The optimum solution must be bistable.

$$a + b > 0 \tag{5.33}$$

The piezoelectric strain must be below its failure strain.

$$\varepsilon_p \leq 2000\ \mu\text{strain} \tag{5.34}$$

Two optima are found (Table 5.4) with layups $[0_P/0/90/90_P]_T$ (global optimum) and $[0_P/90/0/90_P]_T$ (local optimum), where the global solution outperforms the local solution by ~65%. In these cases, the major and minor curvatures are aligned with the polarization direction of the piezoelectric to utilize the d_{33} or d_{31} piezoelectric effect and they are intuitively the optima. The local solution is less optimal due to the reduced curvatures (relative to the global solution) associated with this stacking sequence. What are less obvious are the geometric configurations of piezoelectric area, device aspect ratio, and thickness. These parameters determine the stiffness and actuation behavior of a piezocomposites and govern the energy generated. The optimum ply thickness is 0.626 and 0.619 mm for the global and local solutions, respectively, with significant differences in the piezoelectric areas for the global (72.43%) and local (42.40%) solutions. Despite this significant difference, there is little variation in the actuation force. A close examination reveals that the optima are highly sensitive to the allowable actuation force, i.e., if the available force is less than 3.39 N, the optimum ply thickness and piezoelectric area change. The strain in both designs is well below the

TABLE 5.4

Global and Local Optima

	Global	**Local**
Stacking Sequence	**$[0_P/0/90/90_P]$**	**$[0_P/90/0/90_P]$**
Ply thickness (mm)	0.626	0.619
Piezoelectric area (%)	72.43	42.40
Aspect ratio	1.0	1.0
Maximum strain (μstrain)	1097	1113
Snap-through force (N)	3.41	3.39
Electrical energy (mJ)	33.7	20.4

material failure strain, indicating that the optimum designs are unlikely to suffer from mechanical degradation.

Figure 5.22a shows the energy output for a range of θ values in $[0_P/\theta/\theta + 90/90_P]_T$ laminates. The optima for changing θ (black squares) form a convex hull. The maximum electrical energy is observed when θ is 0°, corresponding to the global solution (white circle). As θ increases from 0° to 45° the major laminate curvature reduces and the alignment of the piezoelectric with laminate curvature becomes less optimal, leading to a decrease in the electrical energy. For θ values from 45° to ~60°, the major curvature continues to decrease but the piezoelectric alignment improves, the net effect being a more gradual decrease in energy generated. At ~60°, this pattern switches and the improved alignment of the piezoelectric with curvature becomes dominant compared with the reduced curvature and the electrical energy increases until the local solution at θ = 90° (white square).

Figure 5.22b shows the effect of varying ply number, *n*, from 1 to 6 for $[0_P/0_n/90_n/90_P]_T$ laminates. The optima for changing *n* are marked by black squares. A consistent pattern is observed for *n* from 1 to 5. The energy generated increases with increasing piezoelectric area to an optimum value with the global optimum at 5 plies (white circle). The optima have larger piezoelectric area as *n* increases from 1 to 5 since the laminate stiffness increases and the piezoelectric stiffness has less influence on curvature. However, for $n = 6$, both the optimum area and electrical energy decrease due to the loss of bistability in these stiff laminates (between 6 and 7 plies), leading to significantly reduced curvatures.

Interestingly, square laminates were consistently found to be the optimum, despite the significant out-of-plane deflections achievable by high aspect

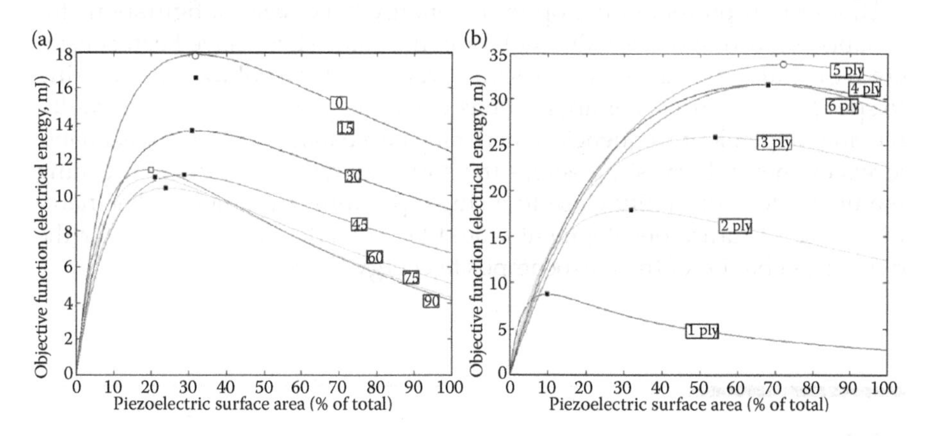

FIGURE 5.22
(a) Variation in electrical energy with θ (shown on each line) and piezoelectric surface area for $[0_P/\theta/\theta + 90/90_P]_T$ laminates, and (b) variation in electrical energy with $n \times 0.125$ mm plies (shown on each line) and piezoelectric surface area for $[0_P/0_n/90_n/90_P]_T$ laminates. Black squares mark optima for each θ or *n*; white circles mark global optima; white squares mark local optima.

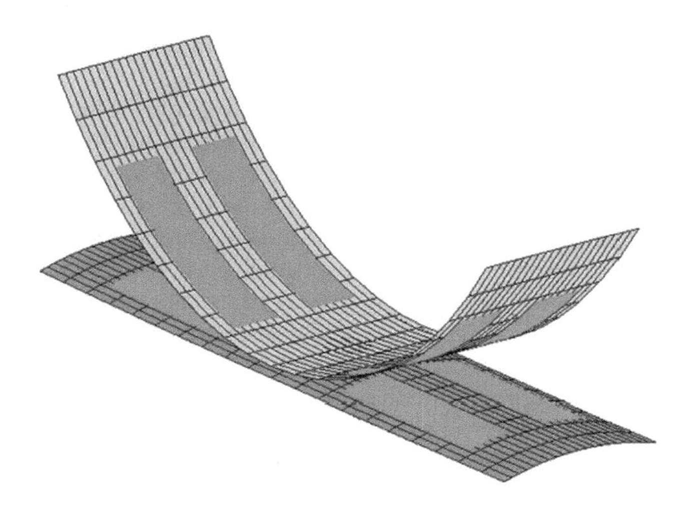

FIGURE 5.23
Two stable states of a $[0_P/0/90/90_P]_T$ laminate, AR = 3.

ratio laminates (Figure 5.23). For a fixed surface area, the thickness at which bistability is lost decreases with increasing aspect ratio and since a reduced thickness compromises the energy output (Figure 5.22b), a low aspect ratio is optimal. The overall sensitivity of objective function to laminate aspect ratio was small since as the laminate size increases, relative to thickness, the curvatures of square and rectangular laminates tend toward the same limiting value. Since the stress in the piezoelectric is directly related to laminate curvature (Equation 5.1), little variation of the objective function is observed.

This section presented the optimum energy harvester configurations for bistable piezocomposites. Although only the static states have been considered, the numerous geometric design variables offer flexibility for tailoring to specific resonances. For applications where the device experiences multiple time-dependent frequencies, there may be a balance between reasonable levels of energy harvesting across the entire vibration pattern, and tuning the device for optimality close to a single prominent frequency. The natural course of future development would be to model and optimize for the dynamic behavior of the piezocomposite energy harvesters.

5.7 Summary

In this chapter, we have considered the mechanics of bistable laminates that are actuated to induce snap-through between stable states. The analytical and finite element modeling approaches described can be used to predict laminate shape, snap-through events during piezoelectric actuation, and harvested energy. However, the modeling of the dynamics of snap-through is limited and more work is needed in this area. The optimization method enables the system to be considered and designed for a variety of potential applications. Energy harvesting is a promising area of application that needs further investigation, potentially used to power the onboard wireless sensors and electronics for air and ground vehicles.

Appendix

The constants used to express the shape coefficients *a*, *b*, and *c* of Equation 5.20 are given below.

$$\gamma_1 = 2\beta_1^2\beta_2^2\beta_3\beta_8$$

$$\gamma_2 = -2\beta_1^2\beta_2^2\beta_3^2 - 8\beta_2^2(\beta_6 + \beta_1\beta_7)^2$$

$$\gamma_3 = (\beta_2\beta_6 + \beta_1\beta_2\beta_4/2)^2$$

$$\gamma_4 = \beta_1\beta_2^2\beta_5(\beta_6 + \beta_1\beta_4/2)$$

$$\gamma_5 = -2(\beta_2\beta_6 + \beta_1\beta_2\beta_4/2)^2 - 4\beta_1^2\beta_2^2\beta_3^2$$

$$\gamma_6 = \beta_1^2\beta_2^2\beta_5^2/4 - 4\beta_1^2\beta_2\beta_3^2\beta_6$$

$$\gamma_7 = \gamma_1 + (\gamma_1\gamma_5 + \gamma_2\gamma_4)/2\gamma_3 - \beta_1^2\beta_2^2\beta_3^2\gamma_4(\gamma_3 + \gamma_5)/\gamma_3^2$$

$$\gamma_8 = \beta_1^2\beta_2^2\beta_8^2 - 4\beta_2\beta_6(\beta_6 + \beta_1\beta_7)^2 + (\beta_1^2\beta_2^2\beta_3^2(\gamma_4^2 - 2\gamma_3\gamma_6) - \gamma_1\gamma_3\gamma_4)/2\gamma_3^2$$

$$\gamma_9 = \beta_1^2\beta_2^2\beta_3^2\gamma_5/2\gamma_3^2 - \gamma_2/2\gamma_3$$

$$\gamma_{10} = -\beta_1^2\beta_2^2\beta_3^2\gamma_4/2\gamma_3^2 + \gamma_1/2\gamma_3$$

$$\gamma_{11} = \gamma_9^2(\gamma_5^2 - 4\gamma_3^2) - ((\beta_1^2\beta_2^2\beta_3^2\gamma_5^2 - \gamma_2\gamma_3\gamma_5)/2\gamma_3^2)^2$$

$$\gamma_{12} = -2\gamma_9^2\gamma_4\gamma_5 - 4\gamma_9^2\gamma_3\gamma_4 + 2\gamma_5^2\gamma_9\gamma_{10} - 8\gamma_3^2\gamma_9\gamma_{10} - 2\gamma_7(\beta_1^2\beta_2^2\beta_3^2\gamma_5^2 - \gamma_2\gamma_3\gamma_5)/2\gamma_3^2$$

$$\gamma_{13} = \gamma_{10}^2(\gamma_5^2 - 4\gamma_3^2) + \gamma_9^2(\gamma_4^2 - 4\gamma_3\gamma_6) - 4\gamma_4\gamma_9\gamma_{10}(2\gamma_3 + \gamma_5) - \gamma_7^2 - 2\gamma_8(\beta_1^2\beta_2^2\beta_3^2\gamma_5^2 - \gamma_2\gamma_3\gamma_5)/2\gamma_3^2$$

$$\gamma_{14} = 2\gamma_9\gamma_{10}(\gamma_4^2 - 8\gamma_3\gamma_6) - 2\gamma_{10}^2\gamma_4(\gamma_5 + 2\gamma_3) - 2\gamma_7\gamma_8$$

$$\gamma_{15} = \gamma_{13}/\gamma_{11} - 3\gamma_{12}^2/8\gamma_{11}^2$$

$$\gamma_{16} = \gamma_{12}^3/8\gamma_{11}^3 - \gamma_{12}\gamma_{13}/2\gamma_{11}^2 + \gamma_{14}/\gamma_{11}$$

$$\gamma_{17} = \gamma_{12}^2\gamma_{13}/16\gamma_{11}^3 - 3\gamma_{12}^4/256\gamma_{11}^4 - \gamma_{12}\gamma_{14}/4\gamma_{11}^2 + (\gamma_{10}^2(\gamma_4^2 - 4\gamma_3\gamma_6) - \gamma_8^2)/\gamma_{11}$$

$$\gamma_{18} = (((\gamma_{15}\gamma_{17}/6 - \gamma_{15}^3/216 - \gamma_{16}^2/16)^2 + (-\gamma_{15}^2/36 - \gamma_{17}/3)^3)^{1/2} - (\gamma_{15}\gamma_{17}/6 - \gamma_{15}^3/216 - \gamma_{16}^2/16))^{1/3}$$

$$\gamma_{19} = \gamma_{18} - 5\gamma_{15}/6 - (-\gamma_{17} - \gamma_{15}^2/12)/3\gamma_{18}$$

$$\gamma_{20} = (\gamma_{15} + 2\gamma_{19})^{1/2}$$

and

$$\beta_1 = 72(A_{11} - A_{12} + 10A_{66})/(L^4(A_{11} + A_{12})(A_{11}A_{66} - A_{12}A_{66} - 2A_{16}^2))$$

$$\beta_2 = (5/2)((A_{12}A_{66} + A_{16}^2)/(A_{11}^2 + 10A_{11}A_{66} - 10A_{16}^2 - A_{12}^2)) + 1/4$$

$$\beta_3 = D_{16}(A_{11} + A_{12}) - B_{11}B_{16}$$

$$\beta_4 = (D_{11} - D_{12})(A_{11} + A_{12}) - B_{11}^2$$

$$\beta_5 = 2(M_x(A_{11} + A_{12}) - B_{11}N_x)$$

$$\beta_6 = \frac{(10A_{66} + A_{11} - A_{12})(36(D_{11} + D_{12})(A_{66}(A_{11} - A_{12}) - 2A_{16}^2) - 72B_{16}^2(A_{11} - A_{12}) + 144A_{16}B_{11}B_{16} - 36A_{66}B_{11}^2)}{(L^4(A_{11}A_{66} - A_{12}A_{66} - 2A_{16}^2)^2)}$$

$$\beta_7 = D_{66}(A_{11} + A_{12}) - 2B_{16}^2$$

$$\beta_8 = (A_{11} + A_{12})M_{xy} - 2B_{16}N_x$$

References

1. Jun, W.J. and Hong, C.S. (1990). "Effect of residual shear strain on the cured shape of unsymmetric cross-ply thin laminates." *Composites Science and Technology* 38(1), 55–67.
2. Smart Material Corp, http://www.smart-material.com.
3. Hyer, M.W. (1981). "Some observations on the cured shape of thin unsymmetric laminates." *Journal of Composite Materials* 15(MAR), 175–194.
4. Dano, M.L. and Hyer, M.W. (1998). "Thermally-induced deformation behavior of unsymmetric laminates." *International Journal of Solids and Structures* 35(17), 2101–2120.
5. Betts, D.N., Salo, A.I.T, Bowen, C.R. et al. (2010). "Characterisation and modelling of the cured shapes of arbitrary layup bistable composite laminates." *Composite Structures* 92(7), 1694–1700.
6. Nigg, B.M., Cole, G.K., and Wright, I.C. (2007). "Optical methods." In: Nigg, B.M. and Herzog, W. editors. *Biomechanics of the Musculo-Skeletal System*. Chichester, UK: John Wiley & Sons, pp. 362–391.
7. Hamamoto, A. and Hyer, M.W. (1987). "Nonlinear temperature-curvature relationships for unsymmetric graphite-epoxy laminates." *International Journal of Solids and Structures* 23(7), 919–935.
8. Kant, T. and Swaminathan, K. (2000). "Estimation of transverse/interlaminar stresses in laminated composites—A selective review and survey of current developments." *Composite Structures* 49(1), 65–75.
9. Nosier, A. and Maleki, M. (2008). "Free-edge stresses in general composite laminates." *International Journal of Mechanical Science* 50(10–11), 1435–1447.
10. Bowen, C.R., Giddings, P.F., Salo, A.I.T. et al. (2011). "Modelling and characterization of piezoelectrically actuated bistable composites." *IEEE Transactions Ultrasonics, Ferroelectrics and Frequency Control* 58(9), 1737–1750.
11. Williams, R.B., Inman, D.J., Schultz, M.R. et al. (2004). "Nonlinear tensile and shear behavior of macro fiber composite actuators." *Journal of Composite Materials* 38(10), 855–869.
12. Deraemaeker, A., Nasser, H., Benjeddou, A. et al. (2009). "Mixing rules for the piezoelectric properties of macro fiber composites." *Journal of Intelligent Materials Systems and Structures* 20(12), 1475–1482.
13. Jaffe, H. and Berlincourt, D.A. (1965) "Piezoelectric transducer materials." *Proceedings of the IEEE* 53(10), 1372–1386.
14. ANSYS Inc. (2007). "Structures with geometric non-linearities. Paper 3." *Ansys Theory Reference V11.0*. Canonsburg, PA: ANSYS Inc.
15. Ren, L.B. (2008). "A theoretical study on shape control of arbitrary lay-up laminates using piezoelectric actuators." *Composite Structures* 83, 110–118.
16. Dano, M.L. and Hyer, M.W. (2003). "SMA-induced snap-through of unsymmetric fiber-reinforced composite laminates." *International Journal of Solids and Structures* 40(22), 5949–5972.
17. Bowen, C.R., Butler, R., Jervis, R. et al. (2007). "Morphing and shape control using unsymmetrical composites." *Journal of Intelligent Material Systems and Structures* 18, 89–98.

18. Schultz, M.R., Wilkie, W.K., and Bryan, R.G. (2007). "Investigation of self-resetting active multistable laminates." *Journal of Aircraft* 44(4), 1069–1076.
19. Kim, H.A., Betts, D.N., Salo, A.I.T. et al. (2010). "Shape memory alloy-piezoelectric active structures for reversible actuation of bistable composites." *AIAA Journal* 48(6), 1265–1268.
20. Schultz, M.R. (2008). "A concept for airfoil-like active bistable twisting structures." *Journal of Intelligent Material Systems and Structures* 19(2), 157–169.
21. Barbarino, S., Ameduri, S., Lecce, L. et al. (2009). "Wing shape control through an SMA-based device." *Journal of Intelligent Material Systems and Structures* 20(3), 283–296.
22. Betts, D.N., Kim, H.A., and Bowen, C.R. (2011). "Modeling and optimization of bistable composite laminates for piezoelectric actuation." *Journal of Intelligent Material Systems and Structures* 22(18), 2181–2191.
23. Arrieta, A.F., Hagedorn, P., Erturk, A. et al. (2010). "A piezoelectric bistable plate for nonlinear broadband energy harvesting." *Applied Physics Letters* 97, 104102.
24. Erturk, A. and Inman, S.J. (2009). "An experimentally validated bimorph cantilever model for piezoelectric energy harvesting from base excitations." *Smart Materials and Structures* 18, 025009.
25. Cottone, F., Vocca, H., and Gammaitoni, L. (2009). "Nonlinear energy harvesting." *Physical Review Letters* 102, 080601.
26. Betts, D.N., Kim, H.A., Bowen, C.R. et al. (2012). "Optimal configurations of bistable piezo-composites for energy harvesting." *Applied Physics Letters* 100, 114104.
27. Priya, S. (2007). "Advances in energy harvesting using low profile piezoelectric transducers." *Journal of Electroceramics* 19, 167–184.

6

Wing Morphing Design Using Macro-Fiber Composites

Onur Bilgen, Kevin B. Kochersberger, and Daniel J. Inman

CONTENTS

6.1 Introduction

The ability of an aircraft wing or any aerodynamic surface to change its geometry or to "morph" during flight has interested aircraft designers over the years, as morphing is almost always observed in nature and results in improved efficiency and control in a wide range of ambient conditions. Morphing is short for metamorphose; however, there is neither an exact definition nor an agreement among the researchers about the type or the extent of the geometrical changes necessary to qualify an aircraft for the title "wing morphing." In general, the geometrical parameters of an aircraft that can be affected by morphing solutions can be categorized into three main areas: planform morphing (span, sweep, and chord), out-of-plane morphing (twist, dihedral/gull, and spanwise bending), and airfoil morphing (camber and thickness). Historically, morphing in man-made aircraft almost always leads to penalties in terms of cost, complexity, and weight, although in certain circumstances these were overcome by system level benefits. The current trend for highly efficient and "green" aircraft makes such compromises less acceptable, calling for innovative and practical morphing designs able to provide more benefits and fewer drawbacks. Recent developments in "smart" materials have been shown to overcome some limitations and enhance the benefits from existing design solutions. The reader is referred to a comprehensive review paper on morphing aircraft by Barbarino et al. (2011) for more background information. As an example of "wing morphing," this chapter will focus on the topic of camber morphing of an airfoil using smart material actuators. In aerodynamics, camber represents the effective curvature of an airfoil. The term camber morphing simply refers to the change of the curvature of the airfoil by means of actuators, in our case, by the means of smart materials.

Piezoelectric materials are the most widely used smart (also referred to as active) materials because they offer actuation and sensing over a wide range of frequencies. A Macro-Fiber Composite (MFC) is a type of piezoelectric device that offers structural flexibility and high actuation authority (up to 0.2% in-plane free strain). A challenge with piezoelectric actuators is that they require high voltage input. In contrast, the current drain is relatively low, which creates small power consumption and requires relatively lightweight electronic components. Piezoelectric materials are already feasible in small platforms due to the continuing developments in the electronic systems. At the same time, recent interest in small unmanned air vehicles (UAVs) and micro air vehicles (MAVs) has driven a need to investigate the use of piezoelectric materials for shape and flow control and for vibration attenuation. For example, field-deployable aircraft have flexible wings that can be folded during transportation, and they can be unfolded for operation. These compliant wings can be realized with the integration of piezocomposite actuators. A major challenge is encountered when operating a relatively compliant, thin structure (desirable for piezocomposite actuators) in situations where there are relatively high external forces. *Establishing a wing configuration that is stiff enough to prevent flutter and divergence but compliant enough to allow the range of available motion is the central challenge in developing a piezocomposite airfoil.* Novel wing morphing designs that incorporate piezocomposite actuators can take advantage of aerodynamic loads to reduce control input moments and increase control effectiveness. This chapter introduces two such designs.

There are several benefits of employing camber morphing via active materials over the discrete trailing edge control using conventional control surfaces in small air vehicles. First, the low Reynolds number flow regime can result in flow separation that reduces the effectiveness of a trailing edge control surface. Camber variation via piezocomposite materials can induce smooth changes in the airfoil curvature, which in return reduce or delay the separation of flow. Second, small UAVs and MAVs cannot afford to lose energy through control surface drag because of their severe power limitations; therefore, airfoil efficiency may be improved by controlling the aircraft via smooth variations in camber. Finally, the opportunity for flow control is inherent in the active material due to its direct effect on circulation and its high operating bandwidth. The bandwidth advantage of a conformal actuator opens the possibility for dynamic actuation that may have significant advantages.

The following section presents a literature survey on the use of piezoelectric materials, including the MFC, in rotary-wing aircraft blades, small fixed-wing aircraft, and rotary-wing exit guide vanes.

6.1.1 Literature Survey

Piezoelectric materials offer relatively high force output in a wide frequency range. Although the strain output is very small for low excitation levels, the response is relatively linear. In the linear regime, the fast response of lead zirconate titanate (PZT) materials caused initial interest in the field of aerodynamic vibration control. Many researchers focused on the application of piezoelectric materials to the blades of rotary-wing aircraft to improve their performance and effectiveness. Steadman et al. (1994) showed an application of a piezoceramic actuator for camber control in helicopter blades. Giurgiutiu et al. (1994) has researched performance improvement on rotor blades using strain-induced actuation methods using PZTs. Wind tunnel experiments proved the control authority of the PZT actuators. Giurgiutiu (2000) presented a comprehensive review on the application of smart material actuation to counteract aeroelastic and vibration effects in helicopters and fixed-wing aircraft.

The "static" control of aerodynamic surfaces using piezoelectric materials started in the early 1990s for large aircraft. Researchers employed both the low-amplitude (mostly linear) and high-amplitude (nonlinear) excitation ranges of the piezoelectric material. Lazarus et al. (1991) examined the feasibility of using representative box wing adaptive structures for static aeroelastic control. Pinkerton and Moses (1997) discussed the feasibility of controlling the wing geometry employing a piezoelectric actuator known as the thin layer composite–unimorph ferroelectric driver and sensor (THUNDER). Hysteresis nonlinearity was observed in the voltage-to-displacement relationship. Barrett and Stutts (1998) proposed an actuator that uses a pair of piezoceramic sheets arranged in a push–pull assembly to turn a spindle of an aerodynamic control surface. Wind tunnel testing showed the feasibility of the concept. Geissler et al. (2000) adopted piezoelectric materials as the actuation element for a morphing leading edge; however, in this case, the leading edge was an independent element able to rotate around an internal hinge. Munday and Jacob (2001) developed a wing with conformal curvature, driven by a THUNDER actuator internally mounted in a position so as to alter the upper surface shape of the airfoil, resulting in a variation of the effective curvature. Wang et al. (2001) presented results from the DARPA/AFRL/NASA Smart Wing Phase 2 Program, which aimed to demonstrate high-rate actuation of hingeless control surfaces using several smart material–based actuators, including piezoelectric materials. Grohmann et al. (2006, 2008) presented the active trailing edge and active twist concepts applied to helicopter rotor blades. The paper presented the optimization of sizing and placement of piezocomposite patches so that desired wing weight and pitching moment output is obtained.

The rapid development and the reduced cost of small electronics in the last decade led to the interest of using piezoelectric materials in small unmanned (or remotely piloted) fixed-wing, rotary-wing, and ducted-fan aircraft. Eggleston et al. (2002) presented the use of piezoceramic materials (THUNDER), shape-memory alloys (SMAs), and conventional servomotors in a small unmanned aircraft. Wind tunnel testing showed the feasibility of the smart material systems. Barrett et al. (2005) employed piezoelectric elements along with elastic elements to magnify control deflections and forces of an aerodynamic surface. The so-called post-buckled precompression (PBP) concept was employed as guide vanes in a small rotary-wing aircraft. Vos et al. (2007, 2008) conducted research to improve the PBP concept for aerodynamic applications. Roll control authority was increased on a 1.4-m wingspan UAV. Kim and Han (2006) and Kim et al. (2009) designed and fabricated a smart flapping wing by using a graphite/epoxy composite material and an MFC actuator. A 20% increase in lift was achieved by changing the camber of the wing at different stages of flapping motion. Bilgen et al. (2009) presented a unique application of piezocomposite actuators on a 0.76-m wingspan morphing wing UAV. In this application, two MFC actuators were bonded to the wings and the camber of the wings was changed asymmetrically with voltage excitation. Adequate roll control authority was demonstrated in the wind tunnel as well as in flight. Paradies and Ciresa (2009) implemented MFCs as actuators into an active composite wing. A scaled prototype wing was manufactured, and models were validated with static and preliminary dynamic tests of the prototype wing. Wickramasinghe et al. (2009) presented the design and verification of a smart wing for a UAV. The proposed smart wing structure consisted of a composite spar and ailerons that have bimorph active ribs with MFC actuators. Butt et al. (2010a,b) and Bilgen et al. (2011a) presented a completely servo-less, wind tunnel and flight-tested remotely piloted aircraft. The MFC actuators were used to create variable-camber, continuous, piezocomposite wings instead of the traditional servomotor controlled, discrete control surfaces. This vehicle became the first fully solid-state piezoelectric material controlled, nontethered, flight-tested, fixed-wing aircraft.

6.1.2 The Macro-Fiber Composite

In many cases, monolithic piezoceramic materials are brittle, making them impractical for direct shape control. The MFC, a type of piezocomposite device, was developed at NASA Langley Research Center (Wilkie et al. 2000). An MFC is a flexible, planar actuation device that employs rectangular cross-section, unidirectional piezoceramic fibers (PZT 5A) embedded in a thermosetting polymer matrix (High and Wilkie 2003). The in-plane poling and subsequent voltage actuation allows the MFC to utilize the *33* piezoelectric effect, which is much stronger than the *31* effect used by traditional PZT actuators with through-the-thickness poling (Hagood et al. 1993). Lloyd (2004), Williams (2004), Sodano et al. (2004, 2006), Bilgen (2010), and Bilgen et al. (2010a) presented detailed linear and nonlinear characterization of the mechanical and piezoelectric behavior of the MFC device. Park and Kim (2005) presented an analytical development of single crystal MFC actuators for active twist rotor blades. Wilkie et al. (2006) presented an experimental study on incorporating single crystal piezoelectric materials into the MFC actuator. Tarazaga et al. (2007) employed the MFC actuator for vibration suppression in a thin-film rigidizable inflatable boom.

6.1.3 Outline of the Chapter

The rest of the chapter is divided into three main parts. Section 6.2 presents the design and analysis of a thin, variable-camber bimorph airfoil that is actuated by MFCs. A unique design is presented and the theoretical analysis of the coupled static fluid–structure interaction problem is shown. A prototype is presented and experimentally analyzed through wind tunnel tests. The structural and aerodynamic response of the variable-camber airfoil is measured and compared with a flat-plate airfoil. Aerodynamic and structural nonlinearities are discussed. Section 6.3 presents a thick variable-camber airfoil employing cascading MFC bimorphs and an internal compliant-box mechanism. First, a parametric examination of the aerodynamic response is employed to optimize the kinematic parameters of the airfoil. A prototype of the concept is presented. Wind tunnel experiments are conducted to quantify the aerodynamic and structural response. Nonlinear effects due to aerodynamic and piezoceramic hysteresis are identified and discussed. Results are compared with conventional symmetric NACA four-series and other airfoils. A brief summary of the chapter is given in the Section 6.4.

6.2 MFC Actuated Simply Supported Thin Airfoil

One of the common goals of recent research in morphing aircraft is to exploit smart material actuation, unique structural boundary conditions, and novel materials to achieve wing morphing. In this section, a thin variable-camber airfoil concept employing MFCs is proposed to take advantage of smooth control surface deformations (see Bilgen et al. 2010b for details). Note that Figures 6.1, 6.3 through 6.9, and 6.11 through 6.17 are reproduced courtesy of IOP Publishing Ltd. A bimorph configuration is used to create the airfoil surface because (1) it can induce large out-of-plane deformations in comparison to a unimorph and (2) it can be actuated with lightweight electronics. Several lightweight high-voltage amplifier designs are presented in Bilgen et al. (2010c).

6.2.1 Thin Bimorph Airfoil Concept

A thin bimorph airfoil with reasonable stiffness and deformation output is possible with an MFC actuator given that the boundary conditions are favorable. Therefore, the support system for a variable-camber device intended for circulation control is proposed in Figure 6.1. The airfoil surface, in its actuated state, gains some camber due to the pressure loading between the supports and has a slight loss of camber on the aft end, also due to the pressure loading. The net result is a change in circulation depending on the boundary conditions, hence the structural response.

In Figure 6.1b, the label "AOA" represents the geometric angle of attack, "β" represents angle of the support condition, and "LEα" represents the leading edge incidence angle with the free stream velocity. The boundary conditions in the design are pinned–pinned (similar to a simply supported beam) for ease of implementation; however, one can choose the second boundary condition (Pin 2) as a slider. The pinned–pinned boundary condition theoretically creates an "artificial" nonlinear stiffness; however, this stiffness is not dominant in an actual implementation of the proposed airfoil geometry and the pin locations. Starting with the baseline design, multiple configurations can be generated by changing the location of the pins as shown in Figure 6.2.

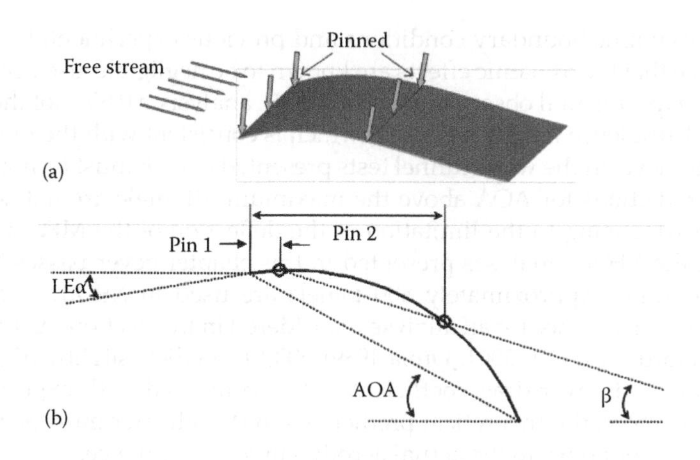

FIGURE 6.1
(a) Actuated shape of the thin variable-camber airfoil concept. (b) Illustration of the possible chord-wise locations of two "pins" and other conventional parameters. (From Bilgen, O., Kochersberger, K. B., Inman, D. J. and Ohanian, O. J., *Smart Materials and Structures*, 19, 055010, 2010. With permission.)

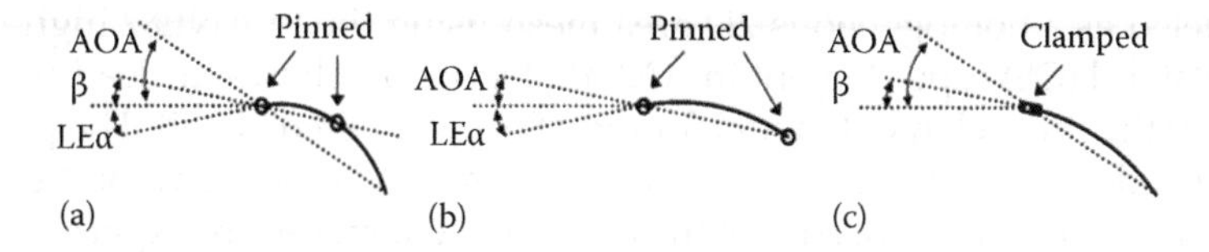

FIGURE 6.2
(a) Baseline configuration; (b) simply supported beam; and (c) cantilevered beam.

The two extreme configurations are as follows: (1) the middle support is moved to the trailing edge, which is similar to a sail or a simply-supported beam; (2) the middle support is moved to the leading edge; hence, the airfoil becomes a cantilevered beam. The "morphing" of the airfoil is achieved by the use of an MFC bimorph actuator that indeed forms most of the surface of the airfoil. A single MFC actuator has a voltage range of –500 to 1500 V. Since the airfoil is in a bimorph configuration, the MFC on the "other" side of the bimorph is actuated with an opposite field and with 3-to-1 fixed ratio. The higher of the two excitation voltages is used in the discussion in this chapter.

6.2.2 Theoretical Static Aeroelastic Analysis

A theoretical analysis is conducted to determine the effect of pin locations on the two-dimensional (2D) aerodynamic response. A MATLAB-based program is developed to solve the static fluid–structure interaction problem by iterating between a panel method software XFOIL (Drela 1989), and a finite element software ANSYS. The analysis considers only chordwise distribution of aerodynamic loads and structural deformations because of the structural and aerodynamic boundary conditions and previous experimental observations. Note that the dynamic effects are known to be negligible also because of previous experimental observations. For XFOIL analysis, a 0.85% (of the mean velocity) turbulence level is assumed, which is consistent with the measured turbulence level in the wind tunnel tests presented later. It must be noted that XFOIL predictions for AOA above the maximum lift angle are not accurate (Drela 1989). Owing to the limitation of the deflection of the MFC bimorph actuator, the XFOIL analyses presented in this chapter never passes beyond this AOA limit. Approximately 800 panels are used in XFOIL to achieve numerical convergence for all analyses considered in this section. As reported in the literature (XFOIL 2001, Drela 1989), XFOIL predicts slightly higher lift coefficients and lower drag coefficients when compared with experimental results; therefore, the theoretical predictions in this chapter must be viewed as an upper boundary to the actual aerodynamic performance.

The airfoil is assumed to be a thin circular arc with 1.00% thickness, have a round leading edge (LE) and tapered trailing edge (TE) with a finite TE thickness of 0.05% chord. A constant curvature is assumed owing to the high chordwise coverage of the active material. The structure of the bimorph airfoil is modeled as a homogeneous 2D area mesh using the PLANE82 high-order quadrilateral (Q8) type element in ANSYS. Figure 6.3 shows the leading-edge region of the finite element model for the 4.00% cambered airfoil. The percent-camber is defined as the percentage of the height over the chord of the circular arc. The material properties of the finite element model are experimentally determined. A 16.5-GPa Young's modulus is used from the experimentally derived bending stiffness, and a 0.267 Poisson's ratio is assumed.

The nonlinear voltage–camber relationship with significant hysteresis is previously quantified through experiments on several specimens of MFC unimorph and bimorph actuators. The change in camber due to excitation voltage is deduced from experiments conducted on a representative MFC bimorph actuator, and this result is used in the theoretical analysis.

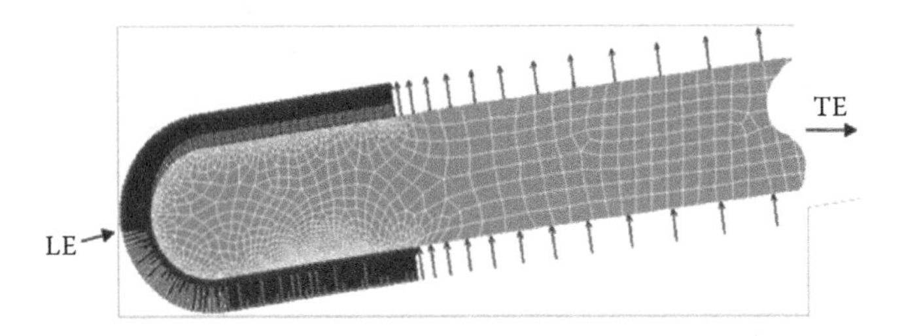

FIGURE 6.3
Two-dimensional structural mesh and pressures loading of the 4.00% cambered 1.00% thick airfoil. LE area is shown with high resolution of pressure distribution and structural elements. Lengths of arrows are not proportional to magnitude of pressure field. (From Bilgen, O., Kochersberger, K. B., Inman, D. J. and Ohanian, O. J., *Smart Materials and Structures*, 19, 055010, 2010. With permission.)

Approximately 4.00% camber is measured at 1400 V excitation for the MFC bimorph employed in this section.

The static aeroelastic model is used in a parametric analysis. The two pin locations are used as independent variables to find a configuration that results in high lift output given that the free stream flow is parallel with the pins (β = 0°). Pin 1 (P1) is varied from 5% chord to 30% chord downstream of the LE. Pin 2 (P2) is varied starting from 5% chord downstream of Pin 1 location to the TE. In the structural model, Pin 1 is constrained in both the streamwise and the perpendicular directions; however, Pin 2 is only constrained in the perpendicular direction. Figure 6.4 shows the theoretical

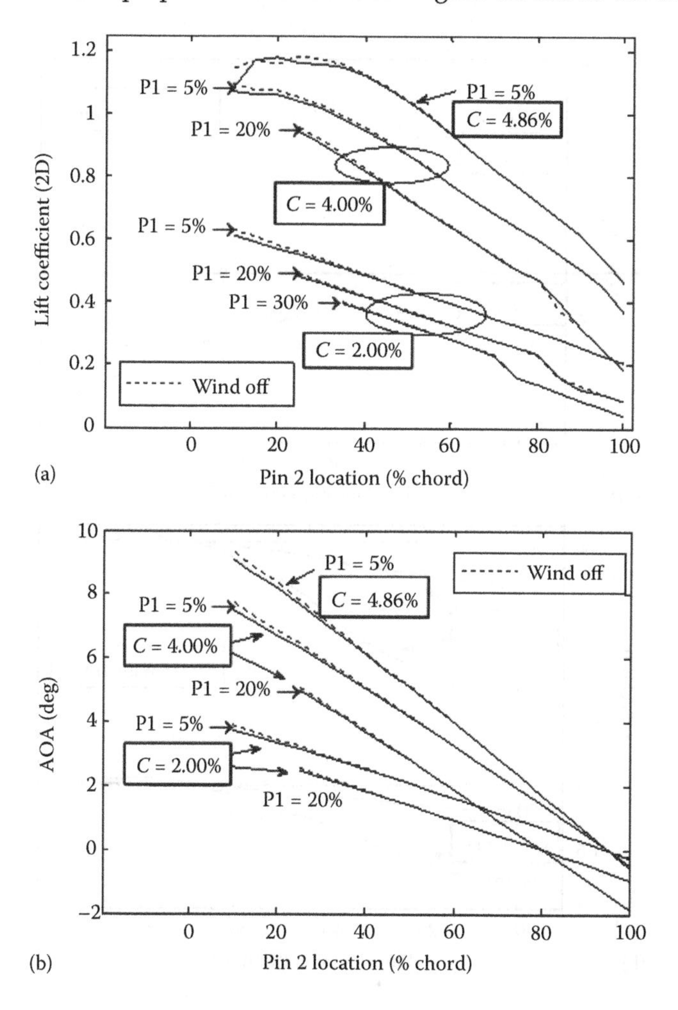

FIGURE 6.4
Theoretical (2D) (a) lift coefficient and (b) angle of attack (AOA) response for a 1.00% thick airfoil at 15 m/s flow. Initial camber is labeled as C. $Re_{chord} = 1.27 \times 10^5$. (From Bilgen, O., Kochersberger, K. B., Inman, D. J. and Ohanian, O. J., *Smart Materials and Structures*, 19, 055010, 2010. With permission.)

change in lift coefficient and AOA for a 1.00% thick circular-arc airfoil with 127-mm chord. The airfoil is actuated to an initial camber (C) of 2.00%, 4.00%, and 4.86% (of chord) and subjected to 15 m/s free stream velocity. The plots present both the "wind-on" (solid lines) and "wind-off" (dotted lines) analyses. Considering the experimentally measured bending stiffness of the bimorph, the flow-induced deformation is negligible at 15 m/s for almost all configurations of the boundary conditions. All results presented in this section are for a constant β angle of zero degrees.

The analysis shows that placing Pin 1 at or close to the LE results in the highest lift output. The remainder of this section considers the cases with

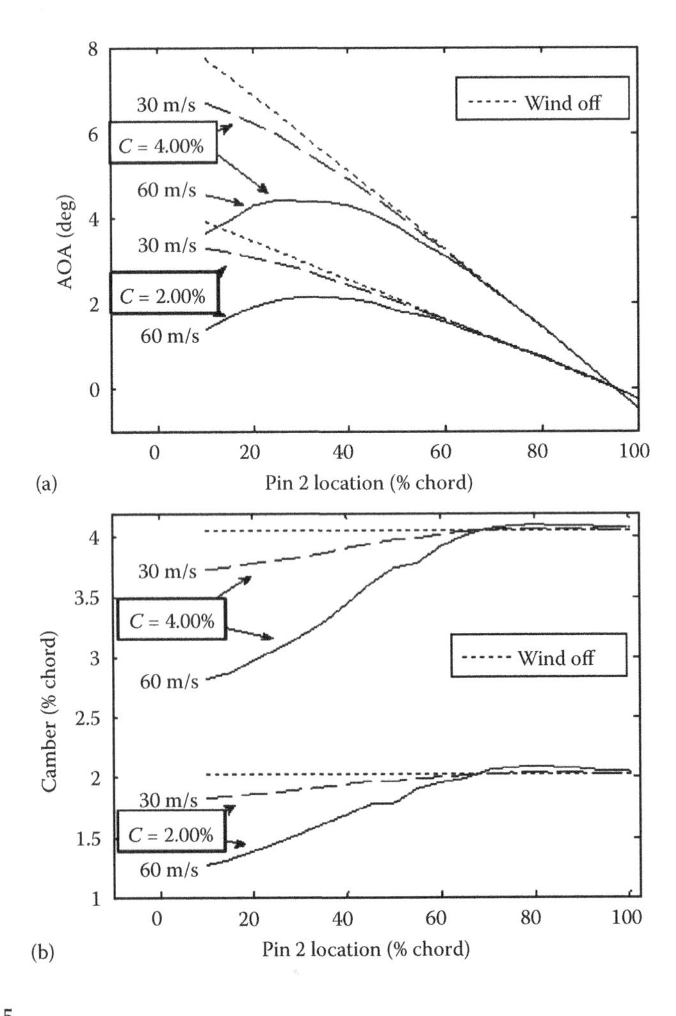

FIGURE 6.5
Theoretical (2D) (a) AOA and (b) effective camber response for a 1.00% thick airfoil with Pin 1 at 5% chord and variable Pin 2 location, subjected to 30 and 60 m/s. (From Bilgen, O., Kochersberger, K. B., Inman, D. J. and Ohanian, O. J., *Smart Materials and Structures*, 19, 055010, 2010. With permission.)

Pin 1 located at 5% chord. The evaluation is extended to higher free stream velocities of 30 and 60 m/s, which can be observed at the exit of the ducted fan. Figure 6.5 presents the effective camber and the AOA.

In Figure 6.5, the initial piezoelectric strain-induced camber (with no aerodynamic loading) is labeled as *C*. As the velocity is increased, a large reduction in camber and AOA is observed for Pin 2 locations up to 60% chord. Pin 2 locations beyond this point results in a small (favorable) increase in camber and AOA. Figure 6.6 presents the variation of lift coefficient and lift-to-drag ratio (*L*/*D*) due to change in Pin 2 location.

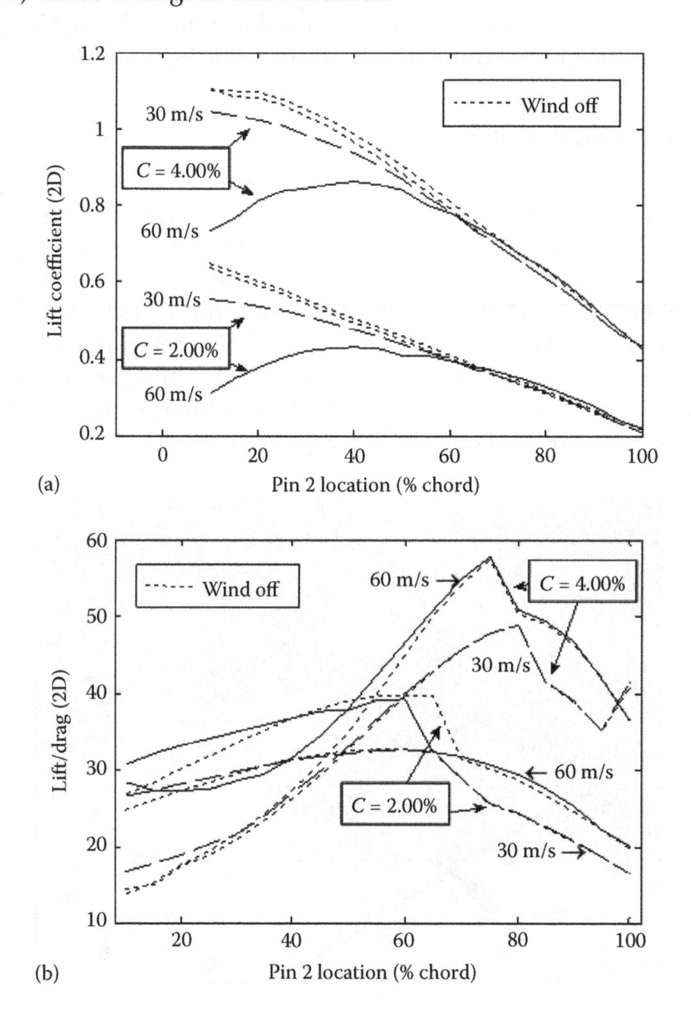

FIGURE 6.6
Theoretical (2D) (a) lift coefficient and (b) lift-to-drag ratio response for a 1.00% thick airfoil with Pin 1 at 5% chord subjected to free stream velocities of 30 and 60 m/s. (From Bilgen, O., Kochersberger, K. B., Inman, D. J. and Ohanian, O. J., *Smart Materials and Structures*, 19, 055010, 2010. With permission.)

The effect of higher dynamic pressures becomes clearly visible at 60 m/s between the wind-on and wind-off analyses. In the ducted-fan aircraft wind tunnel tests (presented later), the airfoil with pinned–pinned boundary condition showed no aerodynamic instability or failure at fan speeds exceeding 45 m/s. In summary, the theoretical analysis shows that a 1.00% thick airfoil presents adequate stiffness at 15 m/s for all analyzed boundary conditions. At higher dynamic pressures, adverse deformations that can reduce circulation (or cause structural failure) can be avoided by more "conservative" pin locations at the expense of reduced aerodynamic output. The configuration with Pin 1 at 5% chord and Pin 2 at 40% chord results in the highest possible lift output for the upper boundary of aerodynamic loading (60 m/s).

6.2.3 Thin Airfoil Prototype

Using the conclusions derived from the theoretical analysis and the geometric constraints of the ducted-fan aircraft (presented later), two MFC actuated thin bimorph airfoils are fabricated. A total of four stainless-steel pins (two on each end) are bonded to the airfoil at 5% and 50% chord from the LE. This configuration is chosen because it is a good compromise between the restrictions of the ducted-fan aircraft geometry and the determined "optimum" pin locations (P1 = 5% and P2 = 40% for high lift output) from the analysis in Section 6.2.2. Figure 6.7 shows one of the two bimorph airfoils employing four (two on each end) MFC M8557-P1-type actuators.

FIGURE 6.7
Prototype variable-camber bimorph airfoil with four MFC M8557-P1 actuators, 127 mm chord, and 133 mm span. (From Bilgen, O., Kochersberger, K. B., Inman, D. J. and Ohanian, O. J., *Smart Materials and Structures*, 19, 055010, 2010. With permission.)

The bimorph is fabricated by sandwiching a 0.027-mm-thick stainless-steel sheet and bonding the laminate under vacuum. The MFCs, each having an 85 mm × 57 mm active area, are aligned at the LE in the chordwise direction. Two layers of 0.027-mm-thick stainless-steel metal (passive material) are bonded to the TE to complete the total chord to 127 mm.

6.2.4 Ducted-Fan Aircraft Wind Tunnel Experiments

The preliminary motivation for the variable-camber airfoil concepts presented in this chapter is to determine the potential effectiveness of MFC actuated bimorph control surfaces for an experimental vertical takeoff and landing (VTOL) ducted-fan aircraft designed by AVID LLC (Blacksburg, VA). This experimental aircraft is developed through a US Air Force Research Laboratory (AFRL) funded Phase II Small Business Innovation Research (SBIR) contract. One of the two thin variable-camber bimorph airfoil prototypes, mounted at the exit of the ducted-fan aircraft, is shown in Figure 6.8. The ducted fan is mounted on its side for wind tunnel testing.

These preliminary and "qualitative" wind tunnel tests are conducted at a maximum fan flow speed of 45 m/s and maximum free stream velocity of 10 m/s. The thin bimorph airfoil demonstrated high lift output and no adverse deformation due to high aerodynamic loads. Most important, a structural failure (e.g., buckling) or an aerodynamic instability is not observed at fan speeds up to 45 m/s.

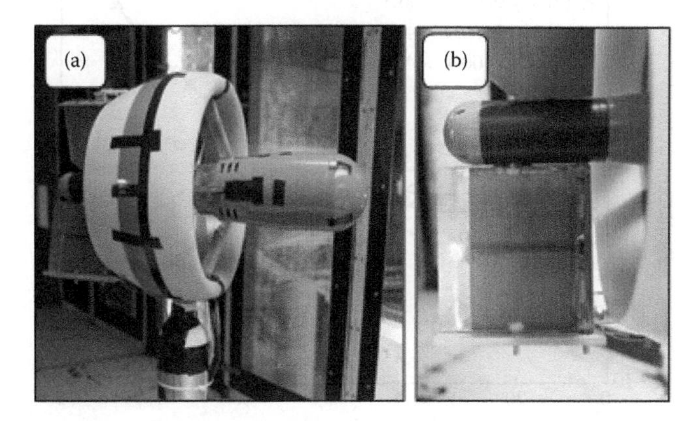

FIGURE 6.8
(a) Ducted-fan vehicle installed in the Virginia Tech 6 ft × 6 ft Stability Tunnel. (b) Close-up of the thin control surface installed in the ducted-fan vehicle. The airflow is from right to left. (From Bilgen, O., Kochersberger, K. B., Inman, D. J. and Ohanian, O. J., *Smart Materials and Structures*, 19, 055010, 2010. With permission.)

6.2.5 Structural and Aerodynamic Experiments

6.2.5.1 Structural Response

The shape of the variable-camber bimorph airfoil at a given condition must be measured accurately owing to the hysteretic nature of the piezocomposite bimorph. The shape measurement is also necessary to observe deformation (if any) due to aerodynamic loading. In the current work, the airfoil shape is approximated by measuring the displacement of the airfoil mid-span section at two locations simultaneously. The measurement is taken using two laser displacement sensors. With the known axis of rotation of the airfoil, a circular-arc shape is fitted to the two measured displacements and the third known axis location. The AOA and the effective camber are calculated for the fitted circular-arc shape.

First, a calibration experiment is conducted to determine the validity of the circular-arc assumption and the accuracy of the laser displacement measurements. Figure 6.9 shows the displacement of the airfoil end-section digitized from a series of photos for a peak-to-peak voltage sweep. A negative sign simply indicates actuation in the reverse direction. In Figure 6.9, the actuation voltage is swept from –1400 to 1400 V and swept back to –1400 V. The effect of piezoelectric hysteresis can be observed in the difference between deflection at zero voltage during the positive and negative voltage sweeps. However, the deflection is very repeatable, as seen at the maximum positive and negative voltage levels. Figure 6.9 also shows that a circular-arc is a

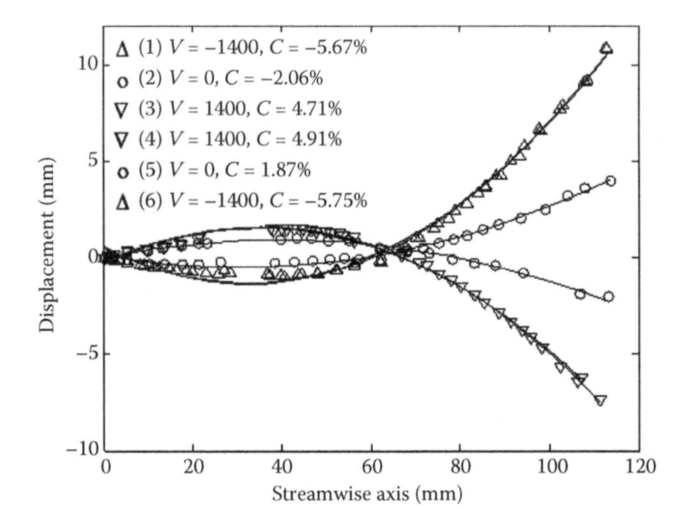

FIGURE 6.9
Airfoil end-section measurements with a digital camera. Legend shows test order, voltage (V), and camber (C) of the circular-arc fit (shown by solid lines). (From Bilgen, O., Kochersberger, K. B., Inman, D. J. and Ohanian, O. J., *Smart Materials and Structures*, 19, 055010, 2010. With permission.)

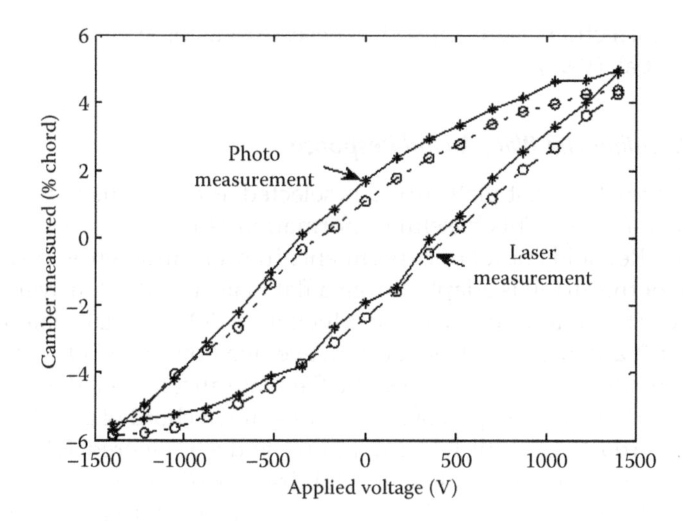

FIGURE 6.10
Comparison of two experimental deflection measurement techniques.

good approximation; however, it is also observed that the actual shape does deviate slightly from this assumption. There is a small and consistent offset between the two measurement methods, which is presented in Figure 6.10. Note that the lasers measure the mid-span section where the photo method measures the end section. No visible spanwise variation of shape is observed; however, such variation is inherent in a composite airfoil due to fabrication limitations and the shape of the MFC bimorph actuator.

6.2.5.2 Wind Tunnel Characteristics

The main wind tunnel tests are performed in a low-speed, open circuit wind tunnel facility with a 2D test-section configuration. A manual beam balance is used to measure the lift and drag forces generated by the airfoils tested in this section. The maximum measured spatial variation of the flow profile is $U_{\text{Velocity}} = \pm 0.54$ m/s at 15.3 m/s; therefore, the flow is assumed spatially uniform for coefficient calculations. A turbulence intensity of 0.60% is measured from 0.1 Hz to 50 kHz bandpass-filtered hot-wire signal for the current average flow speed of 15 m/s. Note that the turbulence of the wind tunnel used in this chapter is relatively large compared with other tunnels used in the research field. The airfoils tested in this chapter have an approximate 1.5-mm gap between the airfoil ends and the tunnel. Mueller and Burns (1982) show that gap sizes around 0.5% of the span are usually acceptable and do not affect the results. For the airfoils tested here, the gap is approximately 1.13%. Although the gap dimension is small, the percentage is still higher than recommended because of the small span of the airfoils. Tunnel wall effects and buoyancy corrections were applied as necessary using the techniques found in Barlow et al. (1999).

6.2.5.3 Baseline Flat Plate Airfoil Response

An aluminum, thin, flat-plate airfoil is selected as a baseline to the variable-camber airfoil tests. This baseline information is necessary because of the lack of low Reynolds number experiments in high turbulence settings. The geometry of the airfoil is adapted from a flat plate airfoil extensively studied by Mueller (1999) and Pelletier and Mueller (2000). The flat plate airfoil is tested for lift and drag coefficients at an average flow speed of 15 m/s and a chord Reynolds number of 127,000. The flat plate airfoil has a thickness of t = 2.54 mm (2.0% of chord), span of b = 133 mm, and a chord of c = 127 mm. The airfoil has a round LE with 1.27 mm radius, and a 3° tapered TE. Figure 6.11 presents comparison of flat plate lift and drag coefficient data from the current test to the ones tested by Mueller (1999), Pelletier and Mueller (2000), and Selig et al. (1989). XFOIL results are also presented. The data from the current evaluation represents the average of AOA *sweep up* and *sweep down* tests. The averaging is done because of the lack of aerodynamic hysteresis.

The lift curve slope matches well for all tests shown, which are slightly lower than the theoretical slope of 2π/rad and the XFOIL prediction. The Reynolds number has negligible effect on the lift coefficient. XFOIL predicts slightly lower lift for AOA above 8°; however, the effect of turbulence is consistent with the experiments. The maximum lift coefficient uncertainty is $U_{CL,max} = \pm 0.055$, which is mostly due to low frequency variation of velocity. The lift curve slope is consistent with the values reported in Hoerner (1975) for a thin flat plate. The drag coefficient from the flat plate airfoil shows a slightly higher trend at lower angles when compared with data reported by Mueller. The zero-degree drag coefficient is 0.024 for the flat plate airfoil. There are several reasons for this difference. First, note that Mueller reports an elliptical LE with ratio of $5t/2$, where the airfoil studied here has a circular LE. At zero-degree AOA, a sharper LE is expected to create less drag. Second, the turbulence levels in different facilities must be noted. Mueller reports 0.05% turbulence intensity over the range of tests conducted. A 0.80-mm gap between the wing and the end plates is also reported. Selig reports 0.358% rms turbulence for 0.01 Hz cutoff frequency, and 0.064% rms turbulence for 1 Hz cutoff frequency, at a Reynolds number of 100,000. Finally, the thin airfoil studied here is supported by four pins (two on each end) that are bonded at 5% and 50% chord. The cylindrical stainless-steel pins are 4.8 mm in diameter and extended 4.8 mm into the flow. The drag coefficient for a finite-length cylinder in a cross-flow is approximately 1.1, as reported by Hoerner (1993). This translates into a corrected drag coefficient of 0.018 at zero AOA for the flat plate. Mueller reports approximately 0.013 for the zero-degree drag coefficient. The observed mismatch in drag coefficient is acceptable considering the maximum drag coefficient uncertainty of $U_{CD,max} = \pm 0.014$.

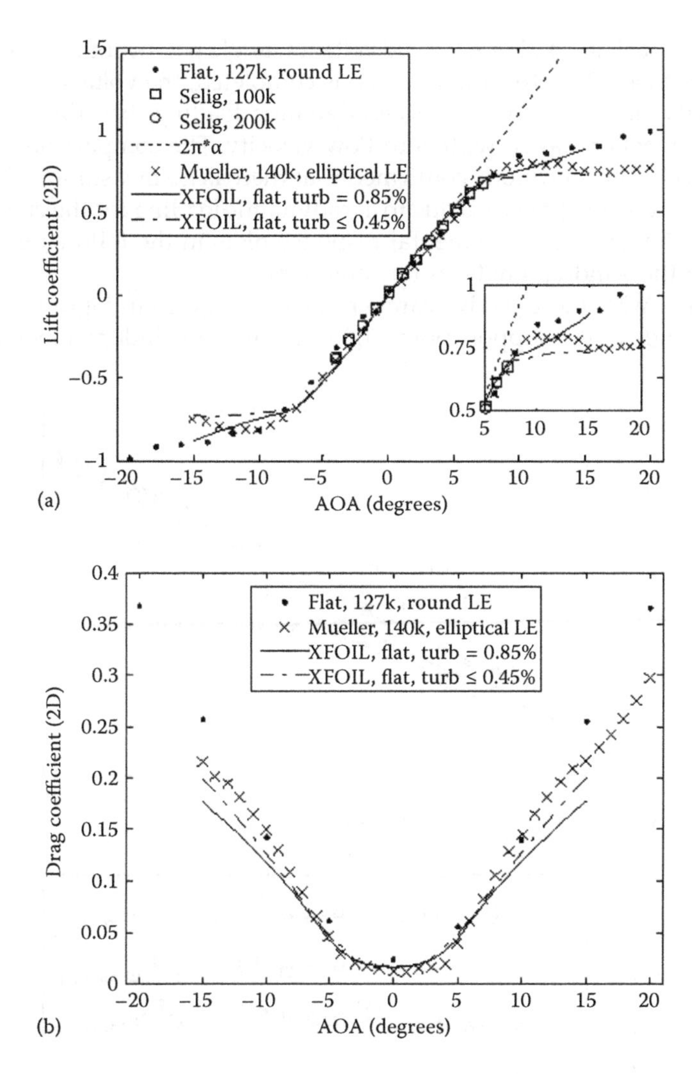

FIGURE 6.11
Experimental (a) lift and (b) drag coefficients versus angle of attack comparison for three thin flat-plate airfoils. (From Bilgen, O., Kochersberger, K. B., Inman, D. J. and Ohanian, O. J., *Smart Materials and Structures*, 19, 055010, 2010. With permission.)

6.2.5.4 Aerodynamic Response: Angle Sweep at Fixed Voltage

The thin variable-camber bimorph airfoil is tested for its lift and drag performance at a flow speed of 15 m/s. Two different test schemes are followed. The first evaluation is performed by setting the voltage of the bimorph at a fixed value, then sweeping the support angle (β) up and down. The complete list of voltages (in order) is as follows: 0 V (+β), 0 V (−β), 700 V (+β), 1400 V (+β), −700 V (−β), and −1400 V (−β). For positive voltages, angle β is swept from −4°

to 20° then back to −4°. For negative voltages, angle β is swept from 4° to −20° then back to 4°. Once the angle sweep is completed, the voltage is changed. To identify deformation due to aerodynamic loading, baseline deflection measurements are taken with zero flow velocity. The comparison of wind-on and wind-off conditions confirmed that there is no measurable deformation (chordwise or spanwise) due to aerodynamic loading for the current test speed of 15 m/s. All experimental response plots in the following sections represent the wind-on analyses for consistency.

First, the AOA and effective camber values are given in Figure 6.12. Note that the angle of the support points (β) is given as the independent variable

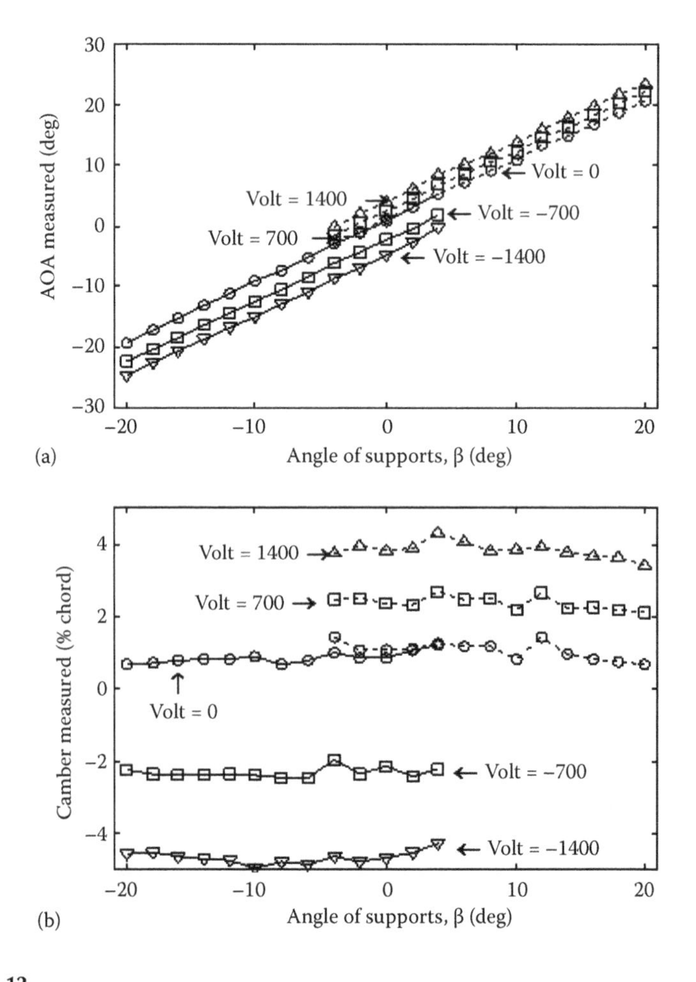

FIGURE 6.12
Experimental (a) angle of attack change due to voltage input and the change in β. (b) Effective camber of the circular-arc fit due to voltage input and the change in β. $Re_{chord} = 1.27 \times 10^5$. (From Bilgen, O., Kochersberger, K. B., Inman, D. J. and Ohanian, O. J., *Smart Materials and Structures*, 19, 055010, 2010. With permission.)

for most of the figures, which is necessary for clear presentation of the data. The angle β is swept up and down for all voltage levels to change the aerodynamic loading on the cambered airfoil. Since no aerodynamic hysteresis is observed in the experiments, all plots related to the "first test scheme" present the average of sweep up and sweep down curves.

The voltage–geometry relationship appears linear (and independent of β) since voltage is changed only when a β sweep is completed. It is observed that the AOA and camber are not zero for the zero degrees β and zero volt condition. This is caused by the residual shape that piezoceramic bimorph holds from previous tests. This is a typical characteristic of a hysteretic piezoceramic actuator that goes through high deformations and therefore builds "memory" of a previous time history of excitation. It is important to note that the peak-to-peak response of the airfoil (e.g., +/–1400 V) will remain the same as long as this range is not exceeded by a "previous" excitation. It is only the intermediate points within this range that are "floating" due to the memory of a time history of excitation. The effects and compensation of hysteresis will be discussed briefly throughout this chapter. The reader is referred to Mayergoyz (2003) for a comprehensive analysis of hysteresis through phenomenological models, such as one presented by Preisach (1935).

The measured change in AOA and camber is consistent with assumptions made in Section 6.2.2. As the support angle is swept up and down, no conclusive deformation is observed because of the change of aerodynamic load distribution on the airfoil. As predicted by the model, the airfoil sustains aerodynamic loading at 15 m/s. Figure 6.13 presents the experimental results for lift and drag coefficients versus angle of supports. The lift coefficient curve for the zero voltage has a slight offset due to the residual camber (and the resulting angle-of-attack) as noted previously. The large voltage-induced peak-to-peak change in lift coefficient at zero degrees should be noted. A lift coefficient of –0.76 at –1400 V and 0.64 at 1400 V is observed, which results in a total lift coefficient change of 1.40 purely through voltage excitation.

6.2.5.5 Aerodynamic Response: Voltage Sweep at Fixed Support Angle

The first test scheme concluded that the aerodynamic hysteresis was negligible for the thin variable-camber airfoil. A second test scheme is designed to identify the hysteresis of the morphing airfoil due to its piezoceramic bimorph nature. The experimental setup and measurements are the same as the previous tests; however, only a fixed support angle of zero degrees is considered, while the applied voltage is gradually increased and decreased. Figure 6.14 presents the camber and the AOA of the airfoil. The test is started at –1400 V. After taking force and displacement measurements, the voltage is incremented by 175 V up to 1400 V, which is labeled as the "Sweep up" curve. The voltage is then swept down from 1400 V to –1400 V in steps

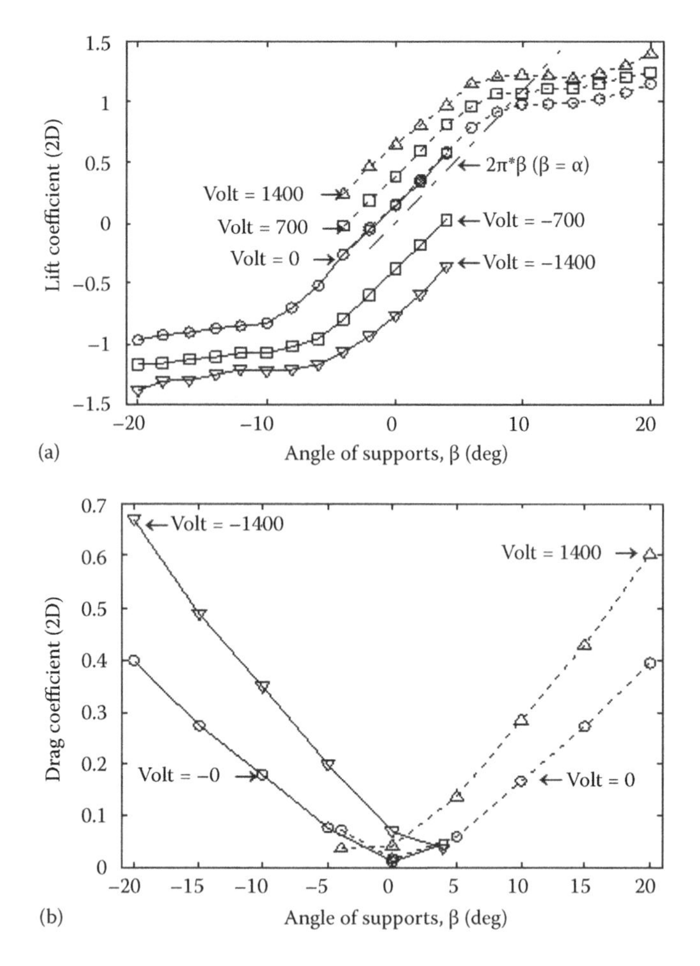

FIGURE 6.13
Experimental (2D) (a) lift and (b) drag coefficients of the variable-camber airfoil at 15 m/s. $Re_{chord} = 1.27 \times 10^5$. (From Bilgen, O., Kochersberger, K. B., Inman, D. J. and Ohanian, O. J., *Smart Materials and Structures*, 19, 055010, 2010. With permission.)

of 175 V, which is labeled as the "Sweep down" curve. As with the previous tests, all plots represent the wind-on analyses. The nonlinear voltage–geometry relationship is due to piezoceramic hysteresis, which can be compensated with a Preisach model (see Bilgen et al. 2011b) and feedback control if desired.

Negative camber values simply indicate that actuation is in the negative direction. Figure 6.15 shows the lift and drag coefficients versus voltage input for the thin variable-camber bimorph airfoil at 15 m/s flow speed. The lift coefficient is measured as 0.657 at +1400 V and −0.807 at the −1400 V. The end slopes of the lift coefficient curves indicate that higher lift can be achieved

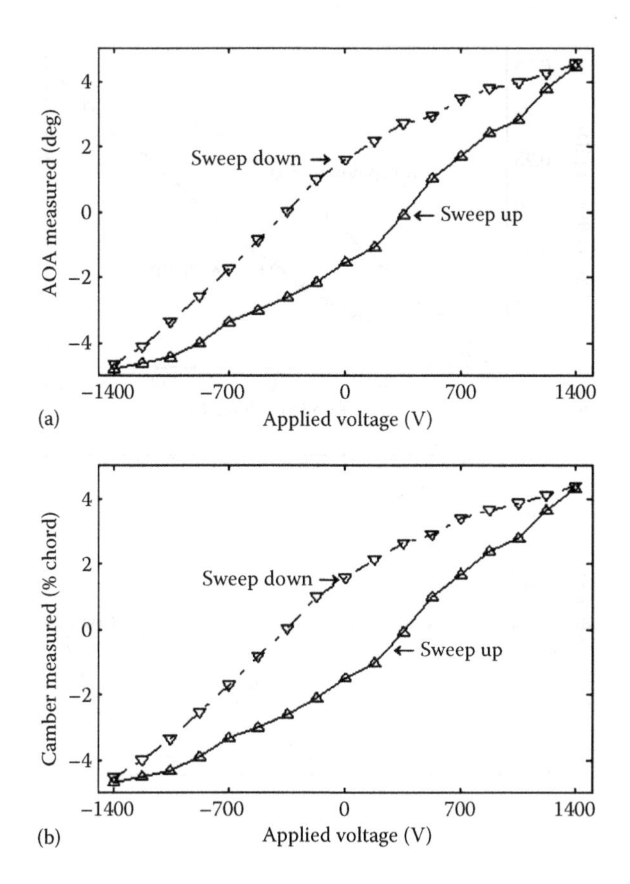

FIGURE 6.14
Experimental (a) AOA of the circular-arc fit due to voltage input. (b) Camber induced by voltage input. $\beta = 0°$, $Re_{chord} = 1.27 \times 10^5$. (From Bilgen, O., Kochersberger, K. B., Inman, D. J. and Ohanian, O. J., *Smart Materials and Structures*, 19, 055010, 2010. With permission.)

if voltage is increased. This is desired since the MFC actuator can safely be actuated up to 1700 V, which will increase operational envelope of the airfoil. A change of 1.46 in lift coefficient is measured for the peak-to-peak voltage input. When the aerodynamic efficiency is considered, a significant change in lift can be achieved for a small drag penalty. Such results confirm an important motivation of using a variable-camber airfoil with continuous curvature.

6.2.5.6 Aerodynamic Comparison

A lift and drag coefficient comparison of different airfoils is given in Figure 6.16. Two airfoils are presented: (1) the thin variable-camber bimorph airfoil and (2) a flat-plate airfoil with elliptical LE and 2.00% thickness. The data for the variable-camber airfoil (from Figure 6.15) are plotted against the

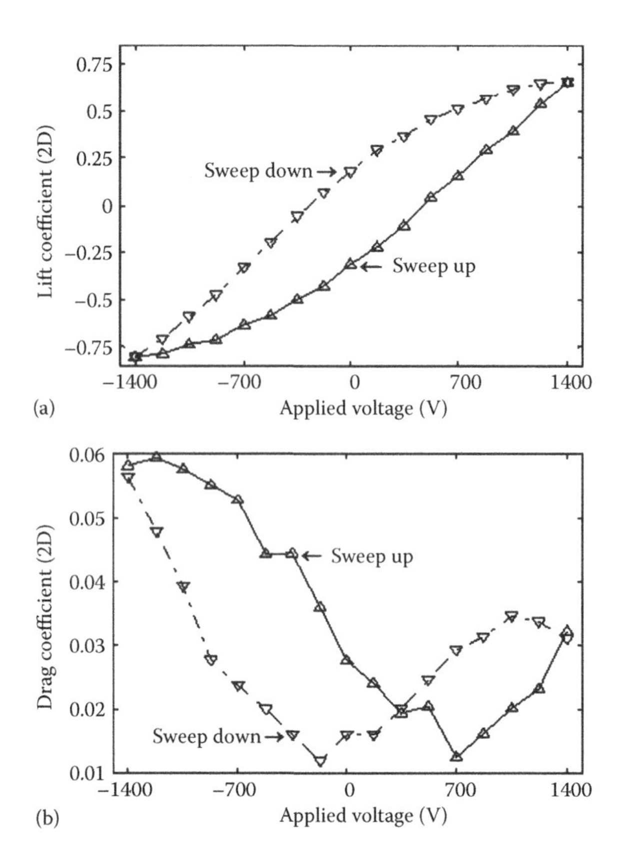

FIGURE 6.15
Experimental (2D) (a) lift and (b) drag coefficients of the variable-camber airfoil from experiments at 15 m/s. $\beta = 0°$, $Re_{chord} = 1.27 \times 10^5$. (From Bilgen, O., Kochersberger, K. B., Inman, D. J. and Ohanian, O. J., *Smart Materials and Structures*, 19, 055010, 2010. With permission.)

measured, voltage-induced AOA (from Figure 6.14). In other words, there is no rotation of the airfoil supports for the morphing airfoil data presented. The comparison is given to allow the reader to evaluate the variable-camber airfoil with respect to a standard airfoil subjected to the same Reynolds number and turbulence intensity. A more desirable comparison would be to an airfoil with a control surface; however, this is not addressed due to the complexity of such an experiment. The tests are conducted at an average flow speed of 15 m/s and a chord Reynolds number of 127,000. Note that the arrows in the legend specify the direction of sweep of voltage (volt) and AOA (α).

There is a significant difference in the variable-camber lift curve slope (0.153 per-degree) when compared with the flat-plate lift slope (0.081 per-degree). The increase of lift slope is due to coupled camber and AOA induced by voltage excitation. The drag coefficient is observed higher for the flat-plate airfoil due to its relatively thick LE when compared with the thin bimorph airfoil.

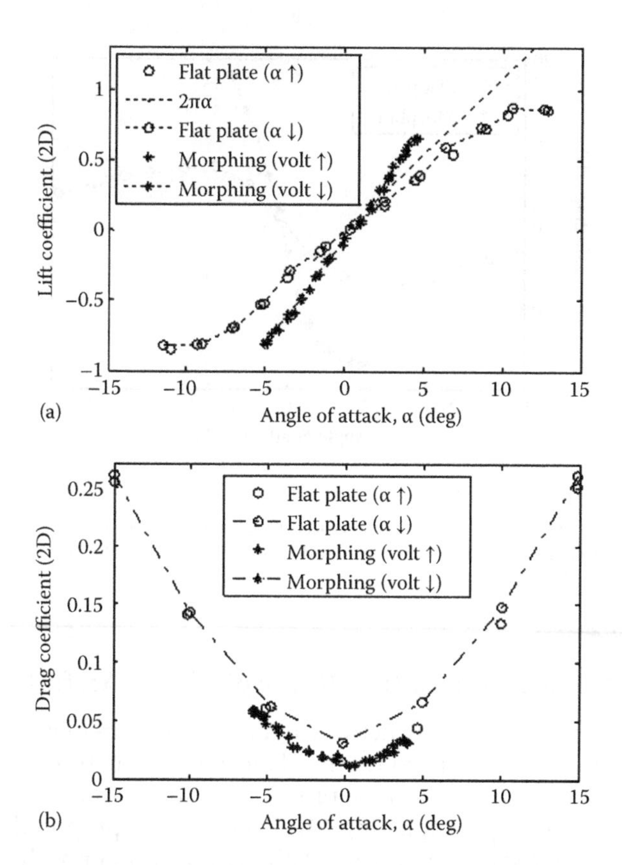

FIGURE 6.16
Experimental (2D) (a) lift and (b) drag coefficient comparisons for two thin airfoils. $Re_{chord} = 1.27 \times 10^5$. (From Bilgen, O., Kochersberger, K. B., Inman, D. J. and Ohanian, O. J., *Smart Materials and Structures*, 19, 055010, 2010. With permission.)

The aerodynamic efficiency of the airfoils is compared by looking at the *L/D* presented in Figure 6.17. Data for both airfoils represent the average of sweep-up and sweep-down curves. The variable-camber airfoil produces an *L/D* of –17.3 at –1225 V ($\alpha = -4.52°$) and an *L/D* of 17.8 at +1400 V ($\alpha = +3.48°$). The flat plate shows a maximum *L/D* of –8.02 at –5.17° AOA. In comparison with the flat-plate airfoil, the variable-camber airfoil generates a significant change in *L/D*; however, even higher *L/D* values are expected in ambient conditions. The current drag measurement from the flat-plate airfoil is significantly higher than those previously published by Mueller (1999) and by Pelletier and Mueller (2000) due to the configuration of the current wind tunnel. The reasons for high drag coefficient measurements are discussed earlier in this chapter.

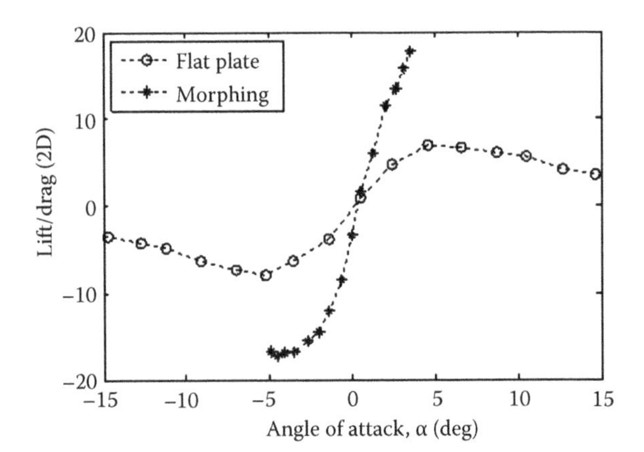

FIGURE 6.17
Experimental (2D) lift-to-drag ratio comparison for two thin airfoils. $Re_{chord} = 1.27 \times 10^5$. (From Bilgen, O., Kochersberger, K. B., Inman, D. J. and Ohanian, O. J., *Smart Materials and Structures*, 19, 055010, 2010. With permission.)

6.3 MFC Actuated Cascading Bimorph Thick Airfoil

The high strain and high structural deflection requirement under aerodynamic loads creates the need for novel mechanisms to be employed along with piezoceramic actuation. In the previous section, a thin variable-camber bimorph airfoil concept employing MFC actuators is proposed to produce relatively large aerodynamic forces. Although a thin bimorph structure is easy to implement, from an aerodynamics point of view, it is well known that thin airfoils are more susceptible to flow separation and therefore have a limited operation envelope. The maximum lift coefficient is typically low for a thin airfoil (e.g., thickness <3% chord) when compared with a thicker counterpart. To address this issue, another variable-camber airfoil concept employing two cascading bimorph surfaces with MFC actuators and a single compliant box mechanism as the internal structure is proposed (see Bilgen et al. 2010d for details). The compliant box allows for a desired thickness between the two cascading bimorph surfaces; hence, the airfoil is significantly thicker than its single bimorph counterpart.

6.3.1 Cascading Bimorph Airfoil Concept

The design proposed here employs two cascading bimorphs that create the top and bottom surfaces of the airfoil that are pinned to each other at the trailing edge. These active surfaces are chosen to be MFC-actuated bimorphs. A compliant box structure is used to create the desired boundary conditions to the "leading edge" of the bimorph surfaces. Figure 6.18 shows the kinematic model and the parameters of the airfoil.

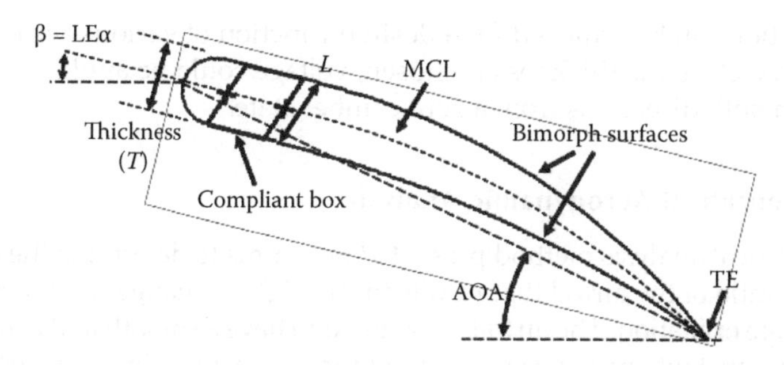

FIGURE 6.18
Cascading bimorph airfoil design and geometric parameters. Morphed state is illustrated.

In the figure, β is the LE incidence angle and MCL is the mean camber line. The compliant box can simply be described as a four-bar mechanism. The two parallel bars that connect top and bottom surfaces have constant length (*L*). The airfoil can be mounted to an aircraft either at the center or at the ends of the vertical bars. The change in camber of the two active bimorph surfaces of the airfoil causes the box to comply and generate a shear-like motion while keeping the end slope of the bimorphs equal to each other. When the airfoil is in the nonactuated state, the link length (*L*) is equivalent to the LE thickness (*T*) of the airfoil. The initial percent thickness (Th = *L*/chord * 100) is calculated at the zero volt state. It is important to note that the concept proposed here consists of (1) the two active unimorph or bimorph surfaces and (2) the three boundary conditions that are necessary to connect the surfaces and create a compliant mechanism. That is to say, the LE geometry is not proposed here, and it can be tailored to the specific aerodynamic application. Here, the LE is designed to be elliptical due to geometric constraints of the ducted-fan aircraft (shown in Figure 6.8).

Figure 6.19 shows a set of possible geometric configurations, with different link lengths and voltage inputs. In Figure 6.19b, only the positive excitation, hence only the positive camber response, is shown since the airfoil is assumed to deform symmetrically in both directions from a zero camber state. The airfoil thickness is determined depending on the application, where an optimum

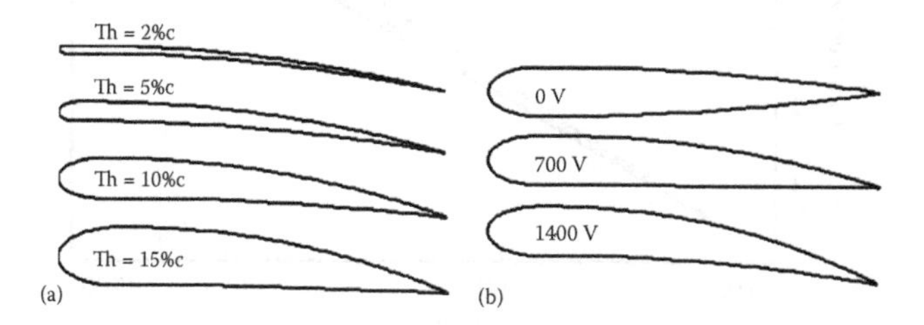

FIGURE 6.19
(a) Airfoil design with different vertical link lengths. (b) Effect of voltage input for an airfoil with 13.0% chord thickness.

configuration can be achieved for a desired function of various aerodynamic coefficients. Once the thickness is chosen, voltage could be applied to induce camber in both directions from a zero camber state.

6.3.2 Theoretical Aerodynamic Analysis

The theoretical analysis method presented here aims to determine the optimal thickness ratio of the airfoil that results in the highest change in lift coefficient with voltage excitation. The airfoil is assumed to have a smooth and continuous surface. A constant curvature is assumed for each active bimorph surface due to the high chordwise coverage of the MFC actuator. An approximate slope of 2.86% camber per kV is assumed for each active surface for the theoretical analysis. The linear assumption simply corresponds to 4.86% camber at 1700 V, 4.00% at 1400 V, and 2.00% at 700 V. A MATLAB-based program is developed to drive the 2D panel method software XFOIL. XFOIL software is employed to calculate

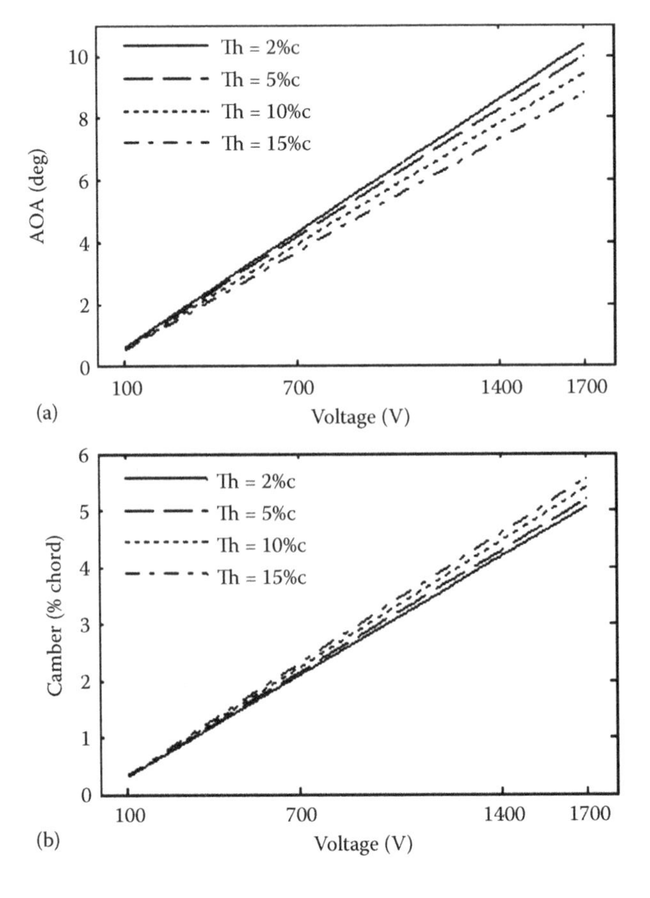

FIGURE 6.20
Theoretical (2D) change in geometry. (a) Angle of attack and (b) effective percent camber.

lift and drag coefficients and the pressure distribution. For XFOIL simulations, a 0.05% chord trailing edge gap and an approximate 0.85% turbulence level is assumed. In this section, the aeroelastic effects are ignored in the theoretical analysis owing to previous theoretical and experimental observations. Figure 6.20 shows the effective geometric parameters of the cascading bimorph airfoil as a function of thickness ratio and actuation voltage. All results presented in this section are for an LE incidence angle of zero degrees ($\beta = 0°$).

Note that both camber and AOA values are zero for zero volt actuation due to the assumption of linear material behavior. Two observations are made. First, the AOA of the airfoil is decreased as the thickness is increased, which is caused by the change in location of the LE with respect to the TE. In contrast, the effective camber of the airfoil increases as the thickness is increased. This is because the mean camber line (the line that is equally distant from the top and bottom surface) forms a higher displacement with the increased thickness. Figure 6.21 shows the theoretical change in lift coefficient (from XFOIL) for the airfoil with 127-mm chord. The airfoil is subjected to 15 m/s free stream velocity.

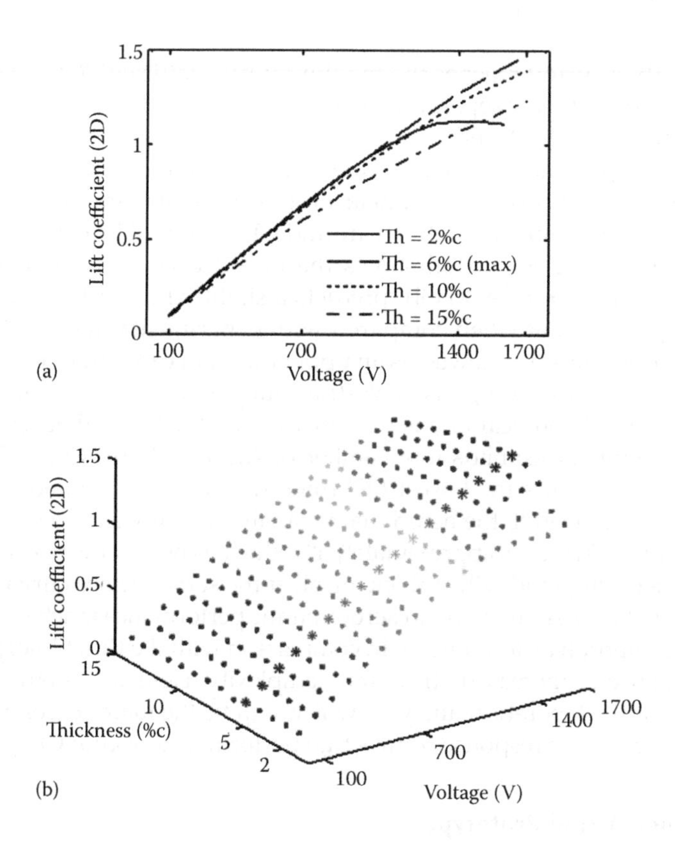

FIGURE 6.21
Theoretical (2D) lift coefficient for the cascading bimorph airfoil subjected to free stream velocity of 15 m/s. (a) 2D and (b) 3D representations. $Re_{chord} = 1.27 \times 10^5$.

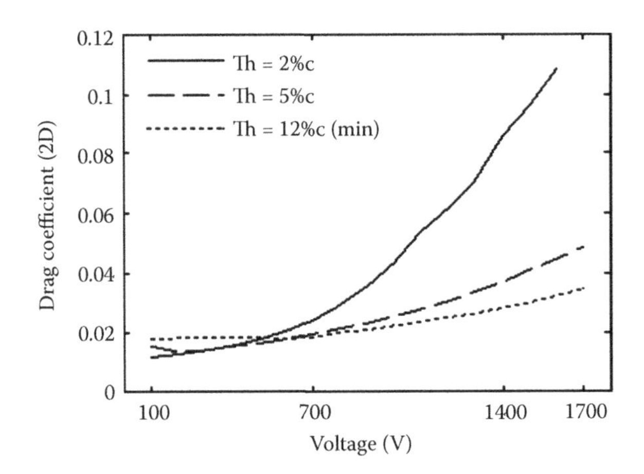

FIGURE 6.22
Theoretical (2D) drag coefficient for the cascading bimorph airfoil subjected to free stream velocity of 15 m/s. $Re_{chord} = 1.27 \times 10^5$.

The 2% thick airfoil reaches the maximum lift coefficient (CL_{max}) of 1.12 at 1400 V due to early flow separation. In contrast, the 6% thick airfoil shows the highest theoretical lift trend.

Figure 6.22 presents the theoretical drag coefficient results. As noted earlier, XFOIL predictions for AOA near or above the maximum lift angle are not accurate; therefore, predictions around CL_{max} should be considered with caution. The 12% thick airfoil shows the lowest drag coefficient trend. As reported in the literature, XFOIL predicts a slightly high lift coefficient and a low drag coefficient when compared with experimental results; therefore, the predictions must be viewed as an upper boundary to actual performance.

Figure 6.23 shows a comparison of lift-to-drag ratio versus actuation voltage. The maximum theoretical *L/D* is 42.6 for the 10% thick airfoil at 1300 V. This operation point corresponds to a camber of 4.12% and an AOA of 7.21°. The 2% thick airfoil shows the lowest *L/D* trend due to early stall and separation around the LE. Overall, the aerodynamic analysis clearly shows that the thin bimorph airfoil (which is approximately 1%*c* thick) is not as effective as generating lift when compared with a thick airfoil at the same surface curvature (and excitation voltage). An increase in aerodynamic performance is achieved by the cascading bimorph airfoil concept in comparison with the thin bimorph airfoil at the expense of increased structural complexity. Here, a smooth surface is assumed for the theoretical analysis, which will be the main reason for deviation of aerodynamic response from actual performance around CL_{max}.

6.3.3 Thick Airfoil Prototype

Using the conclusions derived from the analysis presented above and the geometric constraints of the ducted-fan aircraft, two MFC actuated

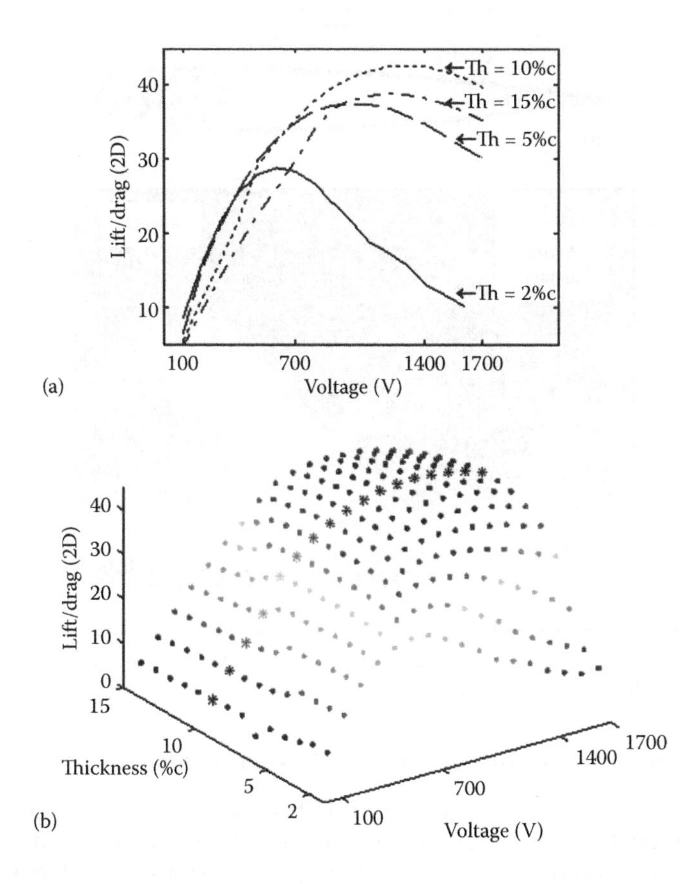

FIGURE 6.23
Theoretical (2D) L/D for the cascading bimorph airfoil at 15 m/s. (a) 2D and (b) 3D representations. $Re_{chord} = 1.27 \times 10^5$.

cascading bimorph airfoils are fabricated. The compliant-box mechanism is designed to keep the thickness to approximately 6% chord and allow free shear-like motion due to piezoceramic actuation. The elliptical leading edge is designed so that it could be attached to a mounting bar directly. The gap between the leading edge and the rest of the airfoil is closed with a narrow and flexible plastic sheet for smooth transition between the two surfaces. Figure 6.24 shows one of the two airfoils employing eight (four on each bimorph) MFC M8557-P1-type actuators. The prototype airfoil has 15 mm thickness, 127 mm chord, and 133 mm span. The airfoil is 12% chord thick (instead of the desired 6% chord thickness for maximum lift authority) due to in-house fabrication limitations. The bimorphs are fabricated by sandwiching a 0.027-mm-thick stainless-steel material with the MFC actuators. The MFCs are aligned at the TE in the chordwise direction for both bimorph surfaces.

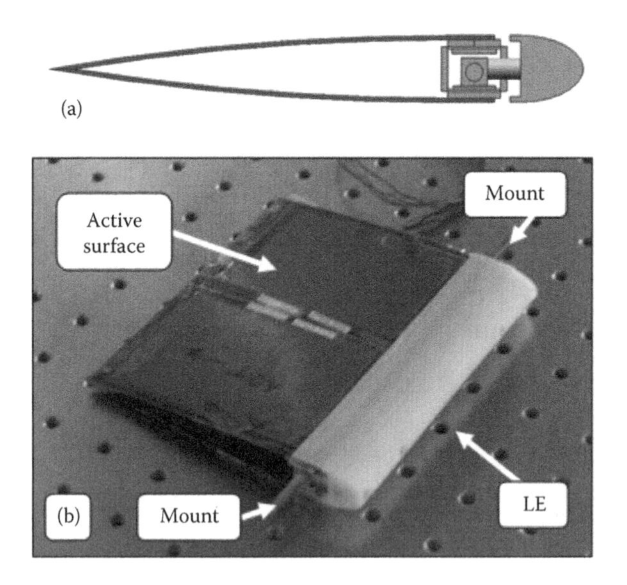

FIGURE 6.24
Cascading bimorph airfoil with eight MFC M8557-P1 actuators, 127 mm chord, and 133 mm span. (a) A simplified illustration and (b) the fabricated airfoil.

6.3.4 Ducted-Fan Aircraft Wind Tunnel Experiments

The two fabricated thick airfoils are initially evaluated through wind tunnel tests of the ducted-fan aircraft. The ducted fan, mounted on its side for wind tunnel testing, and the two fabricated morphing airfoils, mounted at the exit of the ducted fan, is shown in Figure 6.25. The peak-to-peak displacement of the variable-camber airfoils induced by voltage excitation is shown.

These preliminary and qualitative wind tunnel tests are conducted at a maximum fan flow speed of 45 m/s and maximum free stream velocity of

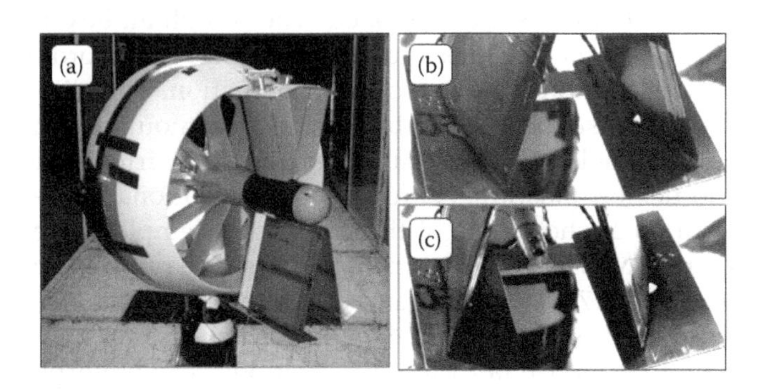

FIGURE 6.25
(a) Ducted-fan vehicle model installed in the Virginia Tech 6 ft × 6 ft Stability Tunnel. (b, c) Close-up of the two morphing control surfaces installed at the exit of the ducted fan.

10 m/s. As with the thin airfoil prototype, the cascading bimorph airfoil demonstrated high force outputs and no adverse deformation due to high aerodynamic loads.

6.3.5 Structural and Aerodynamic Experiments

6.3.5.1 Structural Response

Similar to the analysis of the thin bimorph airfoil, the shape of the cascading bimorph airfoil at a given condition is measured experimentally. Since the airfoil is in a bimorph configuration, the MFC on the "other" side of each bimorph surface is actuated with an opposite field and with a 3-to-1 fixed ratio. The higher of the two excitation voltages is used in the plots. A negative sign simply indicates actuation in the reverse direction. Figure 6.26 shows the displacement of the airfoil end section digitized from a series of photos for a peak-to-peak voltage sweep. The actuation voltage is swept from –1400 to 1400 V and swept back to –1400 V. The effect of piezoelectric hysteresis can be observed from the difference between deflection at 0 V for the "up" and "down" sweeps. However, it should also be noted that the deflection is repeatable, as seen at the maximum positive and negative voltage levels.

6.3.5.2 Baseline NACA Airfoil Response

As in the previous tests of the thin bimorph airfoil, the main wind tunnel tests for the cascading bimorph airfoil are performed in a low-speed, open-circuit wind tunnel facility with a 2D test-section configuration; however, the wind tunnel is upgraded to include a strain-gauge-based balance system. In addition, the inlet of the wind tunnel is upgraded; therefore, a quantitative

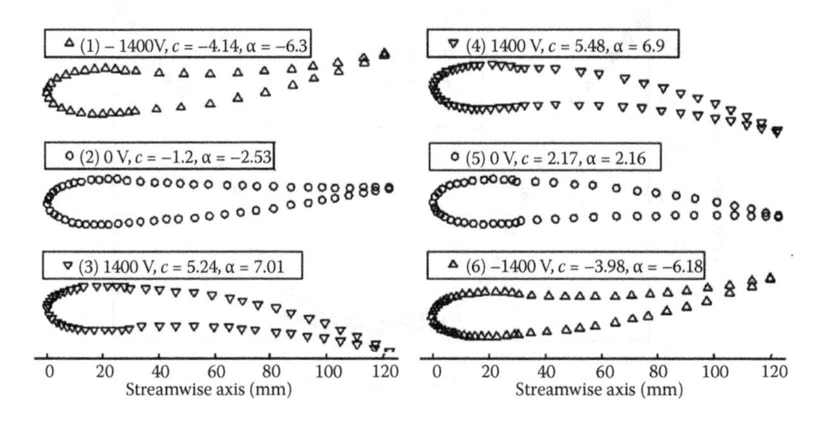

FIGURE 6.26
Composite plot of the airfoil end-section measurements with a digital camera. Legends show test order, voltage (V), and the calculated percent camber (c) and AOA in degrees (α).

comparison of the experimental aerodynamic response of the thin airfoils (presented in Section 6.2) to that of the thick airfoils (presented in the following sections) should be avoided and not presented.

A rapid prototyped (RP) NACA 0009 airfoil is selected as a baseline to the cascading bimorph variable-camber airfoil. The NACA 0009 airfoil is tested for lift and drag coefficients at an average flow speed of 15 m/s and a chord Reynolds number of 127,000. The airfoil has a maximum thickness of 11.3 mm, span b = 133 mm, and a chord c = 127 mm. Figure 6.27 presents the comparison of NACA 0009 lift and drag coefficient data from the current tests to other tests conducted by Selig et al. (1995) labeled as "Selig-1," and by Selig et al. (1989) labeled as "Selig-2." The references and respective Reynolds numbers are presented in the legend of the figures. The data for AOA sweep-up and sweep-down curves are presented for the current tests; however, there was no measurable aerodynamic hysteresis observed.

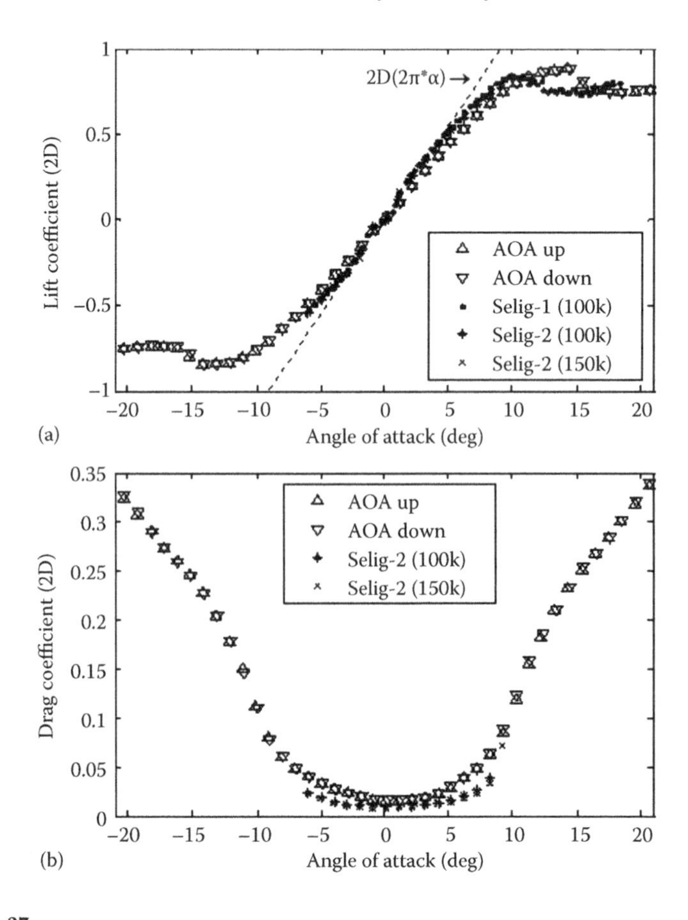

FIGURE 6.27
Experimental (2D) (a) lift and (b) drag coefficients versus angle of attack comparison for NACA 0009 airfoils.

The lift curve slope matches relatively well considering the effect of turbulence. The Reynolds number has negligible effect on the lift coefficient, and there is no aerodynamic hysteresis. The drag coefficient from current evaluation shows a slightly higher trend at lower angles when compared with data reported by Selig. The zero-degree drag coefficient is 0.0158 for the NACA 0009 airfoil tested here. Selig reports approximately 0.0087 at Re = 150,000 and 0.0096 at Re = 100,000 for the zero-degree drag coefficient of the NACA 0009 airfoil.

There are three main reasons for the difference in measured lift and drag between the two facilities compared in this section: (1) The span of the airfoils are designed so that the root and tip is as close to the wind tunnel walls as possible. At high AOA, a small amount of 3D flow exists through the 1.5-mm gap between the airfoil and the test-section walls. This flow is AOA dependent because the pressure gradient between the pressure and suction side increases as the AOA is increased. (2) The relatively large difference of turbulence levels in the two facilities must be noted. Selig et al. (1989, 1995) reports 0.358% rms turbulence for 0.01 Hz cutoff frequency, and 0.064% rms turbulence for 1 Hz cutoff frequency, at a Reynolds number of 100,000. (3) Finally, the airfoil tested in this section is an RP airfoil with some roughness along the spanwise direction. It is known that that NACA 0009 airfoils used by Selig et al. (1989, 1995) have smoother surfaces.

In addition, the experimental measurements are prone to the relative errors induced by uncertainty in setting the airfoil angle, flexibility in the balance system, and a small amount of friction in the balance pivots. The absolute values have uncertainties due to several parameters such as air density, flow velocity measurements, and the theoretical wall and BL corrections. The uncertainty analysis of each measurement is conducted by following the American Institute of Aeronautics and Astronautics (AIAA) Standard (AIAA 1995). The lift coefficient uncertainty is $U_{CL,max} = \pm 0.055$, and the drag coefficient uncertainty is $U_{CD,max} = \pm 0.014$. The major source for the uncertainty is the low frequency variation of the flow velocity.

6.3.5.3 Aerodynamic Response: Angle Sweep at Fixed Voltage

The main wind tunnel tests are performed in a low-speed, open-circuit wind tunnel. The wind tunnel characteristics are discussed earlier in the chapter. This section presents the fixed-voltage aerodynamic experiments on the cascading bimorph airfoil. Overall, two different test schemes are followed. The first evaluation is performed by setting the voltage of the bimorph at a fixed value, then sweeping the support angle (β) up and down. The complete list of voltages (in order) is as follows: (1) –1500 V, (2) –700 V, (3) 0 V, (4) +1500 V, (5) +700 V, and (6) 0 V. The support angle is swept from –20° to 20° and back to –20° in one degree increments. Once the angle sweep is completed, the voltage is changed. Again, note that the zero voltage data shifts depending on the prior voltages applied due to hysteresis in the actuators.

A comparison of wind-on and wind-off conditions confirmed that there is no measurable deformation due to aerodynamic loading. All plots in the following sections represent the wind-on analyses for consistency. As noted before, aerodynamic hysteresis was not observed when the angle β is swept up and down.

The measured AOA and effective camber values are presented in Figure 6.28. The tests are conducted at an average flow speed of 15 m/s. The measured change in AOA and camber are consistent with geometric predictions. The voltage–geometry relationship appears dominantly linear (and independent of β and magnitude of loading) since voltage is changed only when a β sweep is completed. As the support angle is swept up and down, no conclusive deformation is observed.

Figure 6.29 presents the experimental results for lift and drag coefficients versus angle of support. As noted with the thin bimorph airfoil, the large voltage-induced peak-to-peak change in lift coefficient at zero degrees should

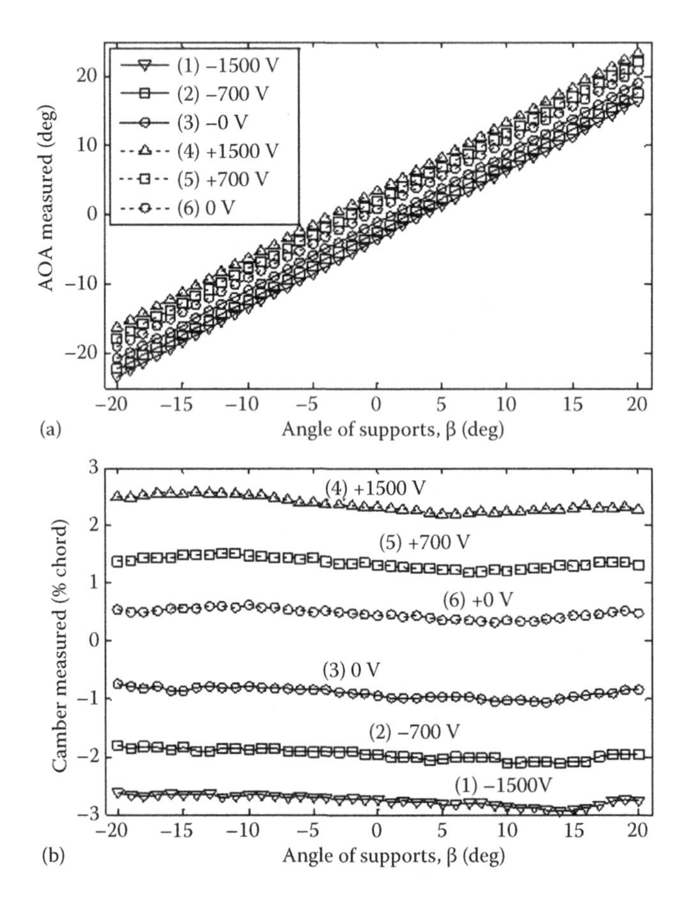

FIGURE 6.28
Experimental (2D) (a) AOA and (b) effective camber variation of the cascading bimorph airfoil due to voltage input and the change in β. $Re_{chord} = 1.27 \times 10^5$.

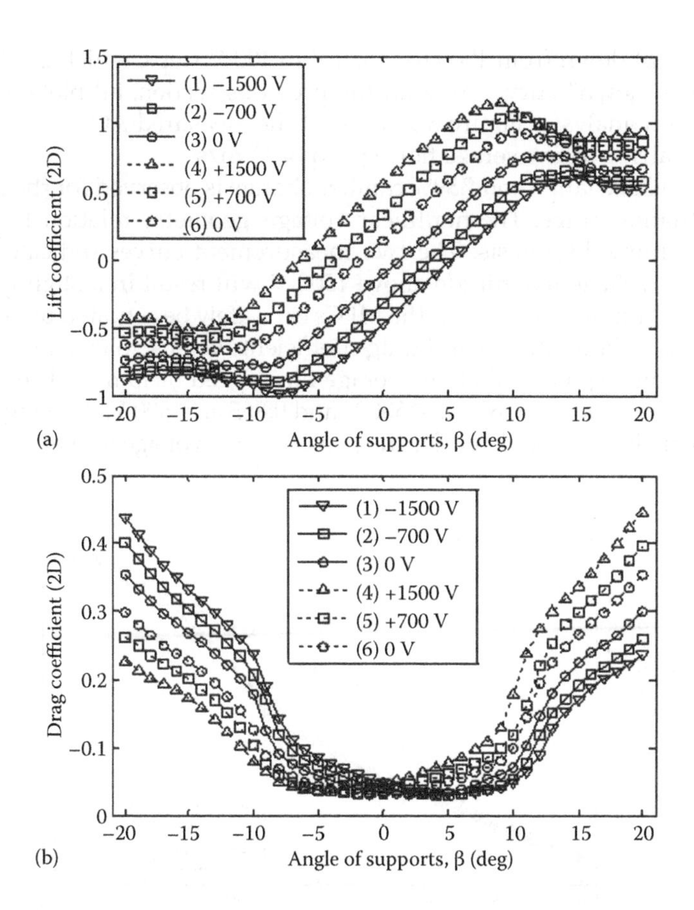

FIGURE 6.29
Experimental (2D) (a) lift and (b) drag coefficient of the cascading bimorph airfoil at 15 m/s. $Re_{chord} = 1.27 \times 10^5$.

be noted. A lift coefficient of –0.47 at –1500 V and 0.56 at 1500 V is observed, which results in a total lift change of 1.03 through voltage excitation at $\beta = 0°$. The minimum drag coefficient is approximately 0.032.

6.3.5.4 Aerodynamic Response: Voltage Sweep at Fixed Support Angle

The first test scheme concluded that the aerodynamic hysteresis was negligible for the morphing airfoil tested. A second test scheme is designed to identify the hysteresis of the morphing airfoil due to its piezoceramic bimorph nature. The experimental setup and measurements are the same as the previous tests; however, only a fixed support angle of zero degrees is considered while the applied voltage is changed. The test is started at –1500 V. After force and displacement measurements are taken, the voltage is incremented by 100 V up to 1500 V, which is labeled as the "Sweep up" curve. The voltage is then swept down from 1500 to –1500 V in 100 V steps, which is labeled as the "Sweep down" curve. As with the previous section, all plots represent the wind-on analyses. Figure 6.30 present the measured AOA and the camber of the airfoil at an average flow speed of 15 m/s.

A 10.7° AOA change and 7.59% camber change is observed for the peak-to-peak actuation range. The nonlinear voltage–geometry relationship is due to piezoceramic hysteresis. The two measurement curves indicate that an increase in voltage magnitude above 1500 V will result in a slightly higher deflection. This is desired since the MFCs can safely be actuated up to 1700 V.

Figure 6.31 shows the lift and drag coefficients versus voltage input for the cascading bimorph airfoil at an average flow speed of 15 m/s. The lift coefficient is measured as –0.677 at –1500 V and 0.865 at +1500 V. A change of 1.54 in lift coefficient is calculated for the peak-to-peak voltage excitation. A high

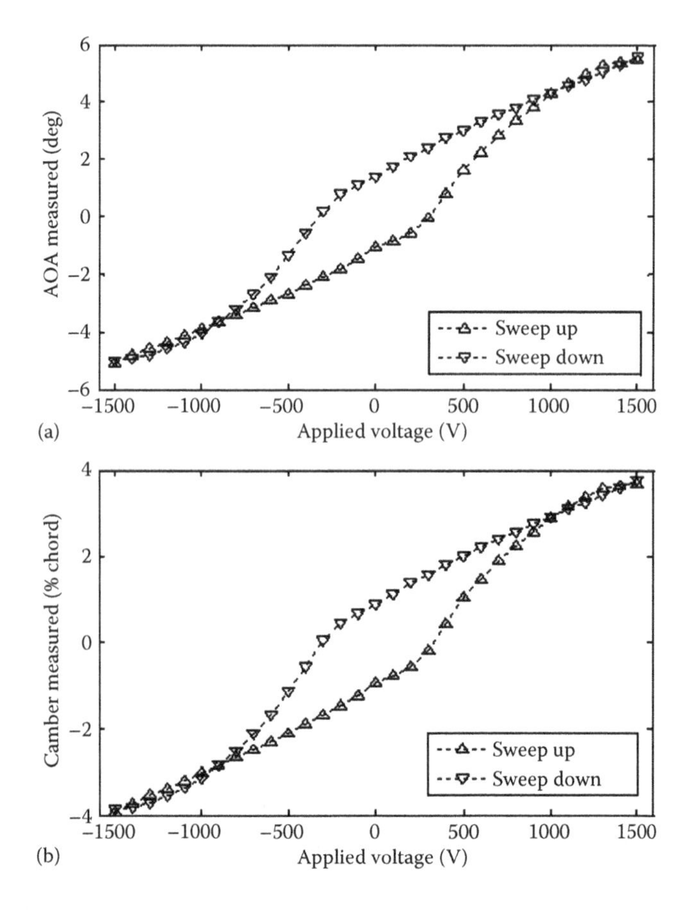

FIGURE 6.30
Experimental (2D) (a) AOA and (b) effective camber of the cascading bimorph airfoil induced by voltage input at $\beta = 0°$. $Re_{chord} = 1.27 \times 10^5$.

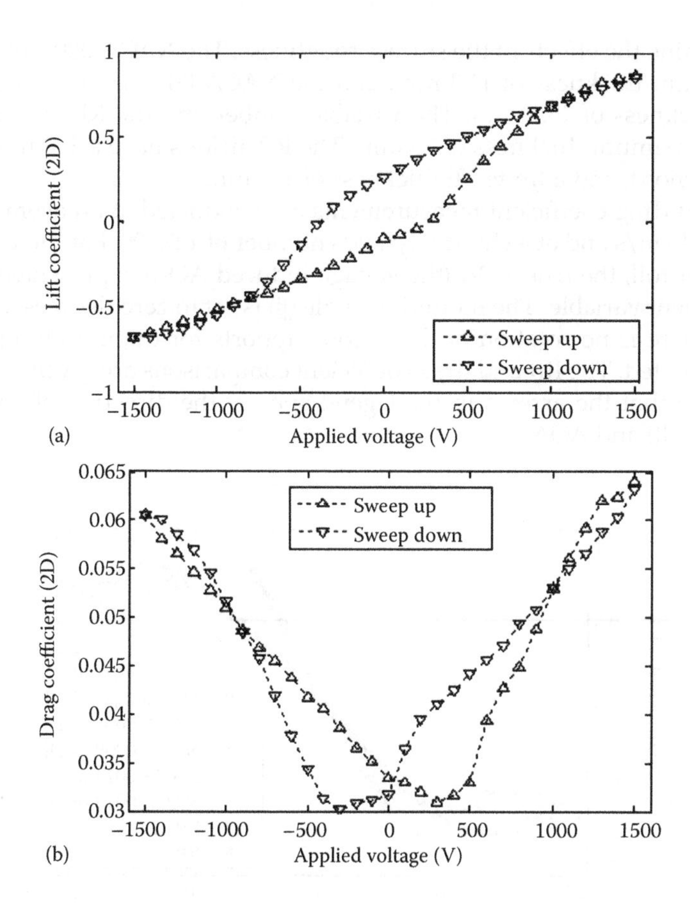

FIGURE 6.31
Experimental (2D) aerodynamic response: (a) lift and (b) drag coefficients of the cascading bimorph airfoil at 15 m/s. $Re_{chord} = 1.27 \times 10^5$.

drag coefficient is observed for the airfoil due to relatively blunt LE and the discontinuities on the surface caused by in-house fabrication limitations. As with the thin airfoil, a significant change in lift can be achieved for a small penalty in drag.

6.3.5.5 Aerodynamic Comparison

An aerodynamic comparison of the cascading bimorph airfoil to other similar (in shape) fixed-camber airfoils is presented in this section. The purpose is to show the advantages of continuously coupled camber–AOA actuation when compared with the mechanical AOA actuation. Four airfoils are presented: (1) cascading bimorph variable-camber airfoil, (2) NACA 0009 airfoil, (3) NACA 0013 airfoil, and (4) an RP airfoil generated from the profile of the variable-camber prototype (at zero camber state). The fourth airfoil is tested to determine the effects of the surface roughness. The NACA 0009 airfoil has a maximum thickness of 11.3 mm, and the NACA 0013 airfoil has a maximum thickness of 16.5 mm. The variable-camber and the RP airfoils both have a maximum thickness of 15 mm. The RP airfoils have a 133 mm span, 127 mm chord, and a finite TE thickness of 1.0 mm.

Lift and drag coefficient measurements are conducted at an average flow speed of 15 m/s and at a chord Reynolds number of 127,000. For the variable-camber airfoil, the true AOA (the voltage-induced AOA) is presented as the independent variable. The mounting angle (β) is set to zero degrees. In other words, there is no rotation of the airfoil supports for the morphing airfoil data presented. The lift and drag coefficient comparisons are given in Figure 6.32. Note that the arrows in the legend specify the direction of sweep of voltage (volt) and AOA (α).

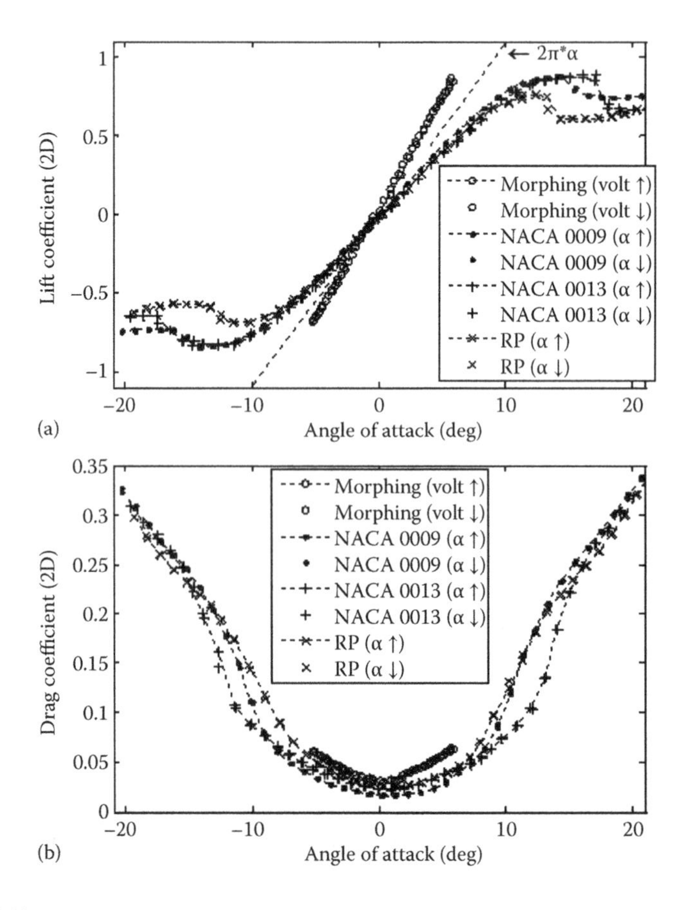

FIGURE 6.32
Experimental (2D) aerodynamic response: (a) lift and (b) drag coefficient comparisons of four airfoils at 15 m/s. $Re_{chord} = 1.27 \times 10^5$.

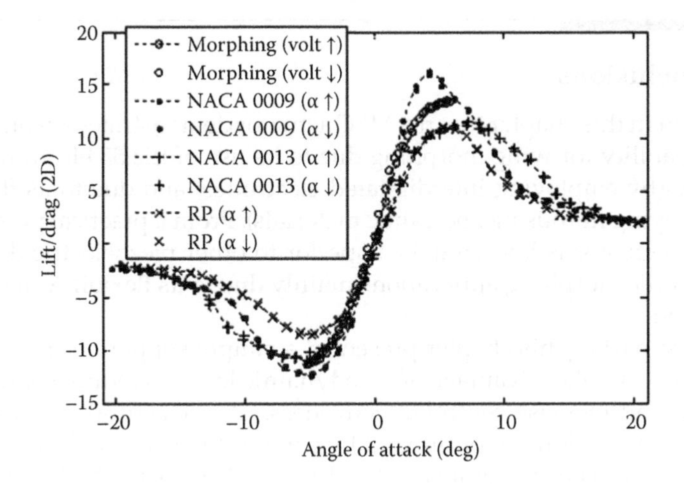

FIGURE 6.33
Experimental lift-to-drag ratio (2D) comparison of four airfoils at 15 m/s. $Re_{chord} = 1.27 \times 10^5$.

There is a significant increase in lift coefficient due to change in camber induced by voltage excitation. A lift curve slope of 0.144 per-degree is measured, which exceeds the NACA 0009 lift slope (0.083 per-degree) by 72%. The plot shows the clear advantage of the lift generation by coupled camber-AOA change induced by voltage. The NACA 0013 and the RP airfoils develop a similar lift curve slope; however, the NACA 0013 achieves a slightly higher CL_{max}. As noted earlier, a high experimental drag is observed for the morphing airfoil due to its relatively blunt (elliptical) LE when compared with the LE of NACA 0009 airfoil. The cascading bimorph airfoil prototype tested here does not have a continuous surface as desired due to in-house fabrication limitations. This is evident when the morphing airfoil at zero AOA is compared to its RP replica airfoil with continuous and relatively smooth surface.

The aerodynamic efficiency is evaluated by comparing the *L/D* presented in Figure 6.33. The variable-camber airfoil produces a maximum *L/D* of 13.4 at 1500 V ($\alpha = 5.78°$) and an *L/D* of −11.2 at −1500 V ($\alpha = -5.20°$). The NACA 0009 airfoil produces a maximum *L/D* of 16.3 at $\alpha = 4.21°$ and an *L/D* of −12.3 at $\alpha = -4.97°$. The cascading bimorph airfoil has higher *L/D* performance when compared with the fixed-camber airfoils with similar thickness (NACA 0013 and RP). In comparison with the conventional airfoils, the morphing airfoil generates a significant change in *L/D*. As noted earlier, owing to the configuration of the wind tunnel facility, the *L/D* performance is significantly lower than expected mainly due to the high drag coefficient. Several reasons along with the wind tunnel characteristics are discussed earlier in this chapter.

6.4 Conclusions

It is shown in this chapter that the MFC actuator demonstrates adequate actuation capability for wing morphing due to its use of the 33 electromechanical mode (by employing interdigitated electrodes) and due to its flexibility (by employing fibrous piezoceramic materials). From a practical perspective, the MFC actuator is known to be superior to other piezoelectric devices in aerodynamic morphing applications mainly due to its flexibility and ease of application.

The first part of this chapter presented a simply supported thin bimorph airfoil that can take advantage of aerodynamic loads to reduce control input moments and increase control effectiveness. The structural boundary conditions of the design are optimized by solving the coupled fluid–structure interaction problem using a structural finite element method and a panel method based on the potential flow theory for fluids. A prototype is tested in the wind tunnel for its 2D aerodynamic coefficients at low Reynolds number flow. The wind tunnel results show comparable effectiveness to conventional actuation systems. The MFC bimorph demonstrated adequate control authority for aerodynamic open-loop shape control and adequate stiffness at a tested flow speed of 15 m/s. An average lift coefficient change of 1.46 is observed purely due to peak-to-peak actuation voltage. A maximum lift-to-drag ratio of 17.8 can be achieved through voltage excitation.

The second part of this chapter presented a variable-camber airfoil employing two cascading bimorph surfaces with MFC actuators and a single compliant box mechanism as the internal structure. The unique choice of boundary conditions induced by the compliant box allows a variable and smooth deformation in both directions from a flat camber line. The compliant box also generates the desired thickness between the two cascading bimorph surfaces; hence, the airfoil is significantly thicker than its single bimorph counterpart. As in the thin airfoil, aerodynamic and structural response is analyzed at a flow rate of 15 m/s and Reynolds number of 127,000. The wind tunnel results showed comparable effectiveness to conventional actuation systems. The MFC actuators demonstrate adequate control authority for aerodynamic shape control and the airfoil concept shows adequate stiffness at a tested flow speed of 15 m/s. An average lift coefficient change of 1.54 is observed purely owing to peak-to-peak actuation voltage. A maximum lift-to-drag ratio of 13.4 can be achieved through voltage excitation. Finally, a 72% increase in lift curve slope is achieved when compared with a NACA 0009 airfoil.

Both concepts show a small increase of drag with actuation voltage, making them efficient variable lift generation devices. Compared with previously published results, the concepts presented in this chapter demonstrate high force outputs and frequency bandwidth (although bandwidth is not discussed in this chapter). In addition, both prototypes are tested and verified to generate high control outputs in the wind tunnel tests of the VTOL ducted-fan vehicle with fan speeds up to 45 m/s. The effect of piezoceramic hysteresis is identified to be the most significant contributor of nonlinearity to both the structural and the aerodynamic response of both concepts.

Overall, the chapter provides sufficient evidence that piezocomposite airfoils can be used in small UAVs. On the other hand, position control of these systems is not addressed. To achieve position control (which is required from a flight control perspective), further linear and nonlinear analysis must be conducted to include (1) piezoelectric hysteresis, (2) structural dynamic, (3) piezoceramic creep, and (4) aerodynamic loading effects. Hysteresis effects can be compensated for by the use of a generalized version of the Preisach model (see Mayergoyz 2003). Once all of the nonlinear effects are compensated, linear feedback controllers can be used to achieve a desired response. In addition, adaptive control laws can achieve further improvement in the response of the piezocomposite airfoils and similar bimorph or unimorph devices.

References

AIAA. 1995. *AIAA Standard on Assessment of Experimental Uncertainty with Application to Wind Tunnel Testing*. S-071A-1995, AIAA, New York.

ANSYS. Computer Software, version 12.0. [http://www.ansys.com/].

Barbarino, S., Bilgen, O., Ajaj, R. M., Friswell, M. I. and Inman, D. J. 2011. A review of morphing aircraft. *Journal of Intelligent Material Systems and Structures* 22(9): 823–877.

Barlow, J. B., Rae, W. H. and Pope, A. 1999. *Low-Speed Wind Tunnel Testing*, 3 Subedition. Wiley-Interscience, New York.

Barrett, R. M., Vos, R., Tiso, P. and De Breuker, R. 2005. Post-buckled precompressed (PBP) actuators: Enhancing VTOL autonomous high speed MAVs. *46th AIAA/ASME/ASCE/AHS/ASC Structure, Structural Dynamics & Materials Conference*, Austin, TX, AIAA 2005-2113.

Barrett, R. M. and Stutts, J. C. 1998. Development of a piezoceramic flight control surface actuator for highly compressed munitions. *39th Structures, Structural Dynamics and Materials Conference*, Long Beach, CA, AIAA-98-2034.

Bilgen, O., Kochersberger, K. B. and Inman, D. J. 2009. Macro-fiber composite actuators for a swept wing unmanned aircraft. *Aeronautical Journal* 113(1144): 385–395.

Bilgen, O. 2010. Aerodynamic and electromechanical design, modeling and implementation of piezocomposite airfoils. Ph.D. Dissertation, Mechanical Engineering Dept., Virginia Tech, Blacksburg, VA. [http://scholar.lib.vt.edu/theses/available/etd-08142010-142319/].

Bilgen, O., Erturk, A. and Inman, D. J. 2010a. Analytical and experimental characterization of macro-fiber composite actuated thin clamped-free unimorph benders. *Journal of Vibration and Acoustics* 132: 051005-1.

Bilgen, O., Kochersberger, K. B., Inman, D. J. and Ohanian, O. J. 2010b. Macro-fiber composite actuated simply-supported thin morphing airfoils. *Smart Materials and Structures* 19: 055010.

Bilgen, O., Kochersberger, K. B., Inman, D. J. and Ohanian, O. J. 2010c. Lightweight high voltage electronic circuits for piezoelectric composite actuators. *Journal of Intelligent Material Systems and Structures* 21(14): 1417–1426, doi: 10.1177/1045389X10381657.

Bilgen, O., Kochersberger, K. B., Inman, D. J. and Ohanian, O. J. 2010d. Novel, bi-directional, variable camber airfoil via macro-fiber composite actuators. *Journal of Aircraft* 47(1): 303–314.

Bilgen, O., Butt, L. M., Day, S. R., Sossi, C. A., Weaver, J. P., Wolek, A., Mason, W. H. and Inman, D. J. 2011a. A novel unmanned aircraft with solid-state control surfaces: Analysis and flight demonstration. *52nd AIAA/ASME/ASCE/AHS/ASC Structures, Structural Dynamics, and Materials*, Denver.

Bilgen, O., Inman, D. J. and Friswell, M. I. 2011b. Theoretical and experimental analysis of hysteresis in piezocomposite airfoils using the classical Preisach model. *Journal of Aircraft* 48(6): 1935–1947, doi: 10.2514/1.56694.

Butt, L., Bilgen, O., Day, S., Sossi, C., Weaver, J., Wolek, A., Inman, D. J. and Mason, W. H. 2010a. Wing morphing design utilizing macro fiber composite. *Society of Allied Weight Engineers (SAWE) 69th Annual Conference*, Virginia Beach, VA, Paper No. 3515-S.

Butt, L. M., Day, S. R., Sossi, C. A., Weaver, J. P., Wolek, A., Bilgen, O., Inman, D. J. and Mason, W. H. 2010b. Wing Morphing Design Team Final Report 2010, Virginia Tech Departments of Mechanical Engineering and Aerospace and Ocean Engineering Senior Design Project, Blacksburg, VA.

Drela, M. 1989. XFOIL: An analysis and design system for low Reynolds number airfoils. *Conference on Low Reynolds Number Airfoil Aerodynamics*, University of Notre Dame, Notre Dame, IN.

Eggleston, G., Hutchison, C., Johnston, C., Koch, B., Wargo, G. and Williams, K. 2002. Morphing Aircraft Design Team, Virginia Tech Aerospace Engineering Senior Design Project, Blacksburg, VA.

Geissler, W., Sobieczky, H. and Trenker, M. 2000. New rotor airfoil design procedure for unsteady flow control. *26th European Rotorcraft Forum*, The Hague, Holland, Paper No. 31.

Giurgiutiu, V. 2000. Review of smart-materials actuation solutions for aeroelastic and vibration control. *Journal of Intelligent Material Systems and Structures* 11(7): 525–544.

Giurgiutiu, V., Chaudhry, Z. and Rogers, C. A. 1994. Engineering feasibility of induced-strain actuators for rotor blade active vibration control. *Smart Structures and Materials '94 Conference*, Orlando, FL, SPIE Paper No. 2190-11: 107–122.

Grohmann, B. A., Maucher, C. K. and Janker, P. 2006. Actuation concepts for morphing helicopter rotor blades. *25th International Congress of the Aeronautical Sciences*, Hamburg, Germany.

Grohmann, B. A., Maucher, C. K., Janker, P. and Wierach, P. 2008. Embedded piezoceramic actuators for smart helicopter rotor blades. *49th AIAA/ASME/ASCE/AHS/ASC Structures, Structural Dynamics, and Materials*, Schaumburg, IL, AIAA-2008-1702.

Hagood, N. W., Kindel, R., Ghandi, K. and Gaudenzi, P. 1993. Improving transverse actuation using interdigitated surface electrodes. *North American Conference on Smart Structures and Materials*, Albuquerque, NM, SPIE Paper No. 1917-25: 341–352.

High, J. W. and Wilkie, W. K. 2003. *Method of Fabricating NASA-Standard Macro-Fiber Composite Piezoelectric Actuators*. NASA/TM-2003-212427, ARL-TR-2833.

Hoerner, S. F. 1993. *Fluid Dynamic Drag*. Hoerner Fluid Dynamics, Brick Town, NJ.
Hoerner, S. F. 1975. *Fluid Dynamic Lift*. L. A. Hoerner, Hoerner Fluid Dynamics, Brick Town, NJ, Chapter 2.
Kim, D. K. and Han, J. H. 2006. Smart flapping wing using macro-fiber composite actuators. *Proceedings of SPIE* 6173 61730F: 1–9.
Kim, D. K., Han, J. H. and Kwon, K. J. 2009. Wind tunnel tests for a flapping wing model with a changeable camber using macro-fiber composite actuators. *Smart Materials and Structures* 18(2): 024008, 8 pp.
Lazarus, K. B., Crawley, E. F. and Bohlmann, J. D. 1991. Static aeroelastic control using strain actuated adaptive structures. *Journal of Intelligent Material Systems and Structures* 2: 386–410.
Lloyd, J. M. 2004. Electrical properties of macro-fiber composite actuators and sensors. MS Thesis, Mechanical Engineering Department, Virginia Tech, Blacksburg, VA.
MATLAB. Computer Software, version R2009b. [http://www.mathworks.com/products/matlab/].
Mayergoyz, I. 2003. *Mathematical Models of Hysteresis and their Applications*, Series in Electromagnetism. Elsevier, New York.
Mueller, T. J. and Burns, T. F. 1982. *Experimental Studies of the Eppler 61 Airfoil at Low Reynolds Numbers*. AIAA Paper 82-0345.
Mueller, T. J. 1999. Aerodynamic measurements at low Reynolds numbers for fixed wing micro-air vehicles. Presented at the Development and Operation of UAVs for Military and Civil Applications course held at the von Karman Institute for Fluid Dynamics, Belgium.
Munday, D. and Jacob, J. 2001. Active control of separation on a wing with conformal camber. *39th AIAA Aerospace Sciences Meeting and Exhibit*, Reno, NV, AIAA-2001-0293.
Paradies, R. and Ciresa, P. 2009. Active wing design with integrated flight control using piezoelectric macro fiber composites. *Smart Materials and Structures* 18: 035010, 9 pp. doi:10.1088/0964-1726/18/3/035010.
Park, J. S. and Kim, J. H. 2005. Analytical development of single crystal macro fiber composite actuators for active twist rotor blades. *Smart Materials and Structures* 14: 745–753.
Pelletier, A. and Mueller, T. J. 2000. Low Reynolds number aerodynamics of low aspect-ratio, thin/flat/cambered-plate wings. *Journal of Aircraft* 37(5): 825–832.
Pinkerton, J. L. and Moses, R. W. 1997. *A Feasibility Study to Control Airfoil Shape Using THUNDER*. NASA Langley Research Center Technical Memorandum, Hampton, VA.
Preisach, F. Z. 1935. Uber die Magnetische Nachwirkung, *Zeitschrift fur Physik* 94: 277–302.
Selig, M. S., Donovan, J. F. and Fraser, D. B. 1989. *Airfoils at Low Speeds, Soartech 8*. Published by H. A. Stokely, Virginia Beach, VA.
Selig, M. S., Guglielmo, J. J., Broeren, A. P. and Giguere, P. 1995. *Summary of Low-Speed Airfoil Data*, Vol. 1. Soartech Publications, Virginia Beach, VA.
Sodano, H. A., Park, G. and Inman, D. J. 2004. An investigation into the performance of macro-fiber composites for sensing and structural vibration applications. *Mechanical Systems and Signal Processing* 18(3): 683–697.
Sodano, H. A., Lloyd, J. and Inman, D. J. 2006. An experimental comparison between several active composite actuators for power generation. *Smart Materials and Structures* 15: 1211–1216.

Steadman, D. L., Griffin S. L. and Hanagud, S. V. 1994. *Structure–Control Interaction and the Design of Piezoceramic Actuated Adaptive Airfoils*. AIAA-1994-1747.
Tarazaga, P. A., Inman, D. J. and Wilkie, W. K. 2007. Control of a space rigidizable inflatable boom using macro-fiber composite actuators. *Journal of Vibration and Control* 13(7): 935–950.
Vos, R., Barrett, R. and Zehr, D. 2008. Magnification of work output in PBP class actuators using buckling/converse buckling techniques. *49th AIAA/ASME/ASCE/AHS/ASC Structures, Structural Dynamics, and Materials Conference*, Schaumburg, IL, AIAA-2008-1705.
Vos, R., De Breuker, R., Barrett, R. and Tiso, P. 2007. Morphing wing flight control via postbuckled precompressed piezoelectric actuators. *Journal of Aircraft* 44(4): 1060–1068.
Wang, D. P., Bartley-Cho, J. D., Martin, C. A. and Hallam, B. J. 2001. *Development of High-Rate, Large Deflection, Hingeless Trailing Edge Control Surface for the Smart Wing Wind Tunnel Model, Smart Structures and Materials 2001: Industrial and Commercial Applications of Smart Structures Technologies*, SPIE Vol. 4332.
Wickramasinghe, V. K., Chen, Y., Martinez, M., Kernaghan, R. and Wong, F. 2009. Design and verification of a smart wing for an extremely-agile micro-air-vehicle. *50th AIAA/ASME/ASCE/AHS/ASC Structures, Structural Dynamics, and Materials Conference*, Palm Springs, CA, AIAA-2009-2132.
Wilkie, W. K., Bryant, G. R. and High, J. W. 2000. Low-cost piezocomposite actuator for structural control applications. *SPIE 7th Annual International Symposium on Smart Structures and Materials*, Newport Beach, CA.
Wilkie, W. K., Inman, D. J., Lloyd, J. M. and High, J. W. 2006. Anisotropic laminar piezocomposite actuator incorporating machined PMN-PT single-crystal fibers. *Journal of Intelligent Material Systems and Structures* 17(1): 15–28.
Williams, R. B. 2004. Nonlinear mechanical and actuation characterization of piezoceramic fiber composites. PhD Dissertation, Mechanical Engineering Dept., Virginia Tech, Blacksburg, VA.
XFOIL. 2001. User primer for XFOIL version 6.9. XFOIL: Subsonic Airfoil Development System, Massachusetts Institute of Technology. [http://web.mit.edu/drela/Public/web/xfoil/xfoil_doc.txt].

7

Analyses of Multifunctional Layered Composite Beams

Sukanya Doshi, Amir Sohrabi, Anastasia Muliana, and J. N. Reddy

CONTENTS

7.1 Introduction

A new generation of engineering materials with predefined mechanical characteristics has shown to be the source for the scientific and technological development of the modern society. For example, the use of multifunctional composites offers great potential for the development of novel adaptive and morphing structures, structural components for extreme environments, energy harvesting and harnessing devices, actuators, sensors, and many other applications. Each of the above applications would subject multifunctional composites to complex mechanical loadings coupled with other non-mechanical effects, such as temperature changes and applied electric fields. Depending on the severity of prescribed external stimuli, the mechanical and physical properties of the multifunctional composites would vary with the prescribed external stimuli, leading to nonlinear response; they would also vary with time and frequencies, termed as time-dependent properties. It is then necessary to understand, predict, and model the properties of these new generation of materials under a combined external influence, e.g., mechanical, thermal, electromagnetic, etc., before analyzing and designing structures with multifunctional capabilities.

To achieve multifunctional capabilities, it is often necessary to combine different materials (constituents) that form composite systems. There have been several types of composites, based on their microstructural arrangements, manufactured and used for various structural applications. Layered composites are some of the commonly used composite systems. Examples of layered composites are fiber-reinforced polymer (FRP) laminated composites with various off-axis stacking sequences for aircraft structural components and retrofitted bridge decks, fiber metal laminate, which is a hybrid layered composite composed of alternating layers of metal and FRP, used for aircraft fuselages, sandwich composites having polymeric or metallic foam cores and FRP skins for naval applications, and piezoelectric stack actuators for high-accuracy positioning applications. In piezoelectric stack actuators, a multilayer arrangement is considered in which the actuators consist of several active layers made of piezoelectric materials and passive (elastic) layers in order to perform axial, bending, and twisting deformations.

Several composite beam theories based on the Euler–Bernoulli, first-order shear deformation, and higher-order shear deformation beams have been proposed to analyze linear elastic deformations in laminated composite beams. Extensive discussion of the above beam theories can be found in the works of Reddy (1990, 2004). Thermal and moisture effects on the deformation of composites have been studied through the thermal and moisture expansion coefficients by Rosen and Hashin (1970) and Shen and Springer (1976). Sai Ram and Sinha (1991) studied the effect of temperature and moisture on the bending characteristics of laminated composite plates. Pipes et al. (1976) derived a method for the hygrothermal response of a laminated composite plate element. The classic lamination theory has also been extended for analyzing deformation in piezoelectric multilayered composites by assuming linear piezoelastic response for each layer. Various such theories have been reviewed by Gopinathan et al. (2000). Robbins and Reddy (1991) developed various layer-wise theories for beams with piezoelectric actuators, while Saravanos et al. (1997) describe theories for laminate composite plates with embedded piezoelectric layers.

The aforementioned laminated composite beam theories, although incorporating coupled mechanical and nonmechanical effects, are derived for linear thermo-electro-elastic response in which superposition and proportionality conditions are applicable among the field variables, i.e., stress, strain, electric field, and electric displacement. Each layer in the laminated composite could exhibit nonlinear and inelastic response, such as viscoelastic, plastic, and viscoplastic, and this nonlinear and inelastic response could also vary significantly with temperatures. There have been several experimental studies aimed at investigating the viscoelastic and viscoplastic behaviors (Guedes et al. 1998, Megnis and Varna 2003) and the effect of elevated temperatures on the time-dependent response (Yeow et al. 1979, Tuttle and Brinson 1986, Tuttle et al. 1995, Muliana et al. 2006, Muddasani et al. 2010) of FRP laminated composites. Piezoelectric materials such as lead zirconate titanate (PZT) and polyvinylidene fluoride exhibit time-dependent electromechanical response, which have been shown by Fett and Thun (1998), Cao and Evans (1993), Schäufele and Heinz Härdtl (1996), Strobl (1997), and Hall (2001). To better understand and predict the overall performance of multifunctional composites, it is then necessary to consider the time-dependent response of these composites.

This chapter presents analyses of multifunctional layered composite beams, focusing on the Euler–Bernoulli beam theory, undergoing coupled mechanical and nonmechanical stimuli, such as thermal and electrical stimuli that can support the design of multifunctional structures. Discussion will be emphasized on a linearized thermo-electro-mechanical response, including the time-dependent effect. Section 7.2 is dedicated to an analysis of linear thermo-electro-elastic response of homogeneous beams followed by several numerical examples. Section 7.3 presents a linear thermo-electro-elastic analysis of multifunctional layered composite beams. An extension of a linear thermo-electro-elastic layered beam theory to a linear time-dependent response of multifunctional layered composite beams, through the use of correspondence principle, is discussed in Section 7.4. Finally, Section 7.5 discusses a limitation of the present beam theory and its implications.

7.2 Linear Thermo-Electro-Elastic Analysis of Homogeneous Beams

7.2.1 Thermoelastic Analysis of the Euler–Bernoulli Beams

Consider an elastic isotropic prismatic beam (Figure 7.1) with a longitudinal axis of the beam lying along the x-axis; the x, y, and z-axes pass through the centroidal point of the cross-section, and the cross-sectional dimension of the beam is relatively small compared with the length of the beam. If the beam is subjected to a temperature change $T(x,y,z)$, which is an arbitrary function of x, y, and z, except that it does not cause twisting, the beam

FIGURE 7.1
Prismatic homogeneous beam.

undergoes bending and axial deformations. Assuming that the plane cross-section remains plane before and after the deformations (i.e., straightness of a transverse normal line), the displacements in the beam are assumed to be of the form

$$u_x = u_x(x, y, z) = A_0\,(x) + A_1(x)y + A_2(x)z,\ u_y = u_y(x),\ u_z = u_z(x) \tag{7.1}$$

The corresponding linearized axial strain is

$$\varepsilon_{xx} = \frac{\partial u_x}{\partial x} = \frac{\partial A_0(x)}{\partial x} + \frac{\partial A_1(x)}{\partial x}y + \frac{\partial A_2(x)}{\partial x}z = A_0' + A_1'y + A_2'z \tag{7.2}$$

In Euler–Bernoulli beams, the only nonzero normal strain is the axial strain. For a linear thermoelastic isotropic beam with E and α as the elastic modulus and coefficient of thermal expansion (CTE),* the axial stress is

$$\sigma_{xx} = E(\varepsilon_{xx} - \alpha T) = E\left(A_0' + A_1'y + A_2'z - \alpha T\right) \tag{7.3}$$

To determine the coefficients A_0', A_1', and A_2' for a statically determinate problem, we need to satisfy the equilibrium conditions, while for a statically indeterminate problem, in addition to the equilibrium conditions we need to satisfy the compatibility conditions. In this chapter, statically determinate problems will be emphasized.

Let us consider a statically determinate beam, in the absence of the prescribed mechanical loads; the equilibrium conditions lead to the following expressions:

$$\begin{aligned} \sum F_x &= 0.0, & \int_A \sigma_{xx}\,\mathrm{d}A &= 0.0 \\ \sum M_y &= 0.0, & \int_A \sigma_{xx} z\,\mathrm{d}A &= 0.0 \\ \sum M_z &= 0.0, & \int_A \sigma_{xx} y\,\mathrm{d}A &= 0.0 \end{aligned} \tag{7.4}$$

where F_x, M_z, and M_y are the axial force and bending moments about the z and y axes, respectively, and A is the cross-sectional area of the beam. Equation 7.4 gives zero net forces, but not necessarily zero tractions at all material points, which will result in violating boundary conditions; in the

* It is noted that E and α can also vary spatially with x, y, and z. In such case, the beam is often considered as heterogeneous beams.

absence of prescribed mechanical loads, all surfaces of the beam should be under traction-free boundary conditions; however, the expressions in Equation 7.4 will satisfy traction-free boundary conditions only if the axial stress σ_{xx} is zero. It is noted that field variables determined from the mechanics of materials approach are valid at the locations away from the boundaries and/or from the applications of external mechanical forces.

Substituting Equation 7.3 into Equation 7.4, the coefficients A_0', A_1', and A_2' can be determined and finally the axial stress is given as

$$\sigma_{xx} = -\alpha ET + \frac{N_x^T}{A} + \left(\frac{-M_z^T I_y - M_y^T I_{yz}}{I_y I_z - I_{yz}^2} \right) y + \left(\frac{M_y^T I_z + M_z^T I_{yz}}{I_y I_z - I_{yz}^2} \right) z \tag{7.5}$$

where the internal axial forces and bending moment due to the thermal effect are

$$N_x^T = \int_A \alpha ET \, \mathrm{d}A, \quad M_y^T = \int_A \alpha ETz \, \mathrm{d}A, \quad M_z^T = -\int_A \alpha ETy \, \mathrm{d}A \tag{7.6}$$

and the second moments of an area are

$$I_y = \int_A z^2 \, \mathrm{d}A, \quad I_z = \int_A y^2 \, \mathrm{d}A, \quad I_{yz} = \int_A yz \, \mathrm{d}A \tag{7.7}$$

Referring to Equation 7.3, the axial strain and deflection are

$$\begin{aligned} \varepsilon_{xx} &= \frac{\sigma_{xx}}{E} + \alpha T = \frac{1}{E} \left\{ \frac{N_x^T}{A} + \left(\frac{-M_z^T I_y - M_y^T I_{yz}}{I_y I_z - I_{yz}^2} \right) y + \left(\frac{M_y^T I_z + M_z^T I_{yz}}{I_y I_z - I_{yz}^2} \right) z \right\} \\ &\equiv \varepsilon_{xx}^0 - \kappa_y y - \kappa_z z \end{aligned} \tag{7.8}$$

$$u_x = u_x(x, y, z) = \frac{1}{E} \int_0^x \left\{ \frac{N_x^T}{A} + \left(\frac{-M_z^T I_y - M_y^T I_{yz}}{I_y I_z - I_{yz}^2} \right) y + \left(\frac{M_y^T I_z + M_z^T I_{yz}}{I_y I_z - I_{yz}^2} \right) z \right\} \mathrm{d}s \tag{7.9}$$

The average axial displacement, which is the displacement along the longitudinal axis of the beam ($y = z = 0.0$), is given as

$$\bar{u}_x = \bar{u}_x(x) = \int_0^x \varepsilon^o \, \mathrm{d}s = \frac{1}{E} \int_0^x \frac{N_x^T}{A} \mathrm{d}s \tag{7.10}$$

The corresponding curvatures are

$$\kappa_y = \frac{\partial^2 u_y}{\partial x^2} = \frac{1}{E}\left(\frac{M_z^T I_y + M_y^T I_{yz}}{I_y I_z - I_{yz}^2}\right) \qquad \kappa_z = \frac{\partial^2 u_z}{\partial x^2} = \frac{1}{E}\left(\frac{-M_y^T I_z - M_z^T I_{yz}}{I_y I_z - I_{yz}^2}\right) \tag{7.11}$$

Finally, the corresponding lateral deflections u_y and u_z are obtained by integrating the above curvatures twice with respect to the x-variable. When the cross section of the beam possesses at least one symmetric line in which I_{yz} = 0.0, then the above problem reduces to a symmetric bending. If the coupling second moment of an area I_{yz} = 0.0, then the y- and z-axes are the principal axes. If in addition to the prescribed temperature change, there are also prescribed mechanical loads, the internal normal force and bending moments are

$$N_x^M + N_x^T, \qquad M_y^M + M_y^T, \qquad M_z^M + M_z^T$$

where the superscript M denotes the mechanical effect, and the internal forces due to the thermal effect are given in Equation 7.6. It is noted that the above formulas are derived with the following sign conventions: tension and elongation are given a positive sign, and bending moments about the y- and z-axes follow the "right hand rule." Detailed derivation and discussion on the strength of material (SOM) approach and thermal stress analysis in the Euler–Bernoulli beams can be found in Cook and Young (1999) and Boley and Weiner (1997).

EXAMPLE 7.2.1

A slender cantilever beam of length L and rectangular cross-section ($2h \times b$) (Figure 7.2) is subjected to a temperature change $T(z) = T_1 + T_1 \frac{z}{h}$ and a concentrated forced P_z applied at the free end along the z-axis. Determine the magnitude and direction of the force P_z in order to minimize the lateral deflection at the free end. Also, determine the corresponding stress and displacement fields.

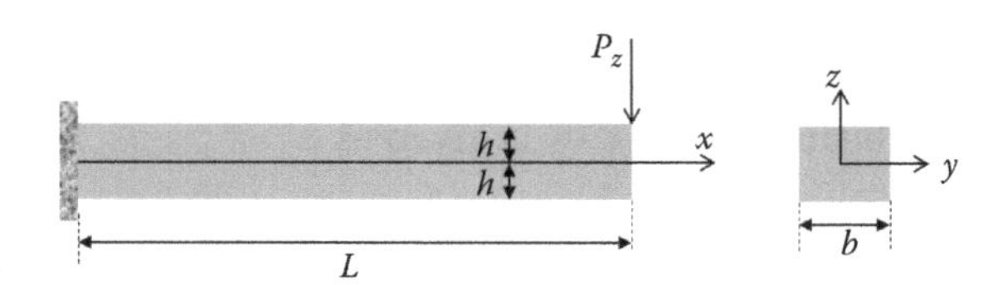

FIGURE 7.2
Homogeneous cantilever beam.

SOLUTION

The beam is subjected to combined mechanical and thermal stimuli. The stress and displacement fields are obtained from Equations 7.5, 7.9, and 7.11; and x, y, and z-axes are the centroidal axes. For this particular problem, the area and second moments of an area of the cross-section are

$$A = 2bh; \quad I_y = \frac{2}{3}bh^3; \quad I_z = \frac{1}{6}hb^3; \quad I_{yz} = 0.0 \tag{a}$$

The corresponding internal normal forces and bending moments are

$$N_x^M = 0.0; \quad M_y^M = P_z L - P_z x; \quad M_z^M = 0.0 \tag{b}$$

$$\begin{aligned} N_x^T &= \int_A \alpha E T(z)\,dA = b\alpha E \int_{-h}^{h} T(z)\,dz = b\alpha E 2T_1 h \\ M_y^T &= \int_A \alpha E T(z) z\,dA = b\alpha E \int_{-h}^{h} T(z) z\,dz = b\alpha E \frac{2}{3} T_1 h^2 \\ M_z^T &= \int_A \alpha E T(z) y\,dA = \alpha E \int_{-h}^{h}\int_{-b/2}^{b/2} T(z) y\,dy\,dz = 0.0 \end{aligned} \tag{c}$$

It is seen that the temperature change produces axial deformation and bending about y-axis with a negative curvature κ_z. The curvature is determined from Equation 7.11:

$$\kappa_z = \frac{\partial^2 u_z}{\partial x^2} = -\frac{M_y^M + M_y^T}{EI_y} = -\frac{P_z L - P_z x + b\alpha E \frac{2}{3} T_1 h^2}{E\frac{2}{3}bh^3} \tag{d}$$

Solving the above differential equation, the lateral displacement is

$$u_z(x) = -\frac{1}{E\frac{2}{3}bh^3}\left\{\left(P_z L + b\alpha E \frac{2}{3} T_1 h^2\right)\frac{x^2}{2} - P_z \frac{x^3}{6} + C_1 x + C_2\right\} \tag{e}$$

The two constants from integrating equation (d) are determined from the prescribed boundary conditions $u_z(0) = \frac{\partial u_z}{\partial x}(0) = 0.0$, which yields $C_1 = C_2 = 0.0$. The lateral deflection at the free end ($x = L$) is minimized by applying the concentrated force P_z:

$$u_z(L) = 0.0, \quad \left(P_z L + b\alpha E \frac{2}{3} T_1 h^2\right)\frac{L^2}{2} - P_z \frac{L^3}{6} = 0.0$$

$$P_z = -\frac{b\alpha E T_1 h^2}{L} \tag{f}$$

Thus, the force P_z should be applied in the opposite direction to the direction illustrated in Figure 7.2. Finally, the corresponding stress and displacement fields are

$$\sigma_{xx} = -\alpha ET + \frac{N_x^T}{A} + \frac{M_y^M + M_y^T}{I_y} z$$
$$= -\alpha E\left(T_1 + T_1 \frac{z}{h}\right) + \alpha ET_1 - \frac{\alpha ET_1 z}{2h} + \alpha ET_1 \frac{3xz}{2hL} = \frac{3}{2}\frac{z}{h}\alpha ET_1\left(\frac{x}{L} - 1\right) \quad \text{(h)}$$

$$u_x = \int_0^x \left\{\frac{\sigma_{xx}}{E} + \alpha T(z)\right\} \mathrm{d}s = \int_0^x \left\{\frac{3}{2}\frac{z}{h}\alpha T_1\left(\frac{s}{L} - 1\right) + \alpha T_1\left(1 + \frac{z}{h}\right)\right\} \mathrm{d}s$$
$$= \alpha T_1\left(\frac{3}{2}\frac{z}{h}\frac{x^2}{2L} - \frac{z}{h}x + x\right) \quad \text{(i)}$$

$$u_z = \frac{\alpha T_1}{4h} x^2 \left(1 - \frac{x}{L}\right) \quad \text{(j)}$$

7.2.2 Piezoelastic Analysis of the Euler–Bernoulli Beams

The previous discussion of the thermoelastic analysis is now extended for analyzing a linear piezoelectric (piezoelastic) beam subjected to couple thermo-electro-mechanical stimuli. The electromechanical coupling effect in a piezoelectric material is observed once the piezoelectric material has been polarized, which can be done by applying high electric fields. The poling direction (or poling axis) is often indicated by polarization component P_3 (Figure 7.3). Once polarized, the piezoelectric materials can be utilized in operating modes 33 and 31 that correspond to the mechanical stress and deflection along and transverse to the poling axis, respectively.

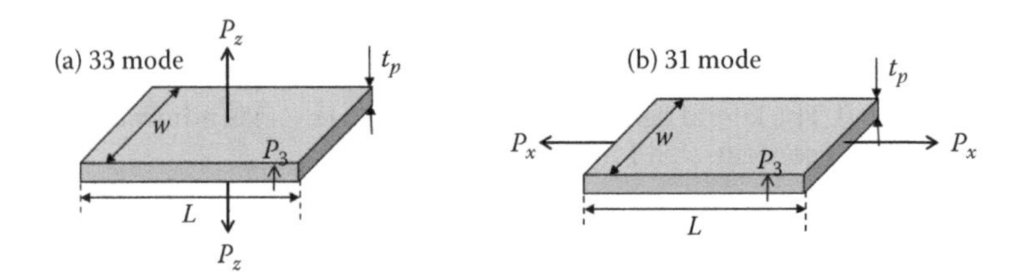

FIGURE 7.3
Piezoelectric layers: (a) 33 mode; (b) 31 mode.

Referring to Figure 7.3a, in the 33 operating mode, the axial stress, axial strain, electric field, and electric displacement in the z-direction are defined as

$$\varepsilon_{zz} = \frac{u_z}{t_p}, \quad \sigma_{zz} = \frac{P_z}{A_z}, \quad E_z = E_3 = \frac{\Delta V}{t_p}, \quad D_z = D_3 = \frac{Q}{A_z} \tag{7.12}$$

where u_z and P_z are the axial displacement and force, respectively, in the z-direction; ΔV and Q are the electric potential difference through the thickness and surface charge, respectively; t_p and $A_z = Lw$ are the thickness of the beam and cross-sectional area with a normal component in the z-direction, respectively. The constitutive relations for the 33 operating mode are

$$\varepsilon_{zz} = \frac{\sigma_{zz}}{E} + d_{333}E_z, \quad D_z = d_{333}\sigma_{zz} + \kappa_{33}E_z \tag{7.13}$$

where the material properties E, d_{333}, and κ_{33} are the elastic modulus* in the through-thickness direction, piezoelectric constant, and permittivity constant. If the piezoelectric material is also subjected to a temperature change $T(x,y,z)$, the thermal strain† αT should be added to the axial strain in Equation 7.2.13, and if the pyroelectric constant p_3 is nonzero, there is also an additional electric displacement p_3T. The axial displacement and surface charge when the piezoelectric material is subjected to an axial force P_z and electric potential difference are

$$u_z = \frac{P_z t_p}{EA_z} + d_{333}\Delta V, \quad Q = d_{333}P_z + \kappa_{33}A_z\frac{\Delta V}{t_p} \tag{7.14}$$

In a similar way, the field variables and constitutive relations in the 31 operating mode, illustrated in Figure 7.3b, are

$$\varepsilon_{xx} = \frac{u_x}{L}, \quad \sigma_{xx} = \frac{P_x}{A_x}, \quad E_z = E_3 = \frac{\Delta V}{t_p}, \quad D_z = D_3 = \frac{Q}{A_z} \tag{7.15}$$

$$\varepsilon_{xx} = \frac{\sigma_{xx}}{E} + d_{311}E_z, \quad D_z = d_{311}\sigma_{xx} + \kappa_{33}E_z \tag{7.16}$$

where $A_x = wt_p$ is the cross-sectional area with a unit-normal vector in the x-direction, E is the elastic modulus in the axial direction, and d_{311} is the piezoelectric constant that relates the electric field in the through-thickness direction (E_z) to the transverse strain component ε_{xx}. The axial displacement and surface charge due to an applied axial force P_x and electric potential difference are

* Once polarized, the piezoelectric material is no longer isotropic. The modulus in the poling direction is not the same as the one in the transverse direction.

† As the polarized piezoelectric material is not isotropic, the CTE in the poling direction might not be the same as the one in the transverse direction.

$$u_x = \frac{P_x L}{EA_x} + d_{311}\Delta V \frac{L}{t_p}, \quad Q = d_{311} P_x \frac{L}{t_p} + \kappa_{33} A_z \frac{\Delta V}{t_p} \tag{7.17}$$

In a single piezoelectric material (beam) polarized through its thickness, an application of electric field along the poling axis can produce only uniform axial elongation and/or contraction while an application of electric field perpendicular to the poling axis creates a uniform shear deformation. When the through-thickness electric field E_z is in the same direction as the poling direction P_3 (Figure 7.3), in which E_z is defined as positive number, the piezoelectric material will experience elongation in the through-thickness direction, 33 mode, and contraction in the transverse direction, 31 mode. Likewise, when E_z is in the opposite direction as the poling direction P_3, in which E_z is defined as a negative number, the piezoelectric material will contract in the through thickness direction and elongate in its transverse direction. Thus, the piezoelectric constant d_{333} is numerically expressed with a positive number, while the piezoelectric constant d_{311} is expressed with a negative number. To produce bending and/or twisting deformations by applying an electric field (an actuator application), it is necessary to stack more than one piezoelectric material or integrate piezoelectric materials with other (elastic) materials, which form a composite system. Analyses of piezoelectric stack actuators are discussed in Section 7.3.

In summary, when a homogeneous piezoelectric prismatic beam is polarized through its thickness and a longitudinal axis of the beam lies on the x-axis, and is subjected to couple mechanical loads, temperature change, and a through-thickness electric field, the constitutive and kinematic relations are

$$\begin{aligned} \sigma_{xx} &= E\left(\varepsilon_{xx} - \varepsilon_{xx}^{T} - \varepsilon_{xx}^{E}\right) = E(\varepsilon_{xx} - \alpha T - d_{311} E_z) \\ \varepsilon_{xx} &= \varepsilon^0 - \kappa_y y - \kappa_z z \end{aligned} \tag{7.18}$$

where κ_y and κ_z are the corresponding curvatures, E is the elastic modulus of the beam in the longitudinal direction, and superscripts T and E denote the thermal and electrical strains, respectively. The three useful equilibrium equations are

$$\int_A \sigma_{xx}\, dA = N_x^M, \quad \int_A \sigma_{xx} z\, dA = M_y^M, \quad \int_A \sigma_{xx} y\, dA = -M_z^M \tag{7.19}$$

Using the axial stress and kinematic relation in Equation 7.18, the equilibrium equations reduce to

$$\begin{aligned} EA\varepsilon^0 - EQ_y\kappa_z - EQ_z\kappa_y &= N_x^M + N_x^T + N_x^E \\ EQ_y\varepsilon^0 - EI_y\kappa_z - EI_{yz}\kappa_y &= M_y^M + M_y^T + M_y^E \\ EQ_z\varepsilon^0 - EI_{yz}\kappa_z - EI_z\kappa_y &= -M_z^M + M_z^T + M_z^E \end{aligned} \tag{7.20}$$

where the internal axial forces and bending moments due to the thermal and electrical effects are

$$N_x^T = \int_A E\varepsilon_{xx}^T \, dA, \quad M_y^T = \int_A E\varepsilon_{xx}^T z \, dA, \quad M_z^T = \int_A E\varepsilon_{xx}^T y \, dA$$
$$N_x^E = \int_A E\varepsilon_{xx}^E \, dA, \quad M_y^E = \int_A E\varepsilon_{xx}^E z \, dA, \quad M_z^E = \int_A E\varepsilon_{xx}^E y \, dA \tag{7.21}$$

The second moments of an area are given in Equation 7.7 and the first moments of an area with respect to the centroidal axes are

$$Q_y = \int_A z \, dA, \quad Q_z = \int_A y \, dA \tag{7.22}$$

If the y- and z-axes are the centroidal axes ($Q_y = Q_z = 0.0$), then the problem reduces to an uncoupled axial-bending problem. It is noted that if the cross-section possesses at least one symmetric line or the y- and z-axes are the principal axes, then $I_{yz} = 0.0$. The uniform axial strain ε^0 and the two curvatures are finally determined by solving Equation 7.2.20 and the corresponding displacement fields are obtained by solving the following differential equations:

$$\varepsilon_{xx} = \frac{\partial u_x}{\partial x}, \quad \kappa_y = \frac{\partial^2 u_y}{\partial x^2}, \quad \kappa_z = \frac{\partial^2 u_z}{\partial x^2} \tag{7.23}$$

7.3 Linear Thermo-Electro-Elastic Analysis of a Layered Composite Beam

We extend our discussion to analyses of multifunctional layered composite beam based on the Euler–Bernoulli beam theory. The studied composites consist of N layers of materials having uniform length L and width b (Figure 7.4),

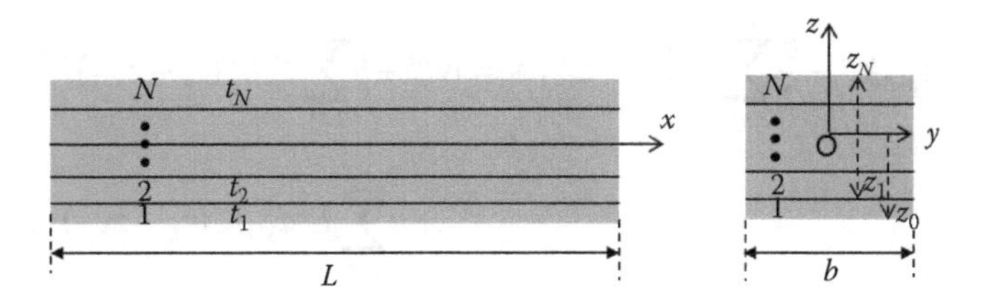

FIGURE 7.4
Multilayered composite beam.

and the thickness of each layer is denoted by t_k ($k = 1 \ldots N$). To reduce complexity, we shall consider the layered composite beams undergoing axial deformation in the x-direction and symmetric bending deformation about the y-axis, leading to $\bar{M}_z = \bar{\kappa}_y = \bar{u}_y = 0.0$. The cross-section of the composite beam is placed on the y–z axes; the y-axis need not be the centroidal axis nor the NA. It is also assumed that all layers are perfectly bonded to each other.

The displacement continuity condition leads to uniform axial strain $\bar{\varepsilon}^0_{xx} = \varepsilon^0_{xx}$ and curvature $\bar{\kappa}_z = \kappa$, and the overall axial strain due to axial and bending deformations is

$$\bar{\varepsilon}_{xx} = \varepsilon^0_{xx} - \kappa z \tag{7.24}$$

It is noted that the axial strain ε^0 and curvature κ are independent on y and z; however, they can vary in the longitudinal x-axis. The axial stress for each layer k is expressed as

$$\sigma^k_{xx} = E^k\left(\varepsilon^0 - \kappa z - \varepsilon^{T,k}_{xx} - \varepsilon^{E,k}_{xx}\right) = E^k\left(\varepsilon^0 - \kappa z - \alpha^k T^k - d^k_{311}E^k_z\right) \tag{7.25}$$

where the superscript k indicates properties or field variables of layer k.

The overall equilibrium equations lead to

$$\int_A \sigma_{xx}\,\mathrm{d}A = b\int_h \sigma_{xx}\,\mathrm{d}z = b\sum_{k=1}^{N}\int_{z_{k-1}}^{z_k} \sigma^k_{xx}\mathrm{d}z = \bar{N}^M_x$$

$$\int_A \sigma_{xx} z\,\mathrm{d}A = b\int_h \sigma_{xx} z\,\mathrm{d}z = b\sum_{k=1}^{N}\int_{z_{k-1}}^{z_k} \sigma^k_{xx} z\,\mathrm{d}z = \bar{M}^M_y \tag{7.26}$$

where $\bar{N}^M_x$ and $\bar{M}^M_y$ are the effective internal axial force and bending moment due to mechanical loads. Substituting Equation 7.25 into Equation 7.26 gives

$$b\sum_{k=1}^{N} E^k \int_{z_{k-1}}^{z_k} \mathrm{d}z\varepsilon^0 - b\sum_{k=1}^{N} E^k \int_{z_{k-1}}^{z_k} z\,\mathrm{d}z\kappa = \bar{N}^M_x + b\sum_{k=1}^{N} E^k\alpha^k \int_{z_{k-1}}^{z_k} T^k\,\mathrm{d}z + b\sum_{k=1}^{N} E^k d^k_{311}E^k_z \int_{z_{k-1}}^{z_k} \mathrm{d}z$$

$$b\sum_{k=1}^{N} E^k(z_k - z_{k-1})\varepsilon^0 - \frac{b}{2}\sum_{k=1}^{N} E^k\left(z_k^2 - z_{k-1}^2\right)\kappa = \bar{N}^M_x + b\sum_{k=1}^{N} E^k\alpha^k(z_k - z_{k-1})T^k + b\sum_{k=1}^{N} E^k d^k_{311}E^k_z(z_k - z_{k-1})$$

$$\overline{EA}\varepsilon^0 - \overline{EQ}\kappa = \bar{N}^M_x + \bar{N}^T_x + \bar{N}^E_x \tag{7.27}$$

$$b\sum_{k=1}^{N}E^k\int_{z_{k-1}}^{z_k} zdz\varepsilon^0 - b\sum_{k=1}^{N}E^k\int_{z_{k-1}}^{z_k} z^2 dz\kappa = \bar{M}_y^M + b\sum_{k=1}^{N}E^k\alpha^k\int_{z_{k-1}}^{z_k} T^k zdz + b\sum_{k=1}^{N}E^k d_{311}^k E_z^k\int_{z_{k-1}}^{z_k} zdz$$

$$\frac{b}{2}\sum_{k=1}^{N}E^k\left(z_k^2 - z_{k-1}^2\right)\varepsilon^0 - \frac{b}{3}\sum_{k=1}^{N}E^k\left(z_k^3 - z_{k-1}^3\right)\kappa = \bar{M}_y^M + \frac{b}{2}\sum_{k=1}^{N}E^k\alpha^k\left(z_k^2 - z_{k-1}^2\right)T^k$$

$$+\frac{b}{2}\sum_{k=1}^{N}E^k d_{311}^k E_z^k\left(z_k^2 - z_{k-1}^2\right)$$

$$\overline{EQ}\varepsilon^0 - \overline{EI}\kappa = \bar{M}_y^M + \bar{M}_y^T + \bar{M}_y^E \tag{7.28}$$

The axial strain ε^0 and curvature κ are obtained by solving Equations 7.27 and 7.28 simultaneously.

Alternatively, the solution can be obtained by first locating the centroidal axis (NA) under pure bending of the composite and place the centroidal (NA) axis on the y-axis so that the z location is measured from the centroidal axis. The centroidal axis (NA) is obtained from a pure bending case by setting $\bar{N}_x^M = 0.0$ and $\varepsilon_{xx}^0 = 0.0$; thus

$$\int_A \sigma_{xx}\,dA = b\int_h \sigma_{xx}\,dz = b\sum_{k=1}^{N}\sigma_{xx}^k dz = b\sum_{k=1}^{N}E^k\int_{z_{k-1}}^{z_k} z\,dz\kappa = 0.0 \tag{7.29}$$

If the y-axis is the centroidal axis (NA) under a pure bending deformation, it can be shown that the overall first moment of an area of the composite $\overline{EQ} = 0.0$, leading to uncoupled axial and bending deformations.

EXAMPLE 7.3.1

Locate the centroidal axis (NA) of the following two-layered composite cross-section of width b. The top layer has thickness 6 mm and elastic modulus of 2000 MPa, and the bottom layer has thickness 4 mm and elastic modulus 500 MPa.

SOLUTION

Let the centroidal axis lie on the y-axis and h_1 is the distance from the centroidal axis to the top surface of the top layer along the z-direction. The distance h_1 is determined from Equation 7.29, which is

$$\sum_{k=1}^{N}E^k\int_{z_{k-1}}^{z_k} z\,dz = 0.0; \quad \frac{E^1}{2}\left(h_1^2 - (h_1-6)^2\right) + \frac{E^2}{2}\left((h_1-6)^2 - (h_1-10)^2\right) = 0.0$$

$$1000(12h_1 - 36) + 250(8h_1 - 64) = 0.0, \quad h_1 = 3.71\text{ mm} \tag{a}$$

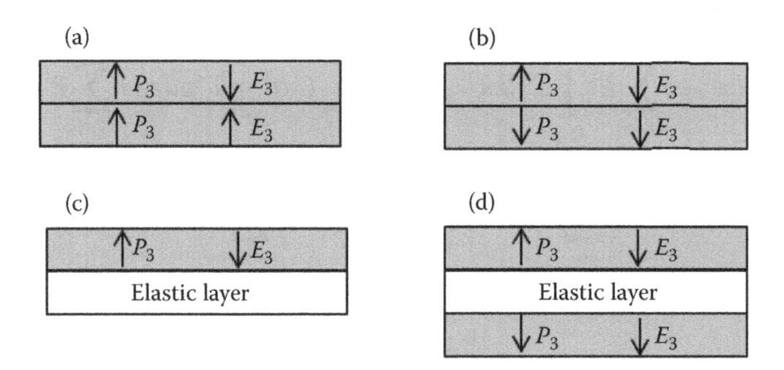

FIGURE 7.5
Examples of stack bending actuators.

Finally, upon solving the axial strain ε^0 and curvature κ, the overall axial and lateral deformations of the composite beam are determined by solving the following differential equations:

$$\bar{\varepsilon}_{xx} = \frac{\partial \bar{u}_x}{\partial x}, \quad \kappa = \frac{\partial^2 \bar{u}_z}{\partial x^2} \tag{7.30}$$

The axial deformation along the neutral surface is determined from $\varepsilon^0 = \dfrac{\partial \bar{u}_x}{\partial x}$.

As briefly mentioned in Section 7.2, bending deformations by applying electric fields in piezoelectric layers, which have been polarized through their thickness, can be achieved by stacking more than one piezoelectric material or stacking piezoelectric materials to other elastic materials, which forms a layered composite. Figure 7.5 illustrates examples of piezoelectric stack actuators for generating bending deformations. For example, when two polarized piezoelectric layers are stacked together, bending can be generated by allowing one layer to contract and another layer to elongate in their transverse direction (Figure 7.5a,b). One can also stack piezoelectric layers with other nonpiezoelectric materials and apply electric fields to form bending deformations, as illustrated in Figure 7.5c,d. For further discussion on several examples and applications of stack actuators, readers may refer to Ballas (2007) and Leo (2007). Bending deformations in a layered composite can also be achieved by prescribing thermal stimuli.

EXAMPLE 7.3.2

A bimetal beam comprising aluminum and tungsten layers is subjected to a uniform temperature change of 100°C. The beam has a length of 100 mm and a width of 6 mm, and each of the aluminum and tungsten layers has a thickness of 2 mm. Determine the thermal stresses and deflections of the bimetal beam due to the temperature change if the elastic moduli and CTEs of aluminum (Al) and tungsten (Ts) are $E^{Al} = 70$ GPa; $E^{Ts} = 350$ GPa; $\alpha^{Al} = 24 \times 10^{-6}$/°C; $\alpha^{Ts} = 4 \times 10^{-6}$/°C.

SOLUTION

Let x-axis be the longitudinal axis and the aluminum layer be the top layer of the bimetal. The NA axis of the beam lies on the y-axis, which is (see Example 7.3.1)

$$\frac{E^{\text{Al}}}{2}\left(h_1^2-(h_1-2)^2\right)+\frac{E^{\text{Ts}}}{2}\left((h_1-2)^2-(h_1-4)^2\right)=0.0$$

$$\frac{70}{2}(4h_1-4)+\frac{350}{2}(4h_1-12)=0.0,\quad h_1=2.67\text{ mm} \tag{a}$$

To determine the corresponding axial strain ε_{xx}^0 and curvature κ, we need to solve Equations 7.27 and 7.28. Since the NA is placed on the y-axis, the problem reduces to a uncouple axial and bending problem, $\overline{EQ}=0.0$. The axial strain ε_{xx}^0 and curvature κ are

$$\varepsilon_{xx}^0=\frac{\bar{N}_x^T}{\overline{EA}}=7.33\times10^{-4}$$

$$\bar{N}_x^T=6\left[(70\times10^3)(24\times10^{-6})2+(350\times10^3)(4\times10^{-6})2\right]100=3696\text{ N}$$

$$\overline{EA}=6\left[(70\times10^3)2+(350\times10^3)2\right]=5.04\times10^6\text{N} \tag{b}$$

$$\kappa=-\frac{\bar{M}_y^T}{\overline{EI}}=-6.5025\times10^{-4}\text{mm}^{-1}$$

$$\bar{M}_y^T=\frac{6}{2}\Big[(70\times10^3)(24\times10^{-6})(2.67^2-0.67^2)$$
$$+(350\times10^3)(4\times10^{-6})(0.67^2-(-1.33)^2)\Big]100=2913.12\text{ Nmm}$$

$$\overline{EI}=\frac{6}{3}\Big[(70\times10^3)(2.67^3-0.67^3)$$
$$+(350\times10^3)(0.67^3-(-1.33)^3)\Big]=4.48\times10^6\text{Nmm}^2 \tag{c}$$

The axial stresses in the aluminum and tungsten layers are determined from Equation 7.25

$$\sigma_{xx}^{\text{Al}}=E^{\text{Al}}\left(\varepsilon_{xx}^0-\kappa z-\alpha^{\text{Al}}T\right)=70\times10^3\left(7.33\times10^{-4}+6.5025\times10^{-4}z-24\times10^{-6}100\right)$$

$$\sigma_{xx}^{\text{Ts}}=E^{\text{Ts}}\left(\varepsilon_{xx}^0-\kappa z-\alpha^{\text{Ts}}T\right)=350\times10^3\left(7.33\times10^{-4}+6.5025\times10^{-4}z-4\times10^{-6}100\right)$$

(d)

The axial and lateral deflections are obtained from Equation 7.30

$$\bar{\varepsilon}_{xx} = 7.33\times10^{-4} + 6.5025\times10^{-4}z, \quad \bar{u}_x = \left(7.33\times10^{-4} + 6.5025\times10^{-4}z\right)x + C_1$$

$$\bar{u}_z = 6.5025\times10^{-4}\frac{x^2}{2} + D_1x + D_2 \tag{e}$$

where the constants C_1, D_1, and D_2 correspond to rigid body motions that can be determined from the boundary conditions.

EXAMPLE 7.3.3

A cantilever beam, comprising an elastic layer sandwiched between PZT layers at the top and bottom parts, is subjected to a concentrated force F_o. The elastic layer has thickness $2t_s$ and each PZT layer has thickness t_p. The beam has uniform width of b and the PZT layers are polarized through the thickness (Figure 7.6). The elastic layer has the elastic modulus, E_{el}, and the PZT layers have the following elastic modulus and piezoelectric constant, E_{PZT}, and d_{311}, respectively. Determine the magnitude and direction of electric fields that should be applied to the PZT layers in order to minimize the lateral deformation at the free end.

SOLUTION

The composite cross-section has a symmetric layer arrangement, and therefore the centroidal axis lies on the symmetric axis. By placing the y-axis on the symmetric axis, the problem reduces to an uncouple axial and bending problem, $\overline{EQ} = 0.0$. To produce bending through the actuators, opposite directions of electric fields should be applied to the top and bottom actuators since they are polarized in the same direction. In the absence of the thermal effect, Equations 7.27 and 7.28 reduce to

$$b\sum_{k=1}^{N} E^k(z_k - z_{k-1})\varepsilon^0 = b\sum_{k=1}^{N} E^k d_{311}^k E_z^k(z_k - z_{k-1})$$

$$2b(E_{PZT}t_p + E_{el}t_s)\varepsilon^0 = b(E_{PZT}d_{311}E_zt_p + E_{PZT}d_{311}(-E_z)t_p) = 0.0$$

$$\overline{EA}\varepsilon^0 = \bar{N}_x^E = 0.0; \quad \varepsilon_{xx}^0 = 0.0 \tag{a}$$

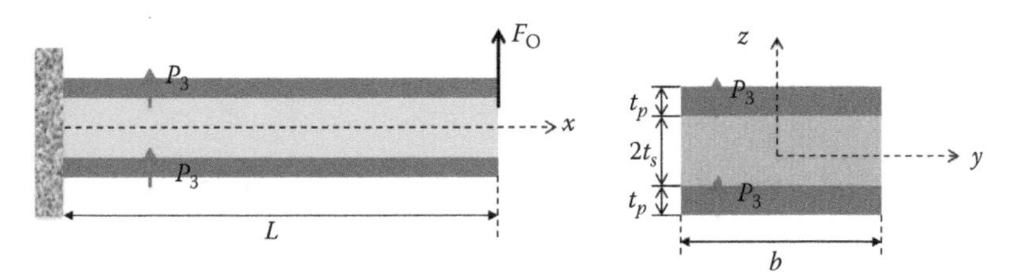

FIGURE 7.6
Active layered composite composed of two piezoelectric and elastic layers.

$$-\frac{b}{3}\sum_{k=1}^{N} E^k\left(z_k^3 - z_{k-1}^3\right)\kappa = F_o x - F_o L + \frac{b}{2}\sum_{k=1}^{N} E^k d_{311}^k E_z^k\left(z_k^2 - z_{k-1}^2\right)$$

$$-\frac{2b}{3}\left(E_{\text{PZT}}\left(t_p^3 + 3t_p^2 t_s + 3t_p t_s^2\right) + E_{\text{el}} t_s^2\right)\kappa = F_o x - F_o L + bE_{\text{PZT}} d_{311} E_z\left(t_p^2 + 2t_p t_s\right)$$

$$-\overline{EI}\kappa = \bar{M}_y^M + \bar{M}_y^E; \quad \kappa = -\frac{F_o x - F_o L + bE_{\text{PZT}} d_{311} E_z\left(t_p^2 + 2t_p t_s\right)}{\overline{EI}} \tag{b}$$

The lateral deflection is now determined by integrating the curvature twice w.r.t x and prescribing boundary conditions at the fixed end, leading to

$$u_z(x) = -\frac{1}{\overline{EI}}\left[F_o\frac{x^3}{6} - F_o L\frac{x^2}{2} + bE_{\text{PZT}} d_{311} E_z\left(t_p^2 + 2t_p t_s\right)\frac{x^2}{2}\right] \tag{c}$$

The lateral deflection at the free end is minimized by setting $u_z(L) = 0.0$. The amount of electric field that should be applied to minimize the lateral deflection at the free end is

$$F_o\frac{L^3}{6} - F_o L\frac{L^2}{2} + bE_{\text{PZT}} d_{311} E_z\left(t_p^2 + 2t_p t_s\right)\frac{L^2}{2} = 0.0$$

$$E_z = \frac{F_o L}{3bE_{\text{PZT}} d_{311}\left(t_p^2 + 2t_p t_s\right)} \tag{d}$$

Since the concentrated load F_o creates a positive curvature, in order to create a negative curvature, the direction of the electric field at the top layer should be opposite to the poling axis, while the electric field at the bottom layer should be in the direction of the poling axis.

Up to now, we have discussed deformations in multilayered composites due to applications of mechanical, electrical, and thermal stimuli. When a polarized piezoelectric material is subject to electromechanical stimuli, depending on the electrical circuit boundary conditions, a surface charge can be also collected (Equations 7.14 and 7.17). In a stack actuator, 33 or 11 operating mode, the total surface charge with N_p piezoelectric layers is given as

$$\bar{Q} = \sum_{k=1}^{N_p} Q^k \tag{7.31}$$

In the 33 operating mode, the total surface charge due to axial force P_z and potential difference ΔV

$$\bar{Q} = \sum_{k=1}^{N_p} d_{333}^k P_z + \sum_{k=1}^{N_p} \kappa_3^k \frac{A_z^k}{t_z^k}\Delta V \tag{7.32}$$

Likewise, when the 31 operating mode is considered, the total surface charge is

$$\bar{Q} = \sum_{k=1}^{N_p} d_{311}^k \sigma_{xx}^k A_z^k + \sum_{k=1}^{N_p} \kappa_3^k \frac{A_z^k}{t_z^k} \Delta V \quad (7.33)$$

with the axial stress for each layer σ_{xx}^k given in Equation 7.25.

7.4 Linear Thermo-Electro-Viscoelastic Analysis

7.4.1 Homogeneous Beams

For a class of problems, it is often convenient to use a correspondence principle in analyzing response of linear viscoelastic materials in that the solution to a linear viscoelastic response of a structure can be constructed from the corresponding linear elastic response (see Wineman and Rajagopal [2000] for a detailed discussion and its applicability). The correspondence principle makes use of the Laplace transform, which is done by integrating with respect to time at a fixed material point. The constitutive relation in a linear viscoelastic response, expressed in terms of the transform parameter, has the same form as the corresponding linear elastic constitutive relation. For example, a linear viscoelastic constitutive relation in its transform form is obtained by replacing the elastic modulus E by a transform parameter s multiplied by the Laplace transform of the elastic modulus $\tilde{E}(s)$ and replacing the field variables by their Laplace transforms

$$\tilde{\sigma}_{xx}(s) = s\tilde{E}(s)\tilde{\varepsilon}_{xx}(s) \quad (7.34)$$

The time-dependent stress $\sigma_{xx}(t)$ is obtained by inverted the Laplace transform $L^{-1}(\tilde{\sigma}_{xx}(s))$. For a linear thermoviscoelastic case, the corresponding thermoelastic relation in Equation 7.3 gives

$$\tilde{\sigma}_{xx}(s) = s\tilde{E}(s)(\tilde{\varepsilon}_{xx}(s) - s\tilde{\alpha}(s)\tilde{T}(s)) \quad (7.35)$$

when the CTE of the viscoelastic material is also time-dependent; if the CTE of the material is fixed (time-independent), then the thermoviscoelastic constitutive relation is reduced to

$$\tilde{\sigma}_{xx}(s) = s\tilde{E}(s)[\tilde{\varepsilon}_{xx}(s) - \alpha\tilde{T}(s)] \quad (7.36)$$

In a similar way for a piezoelectric material with time-dependent elastic modulus, piezoelectric constant, and permittivity constant, the correspondence principle leads to

$$\tilde{\varepsilon}_{xx}(s)=\frac{\tilde{\sigma}_{xx}(s)}{s\tilde{E}(s)}+s\tilde{d}_{311}(s)\tilde{E}_z(s),\quad \tilde{D}_z(s)=s\tilde{d}_{311}(s)\tilde{\sigma}_{xx}(s)+s\tilde{\kappa}_{33}(s)\tilde{E}_z(s) \tag{7.37}$$

The use of a correspondence principle is also applicable beyond the constitutive relation, which can be used for solving boundary value problems involving linear viscoelasticity. For example, referring to Equation 7.17, the time-dependent axial displacement and surface charge due to an applied axial force $P_x(t)$ and electric potential difference $\Delta V(t)$ are obtained as

$$\begin{aligned}
\tilde{u}_x(s)&=\frac{\tilde{P}_x(s)L}{s\tilde{E}(s)A_x}+s\tilde{d}_{311}(s)\Delta\tilde{V}(s)\frac{L}{t_\mathrm{p}},\\
\tilde{Q}(s)&=s\tilde{d}_{311}(s)\tilde{P}_x(s)\frac{L}{t_\mathrm{p}}+s\tilde{\kappa}_{33}(s)A_z\frac{\Delta\tilde{V}(s)}{t_\mathrm{p}}\\
u_x(t)&=L^{-1}(\tilde{u}_x(s)),\quad Q(t)=L^{-1}(\tilde{Q}(s))
\end{aligned} \tag{7.38}$$

Through the use of Laplace transform of derivatives of time-dependent function and Riemann convolution, it can easily be shown that Equation 7.34 can be written as

$$\sigma_{xx}(t)=\int_{0^-}^{t}E(t-\varsigma)\frac{\mathrm{d}\varepsilon_{xx}(\varsigma)}{\mathrm{d}\varsigma}\mathrm{d}\varsigma=\int_{0^-}^{t}\varepsilon_{xx}(t-\varsigma)\frac{\mathrm{d}E(\varsigma)}{\mathrm{d}\varsigma}\mathrm{d}\varsigma \tag{7.39}$$

In a thermoviscoelastic case with a constant (time-independent) CTE, the time-dependent axial stress is given as

$$\begin{aligned}
\sigma_{xx}(t)&=\int_{0^-}^{t}E(t-\varsigma)\left\{\frac{\mathrm{d}\varepsilon_{xx}(\varsigma)}{\mathrm{d}\varsigma}-\alpha\frac{\mathrm{d}T(\varsigma)}{\mathrm{d}\varsigma}\right\}\mathrm{d}\varsigma\\
\sigma_{xx}(t)&=\int_{0^-}^{t}E(t-\varsigma)\frac{\mathrm{d}\varepsilon_{xx}(\varsigma)}{\mathrm{d}\varsigma}\mathrm{d}\varsigma-\alpha\int_{0^-}^{t}E(t-\varsigma)\frac{\mathrm{d}T(\varsigma)}{\mathrm{d}\varsigma}\mathrm{d}\varsigma
\end{aligned} \tag{7.40}$$

Following the convolution integral form, Equation 7.37 is written as

$$\varepsilon_{xx}(t)=\int_{0^-}^{t} S(t-\varsigma)\frac{d\sigma_{xx}(\varsigma)}{d\varsigma}d\varsigma+\int_{0^-}^{t} d_{311}(t-\varsigma)\frac{dE_z(\varsigma)}{d\varsigma}d\varsigma,$$

$$D_z(t)=\int_{0^-}^{t} d_{311}(t-\varsigma)\frac{d\sigma_{xx}(\varsigma)}{d\varsigma}d\varsigma+\int_{0^-}^{t} \kappa_{33}(t-\varsigma)\frac{dE_z(\varsigma)}{d\varsigma}d\varsigma \tag{7.41}$$

where S is the time-dependent compliance, corresponding to $S = 1/E$ in a linear elastic case. In a linear viscoelastic problem, the time-dependent compliance $S(t)$ is obtained from $L^{-1}(\tilde{S}(s))$, where $\tilde{S}(s)=\dfrac{1}{s^2\tilde{E}(s)}$.

For a general thermo-electro-mechanical response, the time-dependent axial stress is

$$\tilde{\sigma}_{xx}(s)=s\tilde{E}(s)\left(\tilde{\varepsilon}^o(s)-\tilde{\kappa}_y(s)y-\tilde{\kappa}_z(s)z-s\tilde{\alpha}(s)\tilde{T}(s)-s\tilde{d}_{311}(s)\tilde{E}_z(s)\right) \tag{7.42}$$

$$\sigma_{xx}(t)=$$

$$\int_{0^-}^{t} E(t-\varsigma)\frac{d\left(\varepsilon^0(\varsigma)-\kappa_y(\varsigma)y-\kappa_z(\varsigma)z-\int_{0^-}^{\varsigma}\alpha(\varsigma-s)\frac{dT}{ds}ds-\int_{0^-}^{\varsigma}d_{311}(\varsigma-s)\frac{dE_z}{ds}ds\right)}{d\varsigma}d\varsigma \tag{7.43}$$

Alternatively, we can use the thermal stress constant $\beta(t)$ and piezoelectric constant $e_{311}(t)$, and the axial stress is written as

$$\sigma_{xx}(t)=\int_{0^-}^{t} E(t-\varsigma)\frac{d\left(\varepsilon^0(\varsigma)-\kappa_y(\varsigma)y-\kappa_z(\varsigma)z\right)}{d\varsigma}d\varsigma$$

$$-\int_{0^-}^{t}\beta(t-\varsigma)\frac{dT(\varsigma)}{d\varsigma}d\varsigma-\int_{0-}^{t}e_{311}(t-\varsigma)\frac{dE_z(\varsigma)}{d\varsigma}d\varsigma \tag{7.44}$$

Like in an elastic beam, it is also necessary to maintain displacement continuity and equilibrium conditions. The kinematic relation in Equation 7.18 is now written as

$$\varepsilon_{xx}(t) = \varepsilon^0(t) - \kappa_y(t)\, y - \kappa_z(t)z \tag{7.45}$$

In a homogeneous viscoelastic beam, the useful equilibrium equations lead to

$$\int_A \sigma_{xx}(t)\,\mathrm{d}A = N_x^M(t), \quad \int_A \sigma_{xx}(t) z\,\mathrm{d}A = M_y^M(t), \quad \int_A \sigma_{xx}(t) y\,\mathrm{d}A = -M_z^M(t) \tag{7.46}$$

Using the axial stress in Equation 7.43 and kinematic relation in Equation 7.45, the equilibrium equations reduce to

$$\begin{aligned}
AE*\mathrm{d}\varepsilon^0 - Q_y E*\mathrm{d}\kappa_z - Q_z E*\mathrm{d}\kappa_y &= N_x^M + N_x^T + N_x^E \\
Q_y E*\mathrm{d}\varepsilon^0 - I_y E*\mathrm{d}\kappa_z - I_{yz} E*\mathrm{d}\kappa_y &= M_y^M + M_y^T + M_y^E \\
Q_z E*\mathrm{d}\varepsilon^0 - I_{yz} E*\mathrm{d}\kappa_z - I_z E*\mathrm{d}\kappa_y &= -M_z^M + M_z^T + M_z^E
\end{aligned} \tag{7.47}$$

where the convolution is $E*\mathrm{d}p = \int_{0^-}^{t} E(t-\varsigma)\frac{\mathrm{d}p(\varsigma)}{\mathrm{d}\varsigma}\mathrm{d}\varsigma$. The time-dependent axial force and bending moments due to the thermal and electrical effects are

$$\begin{aligned}
N_x^T(t) &= \int_A E*\mathrm{d}(\alpha*\mathrm{d}T)\,\mathrm{d}A, \quad N_x^E(t) = \int_A E*\mathrm{d}(d_{311}*\mathrm{d}E_z)\,\mathrm{d}A \\
M_y^T(t) &= \int_A E*\mathrm{d}(\alpha*\mathrm{d}T) z\,\mathrm{d}A, \quad M_y^E(t) = \int_A E*\mathrm{d}(d_{311}*\mathrm{d}E_z) z\,\mathrm{d}A \\
M_z^T(t) &= \int_A E*\mathrm{d}(\alpha*\mathrm{d}T) y\,\mathrm{d}A, \quad M_z^E(t) = \int_A E*\mathrm{d}(d_{311}*\mathrm{d}E_z) y\,\mathrm{d}A
\end{aligned} \tag{7.48}$$

When the y- and z-axes are placed at the centroidal axes and the cross-section possesses at least one symmetric line, then Equation 7.47 reduces to uncouple axial and bending deformations. Furthermore, it is often convenient to express the equilibrium equations in terms of the time-dependent compliance $S(t)$, which for an uncouple axial and bending deformation are

$$\begin{aligned}
A\varepsilon_o(t) &= \int_{0^-}^{t} S(t-\varsigma)\frac{\mathrm{d}\left(N_x^M(\varsigma) + N_x^T(\varsigma) + N_x^E(\varsigma)\right)}{\mathrm{d}\varsigma}\mathrm{d}\varsigma \\
-I_y\kappa_z(t) &= \int_{0^-}^{t} S(t-\varsigma)\frac{\mathrm{d}\left(M_y^M(\varsigma) + M_y^T(\varsigma) + M_y^E(\varsigma)\right)}{\mathrm{d}\varsigma}\mathrm{d}\varsigma \\
-I_z\kappa_y(t) &= \int_{0^-}^{t} S(t-\varsigma)\frac{\mathrm{d}\left(-M_z^M(\varsigma) + M_z^T(\varsigma) + M_z^E(\varsigma)\right)}{\mathrm{d}\varsigma}\mathrm{d}\varsigma
\end{aligned} \tag{7.49}$$

EXAMPLE 7.4.1

A prismatic viscoelastic beam of a rectangular cross section is subjected to a temperature change varying across its thickness and with time $T(z,t)$. The beam has a relaxation modulus $E(t)$ and a constant CTE α. Set up the governing equations of the time-dependent deformations of the viscoelastic beam and determine the thermal stress. You may consider bending moment about the z-axis to be zero.

SOLUTION

The time-dependent axial stress expressed in Equation 7.4.7 is used with the axial strain $\varepsilon_{xx}(z, t) = \varepsilon^0(t) - \kappa(t)z$, which gives

$$\sigma_{xx}(z,t) = \int_{0^-}^{t} E(t-\varsigma)\frac{\mathrm{d}[\varepsilon^0(\varsigma) - \kappa(\varsigma)z]}{\mathrm{d}\varsigma}\mathrm{d}\varsigma - \alpha\int_{0^-}^{t} E(t-\varsigma)\frac{\mathrm{d}T(\varsigma)}{\mathrm{d}\varsigma}\mathrm{d}\varsigma \tag{a}$$

The two useful equilibrium equations given in Equation 7.3.3 become

$$\int_A \sigma_{xx}(t)\mathrm{d}A = N_x^M(t); \quad \int_A \sigma_{xx}(t)z\,\mathrm{d}A = M_y^M(t) \tag{b}$$

In absence of the mechanical loads, the internal axial force $N_x^M(t)$ and bending moment $M_y^M(t)$ are zero. If the y-axis is chosen as the centroidal axis, substituting the axial stress in equation (a) into the equilibrium equations (b) leads to

$$\begin{aligned} &A\int_{0^-}^{t} E(t-\varsigma)\frac{\mathrm{d}\varepsilon^0(\varsigma)}{\mathrm{d}\varsigma}\mathrm{d}\varsigma = N_x^M(t) + \int_{0^-}^{t} E(t-\varsigma)\frac{\mathrm{d}N_x^T(\varsigma)}{\mathrm{d}\varsigma}\mathrm{d}\varsigma \\ &N_x^T(t) = \int_A \alpha T(z,t)\mathrm{d}A \end{aligned} \tag{c}$$

$$\begin{aligned} &-I_y\int_{0^-}^{t} E(t-\varsigma)\frac{\mathrm{d}\kappa(\varsigma)}{\mathrm{d}\varsigma}\mathrm{d}\varsigma = M_y^M(t) + \int_{0^-}^{t} E(t-\varsigma)\frac{\mathrm{d}M_y^T(\varsigma)}{\mathrm{d}\varsigma}\mathrm{d}\varsigma \\ &M_y^T(t) = \int_A \alpha T(z,t)z\,\mathrm{d}A \end{aligned} \tag{d}$$

From equations (a), (c), and (d), the axial stress is

$$\begin{aligned} \sigma_{xx}(z,t) = &\frac{N_x^M(t)}{A} + \frac{1}{A}\int_{0^-}^{t} E(t-\varsigma)\frac{\mathrm{d}N_x^T(\varsigma)}{\mathrm{d}\varsigma}\mathrm{d}\varsigma + z\frac{M_y^M(t)}{I_y} \\ &+ \frac{z}{I_y}\int_{0^-}^{t} E(t-\varsigma)\frac{\mathrm{d}M_y^T(\varsigma)}{\mathrm{d}\varsigma}\mathrm{d}\varsigma - \alpha\int_{0^-}^{t} E(t-\varsigma)\frac{\mathrm{d}T(\varsigma)}{\mathrm{d}\varsigma}\mathrm{d}\varsigma \end{aligned} \tag{e}$$

7.4.2 Layered Composite Beams

An extension of a linearized thermo-electro-elastic response of layered composite (Figure 7.4) to a corresponding linearized thermo-electro-viscoelastic response is done through the use of Laplace transform and/or Riemann convolution integral. The displacement continuity condition in Equation 7.24 leads to the following overall axial strain

$$\bar{\varepsilon}_{xx}(t) = \varepsilon^0(t) - \kappa(t)z \tag{7.50}$$

As in a linearized elastic case, the axial strain ε^0 and curvature κ are independent on y and z, but they can vary in the longitudinal x-axis. The axial stress for each layer k is expressed as

$$\tilde{\sigma}^k_{xx}(s) = s\tilde{E}^k(s)\left(\tilde{\varepsilon}^o(s) - \tilde{\kappa}(s)z - s\tilde{\alpha}^k(s)\tilde{T}^k(s) - s\tilde{d}^k_{311}(s)\tilde{E}^k_z(s)\right) \tag{7.51}$$

$$\sigma^k_{xx}(t) = \int_{0^-}^{t} E^k(t-\varsigma)\frac{d\left(\varepsilon^o(\varsigma) - \kappa(\varsigma)z - \int_{0^-}^{\varsigma}\alpha^k(\varsigma - s)\frac{dT^k}{ds}ds - \int_{0^-}^{\varsigma} d^k_{311}(\varsigma - s)\frac{dE^k_z}{ds}ds\right)}{d\varsigma}d\varsigma \tag{7.52}$$

The overall equilibrium conditions in Equation 7.26 must also be satisfied at all time, leading to

$$\int_A \sigma_{xx}\,dA = b\sum_{k=1}^{N}\int_{z_{k-1}}^{z_k}\sigma^k_{xx}(t)dz = \bar{N}^M_x(t)$$
$$\int_A \sigma_{xx}z\,dA = b\sum_{k=1}^{N}\int_{z_{k-1}}^{z_k}\sigma^k_{xx}(t)z\,dz = \bar{M}^M_y(t) \tag{7.53}$$

NOTE: It can be shown that in a homogeneous beam subject to a pure bending $M^M_y(t)$, the NA is at the centroid of the cross-section for all times t. In a layered composite where each layer has different relaxation modulus, the centroidal axis (or the NA) would not be constant, and instead it would vary with time.

Substituting the axial stress in Equation 7.52 into the equilibrium equations (Equation 7.53) gives algebraic equations involving the time-dependent axial strain $\varepsilon^0(t)$ and curvature $\kappa(t)$. As mentioned above, in layered composites the centroidal axis of the cross section would vary with time, unless the layered composites have symmetric layer arrangements with regard to its mechanical properties, which results in axial and bending coupling. In such a case, obtaining exact closed-form solutions of the time-dependent axial strain $\varepsilon^0(t)$ and curvature $\kappa(t)$ would be challenging, if not possible, and therefore numerical solutions are often sought.

EXAMPLE 7.4.2

A bimorph piezoelectric actuator (Figure 7.7) comprising two piezoelectric layers has length L and width b, and each layer has thickness t_p. The piezoelectric layers are polarized through their thickness. Opposite directions of electric field $E_z(t)$ are applied to the piezoelectric layers in order to generate bending deformation. The piezoelectric layers have the relaxation modulus $E(t)$ and time-dependent piezoelectric constant $e_{311}(t)$. Determine the deformation and stress fields in the actuator.

SOLUTION

This is a pure bending deformation on a symmetric layered composite. The y-axis is placed at the centroidal axis so that the axial strain depends only on the curvature $\varepsilon_{xx}(t) = -\kappa(t)z$. Using the axial stress in Equation 7.4.11, the curvature is

$$-b\sum_{k=1}^{2}\int_{0^-}^{t} E^k(t-\varsigma)\frac{d\kappa(\varsigma)}{d\varsigma}d\varsigma\int_{z_{k-1}}^{z_k} z^2\,dz = b\sum_{k=1}^{2}\int_{0^-}^{t} e_{311}^k(t-\varsigma)\frac{dE_z^k(\varsigma)}{d\varsigma}d\varsigma\int_{z_{k-1}}^{z_k} z\,dz \quad \text{(a)}$$

Since the electric fields on the top and bottom layers are in opposite direction, the two electric fields will have opposite signs. Assuming a positive electric field for the top layer, equation (a) is rewritten as

$$-\frac{2b}{3}t_p^3\int_{0^-}^{t} E(t-\varsigma)\frac{d\kappa(\varsigma)}{d\varsigma}d\varsigma = \frac{2b}{2}t_p^2\int_{0^-}^{t} e_{311}(t-\varsigma)\frac{dE_z(\varsigma)}{d\varsigma}d\varsigma$$

$$\int_{0^-}^{t} E(t-\varsigma)\frac{d\kappa(\varsigma)}{d\varsigma}d\varsigma = -\frac{3}{2t_p}\int_{0^-}^{t} e_{311}(t-\varsigma)\frac{dE_z(\varsigma)}{d\varsigma}d\varsigma \quad \text{(b)}$$

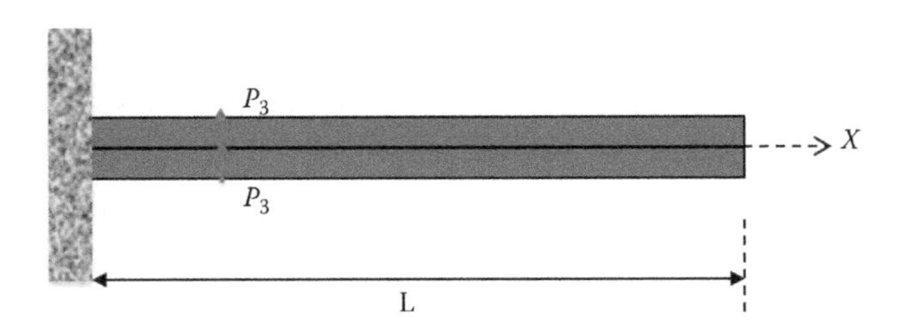

FIGURE 7.7
Bimorph stack piezoelectric actuator.

Furthermore, the time-dependent curvature can be easily solved by expressing the above equation in terms of the time-dependent compliance $S(t)$

$$\frac{\partial^2 u_z}{\partial x^2}(t) = \kappa(t) = -\frac{3}{2t_p}\int_{0^-}^{t} S(t-\varsigma)\frac{\mathrm{d}\left(\int_{0^-}^{\varsigma} e_{311}(\varsigma - s)\frac{\mathrm{d}E_z(s)}{\mathrm{d}s}\mathrm{d}s\right)}{\mathrm{d}\varsigma}\mathrm{d}\varsigma \qquad \text{(c)}$$

The lateral deformation for the cantilever bimorph beam is

$$u_z(x,t) = -\frac{3}{2t_p}\int_{0^-}^{t} S(t-\varsigma)\frac{\mathrm{d}\left(\int_{0^-}^{\varsigma} e_{311}(\varsigma - s)\frac{\mathrm{d}E_z(s)}{\mathrm{d}s}\mathrm{d}s\right)}{\mathrm{d}\varsigma}\mathrm{d}\varsigma\frac{x^2}{2} \qquad \text{(d)}$$

The stress fields for the top and bottom layers are

$$\begin{aligned}
\sigma_{xx}^{\text{top}}(t) &= -z\int_{0^-}^{t} E(t-\varsigma)\frac{\mathrm{d}\kappa(\varsigma)}{\mathrm{d}\varsigma}\mathrm{d}\varsigma - \int_{0^-}^{t} e_{311}(t-\varsigma)\frac{\mathrm{d}E_z(\varsigma)}{\mathrm{d}\varsigma}\mathrm{d}\varsigma; \quad 0 \le z \le t_p \\
\sigma_{xx}^{\text{bottom}}(t) &= -z\int_{0^-}^{t} E(t-\varsigma)\frac{\mathrm{d}\kappa(\varsigma)}{\mathrm{d}\varsigma}\mathrm{d}\varsigma + \int_{0^-}^{t} e_{311}(t-\varsigma)\frac{\mathrm{d}E_z(\varsigma)}{\mathrm{d}\varsigma}\mathrm{d}\varsigma; \quad -t_p \le z \le 0
\end{aligned} \qquad \text{(e)}$$

It can be seen from the lateral deflection in equation (d) and stress fields in equation (e) that the applied electric field produces bending with positive curvature, compression stress at the top layer, and tension stress at the bottom layer.

EXAMPLE 7.4.3

A cantilever beam, comprising a viscoelastic layer sandwiched between PZT layers at the top and bottom parts (Figure 7.8), is subjected to a mechanical bending moment of $M_y = 200\sin(\pi t)$ Nmm about the y-axis. The viscoelastic layer has thickness h_s and each PZT layer has thickness $0.5h_p$. The beam has uniform width of b and the PZT layers are polarized through the thickness along z-axis. The relaxation modulus of the viscoelastic layer is $E_{ve}(t) = 2000 + 2000e^{-0.3t}$ MPa, the relaxation modulus and piezoelectric constant of the PZT layer are $E_{PZT}(t) = 5000 + 5000e^{-t}$ MPa and $e_{311}(t) = 5 + 5e^{-0.2t}$ NV/mm, respectively. Determine the magnitude and direction of electric fields that should be applied to the PZT layers in order to minimize the lateral deformation of the beam ($h = b = 1$ mm and $h_s = 5$ mm).

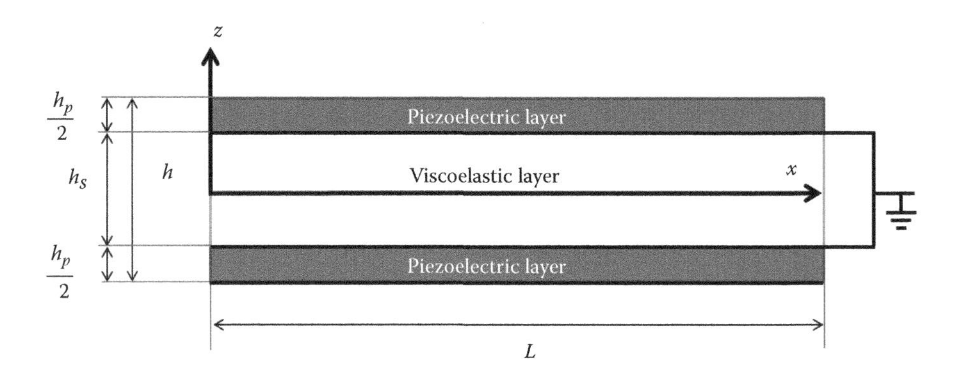

FIGURE 7.8
Piezoelectric sandwich beam.

SOLUTION

If we apply an electric field with opposite directions to the active layers, it will cause one of the layers to expand and the other one to shorten. This mode for applying the electric field finally will lead into the bending of the beam (see Example 7.4.2) if there is no other form of loading.

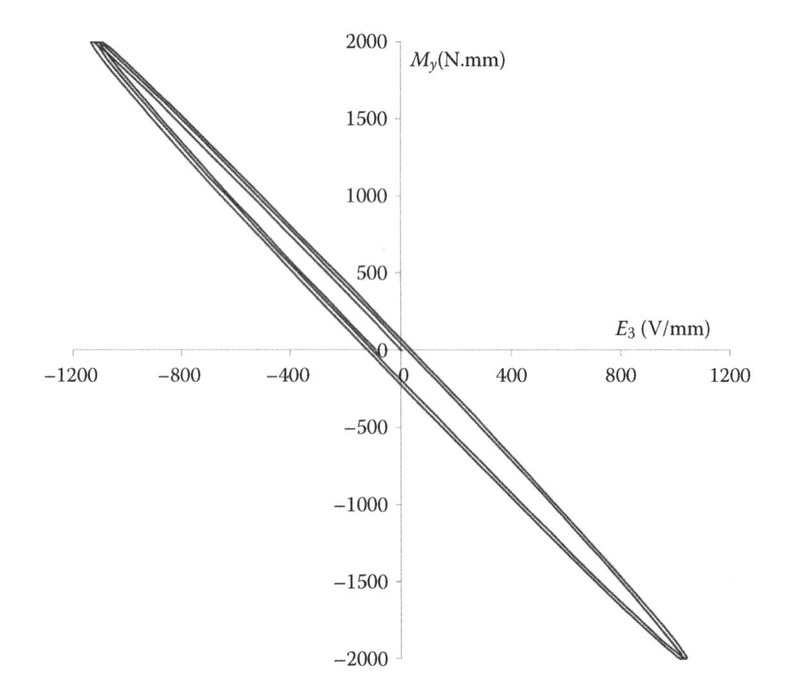

FIGURE 7.9
Corresponding electric field applied to compensate the lateral deformation due to sinusoidal bending moment.

Following the governing equations in Example 7.4.2 with the applied external bending moment $\bar{M}_y^M(t) = 200\sin \pi t$ and writing the equations in the forms of their Laplace transforms, the expression for the electric field, in Laplace transform, can be obtained. The expression of the applied electric field necessary to keep the lateral deflection zero (or curvature zero) is obtained from the inverse transform:

$$E_z(t) = [-67.32e^{-0.1t} + 67.32\cos(1.571t) - 1071\sin(1.57t)]\ \text{V/mm} \qquad \text{(a)}$$

Figure 7.9 illustrates the plot of the electric field and bending moment, which show a hysteretic response due to the time-dependent effect in the elastic and piezoelectric material properties.

7.5 Discussion

With a vast development of engineering materials having multifunctional capabilities, there is a need to understand, predict, and model the properties of these new generations of materials under a combined external influence, e.g., mechanical, thermal, electromagnetic, etc., before analyzing and designing structures with multifunctional capabilities. To achieve such multifunctional capabilities, engineering materials often comprise several constituents (or phases) leading to multifunctional composite systems. This chapter presents analyses of homogeneous and heterogeneous multifunctional beams subject to external mechanical and nonmechanical stimuli. The constitutive models discussed in this chapter were derived on the basis of the classic mechanics approach for linear thermo-electro-mechanical response. The presented analyses were based on the Euler–Bernoulli beam theory and SOM approach. The applicability and limitation of the presented analyses shall be discussed as follows:

1. The kinematic assumption in the Euler–Bernoulli beam is limited to slender beams in which the longitudinal dimension is much larger than the cross-sectional dimensions of the beam. In such a case, the contribution of the transverse shear effect on the lateral deformations and the spatial variations in the lateral deformations of the beam are insignificant and can be neglected. The SOM approach considers only one-dimensional constitutive relation for the axial stress–strain relation, neglecting the contribution of the transverse normal strains (or stresses) on the axial stress–strain response.
2. The linear thermo-electro-elastic constitutive relation assumes that the response due to the thermal, electrical, and mechanical stimuli

can be superposed and the corresponding responses are proportional to the external stimuli (input variables). This assumption can give a reasonable prediction when the magnitudes of the external stimuli are relatively small in that the properties of the materials do not vary significantly with the external stimuli. In many applications, the properties of materials could vary significantly with the prescribed external stimuli, leading to nonlinear responses. For example, mechanical and electrical properties and also thermal properties of materials change significantly with severe temperature changes; electromechanical (piezoelectric) properties vary with prescribed external stresses; a continuous application of an electric field generates significant amount of heat due to the energy dissipation effect, which could significantly increase the body temperature.

3. With the use of the correspondence principle, the thermo-electro-elastic analyses are extended for analyzing linear thermo-electro-viscoelastic response in that the solution to a linear time-dependent response of a structure can be constructed from the corresponding linear thermo-electro-elastic response. In viscoelastic materials temperature changes significantly influence the time-dependent deformation, i.e., creep, relaxation, and hysteresis; elevated temperatures accelerate the creep and relaxation processes and low temperatures slow down the creep and relaxation processes. The linear thermo-electro-viscoelastic analyses based on the correspondence principle are not capable in simulating the accelerated/decelerated time-dependent response due to time-varying temperature changes. Viscoelastic materials are dissipated materials that could generate significant amount of heat during the deformation. Electric currents pass through a material could also generate a nonnegligible amount of heat. Thus, in multifunctional viscoelastic materials, application of mechanical loads and electric fields could dissipate a significant amount of energy, thus increasing the overall body temperature. The presented analyses neglect the dissipation effect on the overall thermo-electro-mechanical deformation of the multifunctional beams.

Acknowledgments

This work was sponsored by the Air Force Office of Scientific Research (AFOSR) under grant FA9550-09-1-0145 and National Science Foundation (NSF) under grant CMMI-1030836.

References

Ballas, R. G. (2007). *Piezoelectric Multilayer Beam Bending Actuators: Static and Dynamic Behavior and Aspects of Sensor Integration*. Berlin, New York, Springer.

Boley, B. A. and J. H. Weiner (1997). *Theory of Thermal Stresses*. Mineola, NY, Dover Pubns.

Cao, H. and A. G. Evans (1993). "Nonlinear deformation of ferroelectric ceramics." *Journal of the American Ceramic Society* **76**(4): 890–896.

Cook, R. D. and W. C. Young (1999). *Advanced Mechanics of Materials*. Upper Saddle River, NJ, Prentice Hall.

Fett, T. and G. Thun (1998). "Determination of room-temperature tensile creep of PZT." *Journal of Materials Science Letters* **17**(22): 1929–1931.

Gopinathan, S. V., V. V. Varadan et al. (2000). "A review and critique of theories for piezoelectric laminates." *Smart Materials and Structures* **9**: 24.

Guedes, R. M., A. T. Marques et al. (1998). "Analytical and experimental evaluation of nonlinear viscoelastic-viscoplastic composite laminates under creep, creep-recovery, relaxation and ramp loading." *Mechanics of Time-Dependent Materials* **2**(2): 113–128.

Hall, D. (2001). "Review nonlinearity in piezoelectric ceramics." *Journal of Materials Science* **36**(19): 4575–4601.

Leo, D. J. (2007). *Engineering Analysis of Smart Material Systems*. Hoboken, NJ, Wiley Online Library.

Megnis, M. and J. Varna (2003). "Micromechanics based modeling of nonlinear visco-plastic response of unidirectional composite." *Composites Science and Technology* **63**(1): 19–31.

Muddasani, M., S. Sawant et al. (2010). "Thermo-viscoelastic responses of multi-layered polymer composites: Experimental and numerical studies." *Composite Structures* **92**(11): 2641–2652.

Muliana, A., A. Nair et al. (2006). "Characterization of thermo-mechanical and long-term behaviors of multi-layered composite materials." *Composites Science and Technology* **66**(15): 2907–2924.

Pipes, R. B., J. R. Vinson et al. (1976). "On the hygrothermal response of laminated composite systems." *Journal of Composite Materials* **10**(2): 129.

Reddy, J. N. (1990). "On refined theories of composite laminates." *Meccanica* **25**(4): 230–238.

Reddy, J. N. (2004). *Mechanics of Laminated Composite Plates and Shells: Theory and Analysis*. Boca Raton, FL, CRC Press.

Robbins, D. and J. N. Reddy (1991). "Analysis of piezoelectrically actuated beams using a layer-wise displacement theory." *Computers & Structures* **41**(2): 265–279.

Rosen, B. W. and Z. Hashin (1970). "Effective thermal expansion coefficients and specific heats of composite materials." *International Journal of Engineering Science* **8**(2): 157–173.

Sai Ram, K. and P. Sinha (1991). "Hygrothermal effects on the bending characteristics of laminated composite plates." *Computers & Structures* **40**(4): 1009–1015.

Saravanos, D. A., P. R. Heyliger et al. (1997). "Layerwise mechanics and finite element for the dynamic analysis of piezoelectric composite plates." *International Journal of Solids and Structures* **34**(3): 359–378.

Schäufele, A. B. and K. Heinz Härdtl (1996). "Ferroelastic properties of lead zirconate titanate ceramics." *Journal of the American Ceramic Society* **79**(10): 2637–2640.

Shen, C. H. and G. S. Springer (1976). "Moisture absorption and desorption of composite materials." *Journal of Composite Materials* **10**(1): 2.

Strobl, G. R. (1997). *The Physics of Polymers: Concepts for Understanding Their Structures and Behavior*. Berlin, New York, Springer Verlag.

Tuttle, M. E. and H. F. Brinson (1986). "Prediction of the long-term creep compliance of general composite laminates." *Experimental Mechanics* **26**(1): 89–102.

Tuttle, M. E., A. Pasricha et al. (1995). "The nonlinear viscoelastic-viscoplastic behavior of IM7/5260 composites subjected to cyclic loading." *Journal of Composite Materials* **29**(15): 2025.

Wineman, A. S. and K. R. Rajagopal (2000). *Mechanical Response of Polymers: An Introduction*. Cambridge [England], New York, Cambridge Univ Press.

Yeow, Y., D. Morris et al. (1979). "Time-temperature behavior of a unidirectional graphite/epoxy composite." *Composite Material: Testing and Design (Fifth Conference)*, ASTM STP 674. Tsai Ed., ASTM: 263–281.

Section III

Sensing

8

Wireless Health Monitoring and Sensing of Smart Structures

R. Andrew Swartz

CONTENTS

Wireless sensors can be an extremely valuable enabling technology for structural health monitoring (SHM) applications. In the correct context, they can be used to dramatically decrease installation cost, reduce system weight, and even improve robustness. Wireless systems are inexpensive, easy to reconfigure, and contain the inherent capability to perform autonomous and embedded data processing. However, when considering implementation of a wireless component in an SHM system, it is also important to consider the limitations and relative disadvantages of the technology as well. In the incorrect context, wireless sensors will degrade performance, lose information, increase long-term costs, or even cause complete failure of the monitoring system. Wireless sensors can suffer from lower data transmission rates than their wired counterparts, are more prone to loss of information, and power supply is very frequently a critical limiting factor to their application. In this chapter, the relative advantages and disadvantages of wireless sensor technology for SHM applications in composite structures are discussed. As part of this discussion, the important core components of a typical wireless sensor are presented. In addition, strategies for overcoming the inherent disadvantages of the wireless approach are presented with special attention given to embedded computation and passive sensing. The goal of the chapter is not to provide an in-depth summary of all wireless sensing approaches to date (the approaches are many and varied and would warrant their own book), but to provide the reader with sufficient information to decide which wireless sensor technology, if any, might be most fitting for their application and what strategies they might consider to improve the performance and reliability of their monitoring application.

8.1 Advantages of Wireless Sensors

All sensors have at their heart, a transducer, which is a device that couples some physical phenomenon of interest (e.g., acceleration, temperature, strain, force, pressure, humidity) with an electrical property that can be measured and detected by the electronic components within the sensor. Wireless sensors differ from their traditional cable-based counterparts in that, once the physical phenomenon of interest is converted into an electrical signal, rather than transmitting the signal to a central data acquisition (DAQ) device via shielded cables, the wireless sensor encodes the signal into a digital radio-frequency (RF) packet and broadcasts it to other sensors in the network (or to a base station). In large structures, the signal cables required for the traditional cable-based sensors may represent a significant source of unwanted cost and weight. For instance, many aerospace structures have severe limitations in terms of space and weight allowances. High-fidelity shielded cables capable of transmitting uncorrupted analog signals over long distances are expensive to purchase, bulky, and heavy. In addition, installation of signal cables can be very expensive as well, particularly in existing structures that may need to be dismantled in order to install new cables. High installation costs create large disincentives to install the kind of informationally rich dense sensor networks that are most effective in SHM applications. Because wireless sensors do not require the installation of large amounts of shielded cables in the structure, these costs, bulk, and weight can be avoided when wireless sensors are used.

Owing to their low cost and relative ease of installation, use of wireless sensors promotes dense sensor installations; however, they do carry additional advantages as well. Analog signal cables are susceptible to physical damage as well as to electrical noise corruption. Because they lack these cables, wireless sensors are immune to these kinds of corruptive influences. Wireless sensors digitize signals at the source, creating digital wireless data packets that are transmitted from one wireless modem to another. Digital data packets may be lost or corrupted; however, transmission error detection protocols identify and reject corrupted packets, allowing them to be retransmitted. In this way, once the data is digitized, it is freed from the possibility of analog-domain signal corruption. In addition, when cables in traditional tethered sensor networks are damaged or broken, the sensors attached to them are compromised until the cables can be repaired or replaced. Wireless sensor networks, lacking cables, are more robust in this regard. In fact, particularly in the case of ad hoc and multi-hop wireless sensor networks (see Section 8.3), multiple wireless transmission paths may exist within the sensor network, enhancing robustness. In addition, with many sensor nodes distributed throughout a structure, a properly designed wireless sensor network enjoys additional robustness because the DAQ system is not limited to a single component (e.g., one DAQ computer) that can disable the entire network upon failure. Large wireless sensor networks have some redundancy and can eliminate the single point-of-failure issue using an adaptable network design.

With respect to composite structures, connecting wired transducers to the DAQ system can be particularly troublesome. The intrusions of wires into portions of the structure may compromise the strength of the structure rather than enhance its performance. Wireless sensors provide a possible means to avoid this problem provided that the sensor is small enough to be embedded inside of the structure. Composite structures that move or rotate (e.g., wind turbine blades) are important candidates for SHM largely because of concerns about fatigue or impact damage. However, moving and rotating structures are particularly troublesome for signal cables. For example, while it is certainly possible to make electrical connections through a rotating assembly using a slip ring, providing dozens or hundreds of channels worth of connections for high-quality analog sensor output across that slip ring is prohibitive. In such an application, wireless transmission of data from the rotating portion of the structure to its stationary part is extremely beneficial.

The advantages of wireless sensor networks presented thus far focus on what the wireless sensor networks lack versus the cable-based networks, namely analog signal cables. However, perhaps the chief advantage that a wireless sensor node has over a traditional sensor is the presence of a microcontroller within the device itself. A microcontroller is required in the sensor node in order to convert the transducer output into a digital signal that can be communicated via the wireless modem to the outside world. The microcontroller is also necessary to operate the various components of the sensing node, including the sensor interface, memory, and wireless modem. However, that microcontroller represents something unique to wireless sensors, an inherent on-board data interrogation platform. Because the microcontroller is necessary for basic operation, each wireless sensor must have one; however, the microcontroller can do much more for the user than simply buffer and transmit data. It can also perform basic engineering algorithms on the data as soon as it is collected. By doing so, a major pitfall that plagues many sensor network installations may be avoided—that of data usage. There is very little benefit to large data network installations that generate copious amounts of operational data if that data then sits in a repository, unprocessed and unexamined. However, this is exactly the result of many such installations when resources are allocated to install an SHM system, but are not allocated to allow qualified engineers or technicians to review and process the data for the remaining lifetime of the structure. By including automated and embedded data processing resources within the sensor, the designer of a wireless sensor network can leverage these resources to ensure that all data collected by the SHM sensor network undergoes some minimum level of interrogation. In addition, embedded data processing is an important tool for overcoming some of the inherent limitations of wireless sensing networks as well. However, a discussion of the constituent components of a typical wireless sensor is warranted to reveal the source of these limitations as well as why embedded data processing can be so useful.

8.2 Composition of a Wireless Sensing Node

A wireless sensing node has some additional components not found on a traditional sensor. While the basic transducer elements are the same (these depend on the exact application and physical phenomenon of interest), wireless sensors will have some basic components that are unique and necessary in order to accomplish their basic operation. Each wireless sensing unit (WSU) must be able to convert basic sensor output into digital data that is suitable for RF transmission to its destination, thus requiring an analog-to-digital converter (ADC). In addition, the RF transmitter itself, usually in the form of a wireless modem, is a necessary component. Data must be digitized and buffered until it can be transmitted, requiring memory resources and a microcontroller as well. Active sensor networks (those that generate their own structural excitation) and even control sensor networks are possible as well, if an actuation interface is present. In this section, the hardware and embedded software (firmware) components of a typical WSU are presented and discussed.

8.2.1 Wireless Sensing Unit Hardware

The hardware components of a WSU consist of four basic building blocks (Figure 8.1). These building blocks include the sensing interface, the computational core, the wireless transceiver, and an actuation interface (if used). A sample WSU (Swartz et al. 2005), minus its protective case, is shown in Figure 8.2 with some of the operational components highlighted. A WSU will also require some kind of power source (usually a battery) as well as an antenna. Each block has its own defined role and offers the designer of the wireless sensing network a set of engineering tradeoffs to consider. The features of the components that make up these building blocks, as well as their respective duties and design considerations, are described below.

8.2.1.1 Sensing Interface

Given the very wide variety of transducers and sensors used for monitoring of composite structures, it is impossible to cover every conceivable sensing modality and sensor output that the reader may be interested in applying for

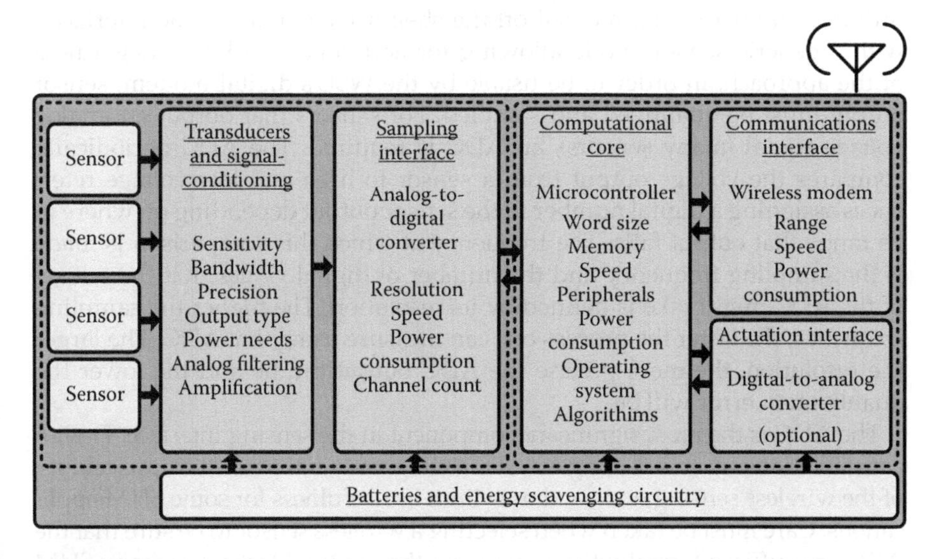

FIGURE 8.1
Functional diagram for a wireless sensing unit.

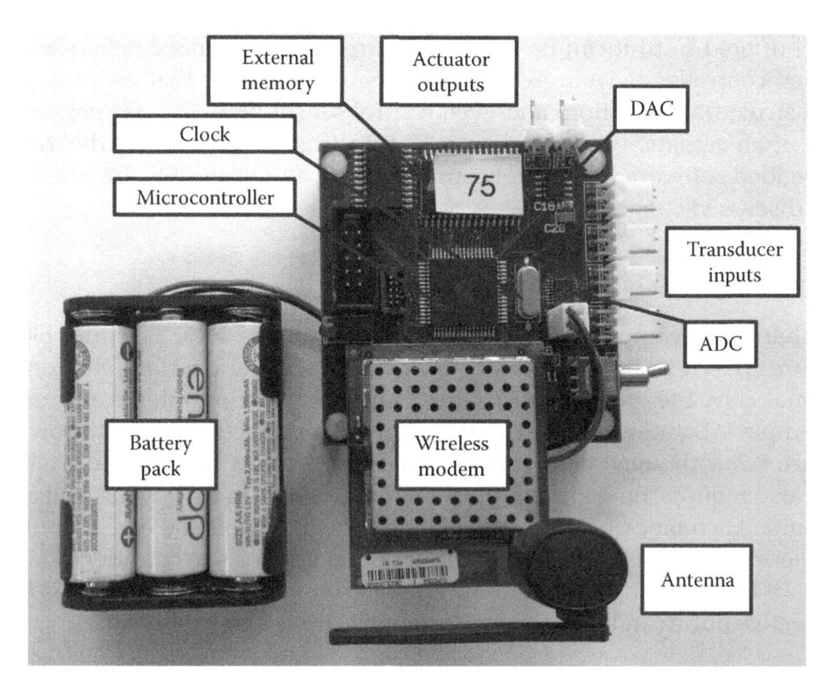

FIGURE 8.2
Wireless sensing prototype with case removed.

his or her own application. For some applications, a wireless sensor node may have the transducers completely integrated within the node itself; others may utilize a number of commercial off-the-shelf sensors that can be interfaced with a generic sensing node allowing for additional flexibility. Regardless of the approach, in order to be usable by the WSU's digital modem, sensor output must be quantized and sampled. For sensors that output an analog voltage signal (many sensors), an ADC is required. The ADC periodically compares the voltage output from a sensor to high and low voltage references assigning a digital number to the sensor output depending on where in its range that output falls. The frequency at which this comparison is made is the sampling frequency and the number of digital values that the output of the ADC might take is defined by its resolution. The higher the sampling frequency, the faster the process one can measure using that ADC. The larger the resolution, the more precise the ADC output can be and the lower the quantization error will be.

The ADC is the most significant component in the sensing interface. Design considerations when selecting an ADC will greatly influence the applicability of the wireless sensing node and may limit its usefulness for some SHM applications. Care must be taken when selecting a wireless sensor to be sure that the ADC has sufficient capabilities to perform the required tasks. For many SHM applications, damage-sensitive features are subtle enough that high-resolution data is required to detect them. In those cases, resolution is a critical design consideration of the ADC (i.e., 16-bit or even 24-bit would be preferred over 12-bit resolution). Guided-wave or other modal-based SHM algorithms rely on high-frequency data that requires high sampling rates to record. Also, if some sensors are located in close proximity to each other, it will be beneficial to provide multiple sensing channels on a single ADC. In addition, wireless sensors often rely on batteries for power; in that case, low-power consumption is an important consideration for the ADC as well. Finally, low cost is always a plus.

If the input voltage range of the ADC is not well matched to the sensor used, then it is useful to add additional signal conditioning circuitry to the sensing interface, possibly including filters for amplification, mean shifting, anti-aliasing, and de-noising. In addition, sensors that do not output analog voltage signals (e.g., analog current or piezoresistive sensors) will require additional signal conditioning circuitry between the sensor and the ADC. These components may either be integrated with the wireless sensor (e.g., placed on the sensor's printed circuit board, or PCB) or designed as modular, external components. If they are integrated, great care must be taken to properly design the analog domain of the PCB to avoid noise contamination from the digital components. Digital noise corruption of the analog domain of the ADC is also an important challenge in the design of any WSU, regardless of the presence or absence of integrated signal conditioning circuitry.

8.2.1.2 Computational Core

Once the signal from the sensor interface is sampled and digitized, it is passed to the computational core. The computational core is not only responsible for buffering sensor data and preparing it for transmission but it also controls all aspects of the WSU operation. The operating system (OS) responsible for operation of the WSU resides within the computational core. The OS coordinates sampling, data buffering, and communication activities within the WSU. Engineering applications, if used, are also performed here. The computational core consists of a microcontroller and any additional external memory resources that might be provided.

The abilities of the microcontroller used in the computational core will largely determine the capabilities and energy consumption properties of the WSU. The primary design tradeoffs in microcontroller selection are processing speed and power versus energy consumption. A microcontroller has a number of features that can lead to improvements in processing power and speed, but most of these features come at the expense of additional energy consumption (which can be a major problem in battery-powered networks). The most basic parameter of a microcontroller is its word size architecture, or the number of bits that the microcontroller can store in a single memory register. Larger registers (e.g., 32 or 64 bits) are more costly and usually require more power than smaller registers (e.g., 8 bits), but are more computationally powerful. Faster processor speed is another major contributor to power consumption, but may be necessary for high-frequency applications, computationally intensive algorithms, or real-time applications. The microcontroller must also have sufficient memory for all of its operations, including instructions, algorithms, and data storage. If large data sets are anticipated, external memory may be added to the computational core, usually in the form of static random-access memory (SRAM) or flash memory. Flash memory is cheaper than SRAM, but has a limited number of erase cycles, effectively limiting the lifespan of the WSU.

Many microcontrollers incorporate a number of useful features that may also factor into selection decisions such as built-in ADCs, digital-to-analog converters (DACs), peripheral and network communication busses (e.g., UART, SPI, I^{2}C, CAN), quadrature decoders, pulse-width modulators, and others. These features may be critical in some wireless SHM applications and less useful in others. Typically, communication busses are necessary for WSU operation, but not every standard will be needed. Likewise, if the microcontroller has a built-in 12-bit ADC, but the SHM application requires higher precision, then an external ADC will still be needed. The best practice in many applications is to use the microcontroller that meets the minimum requirements for the SHM application under consideration; higher capabilities usually come at a cost in terms of both dollars and energy expenditures.

8.2.1.3 Communications Interface

Without some link to the outside world, the WSU is essentially useless. The communications interface provides a means by which the WSU can transmit its data and information and receive additional instructions. The communications interface consists of a wireless modem (or radio), an antenna, and any RF amplification circuitry that may be necessary for quality communications. The modem receives information from the computational core of the sensor, converts it into appropriately formatted wireless data packets, and transmits it over a predefined carrier frequency. Likewise, the communications interface also receives commands and data from the network and relays these to the computational core. These packets must conform to a wireless communication standard (e.g., Bluetooth, Zigbee, WiFi). Wireless modems are typically designed around one of these standards and that standard will have a large bearing on the performance of the modem for SHM applications. These standards define all aspects of the behavior of the wireless modem, including carrier frequency channel allocation, medium access control (MAC) rules, transmission power, packet structure, addressing, and error checking, and others.

A wealth of communication standards have been developed for different applications. For instance, WiFi has been developed for personal computer (PC) networks, requires relatively larger energy and communication bandwidth resources, but can support data-intensive communication activities such as video streaming or large file downloads. Bluetooth is a lower-power standard designed for less data-intensive activities and transmissions over relatively shorter ranges. Zigbee is a popular standard for SHM applications owing to the fact that it is designed for low-power, ad hoc networks (see Section 8.3) but has long enough range to be useful in many SHM applications (Lynch and Loh 2006). The communication standard to which a wireless modem has been designed will have the largest impact of any factor on the selection of the communication interface. Ultimately, the selection is a tradeoff between power consumption and data throughput, and/or power consumption and communication range (though amplification circuitry and high-gain or directional antennas may be used to improve range). Amplification options should only be considered when transmission power will remain within the Federal Communications Commission (FCC)-allowed maximums, for example, 1000 mW for devices using the popular unlicensed industrial, scientific, and medical (ISM) communication bands. Antenna selection and placement are also important aspects to WSU design and deployment. Depending on the size and geometry of the network, a directional antenna (one that radiates most of its energy in one or two directions) may be more advantageous than an omni-directional antenna. Also, in structures defined by many complex reflective surfaces, many communication issues can be solved by (often empirical) antenna placement strategies rather than using more powerful antennas or additional transmission energy.

8.2.1.4 Actuation Interface

WSUs can participate in active sensing applications as well as function as control sensing nodes if they are provided with an actuation interface. An actuation interface gives the microcontroller command of actuators that are collocated with the WSU. In active sensing units, these actuators may be used to excite a dynamic vibrational response in the structure where ambient excitations are not expected to produce useful signals for SHM (e.g., using piezoelectric transducers to excite guided waves). If the SHM network is expected to perform control actions to mitigate the effects of damage (as opposed to simply sensing and reporting damage), then an actuation interface will give the WSU command of control actuators as well. The actuation interface may consist of a DAC and can have one or more channels. Speed, resolution, and power consumption are important design parameters of the actuation interface. Depending on the actuator, significant signal conditioning and amplification circuitry may be required. Large actuators may also require dedicated power sources as well, beyond what can be supplied by the WSU.

8.2.2 Firmware

Besides the hardware components of the WSU, embedded software, termed "firmware," is necessary for data collection and SHM. The firmware residing in the computational core of the sensor is responsible for all aspects of the operation of the WSU, including OS components: basic device drivers for the peripherals (i.e., the wireless modem, external ADC and DAC, and external memory), network and MAC information, memory management rules, and the state machine for the WSU; as well as an application layer that contains the data collection rules and any engineering algorithms necessary for embedded data processing. One view of these layers, based on the open systems interconnection model for communication systems (ISO/IEC 1996), is depicted in Figure 8.3 with OS components near the bottom (closest to the hardware) and the application activities on top.

For engineers new to wireless sensing, commercially available wireless sensor platforms provide many advantages over self-designed solutions to help reduce development time and effort. Commercially available WSUs vary greatly in terms of both hardware and firmware and both must be considered when selecting a wireless sensing platform for SHM applications. The sensor platform must have sufficiently capable hardware for the desired application, but the firmware is just as important. The OS must have adequate device drivers to allow the user to access the full functionality of the hardware. The state machine must support the SHM activities required by the algorithm. Active sensing and control nodes must be able to provide deterministic, real-time performance. Necessary support functions for SHM such as time synchronization and multi-hopping communication must be available. Most important, the user must be allowed sufficient programming permissions to write and embed SHM algorithms that will run in an embedded fashion. Without this ability, many of the inherent advantages of the wireless sensors will be squandered. Furthermore, communication network topographies (see Section 8.3) that promote effective use of SHM algorithms must be supported.

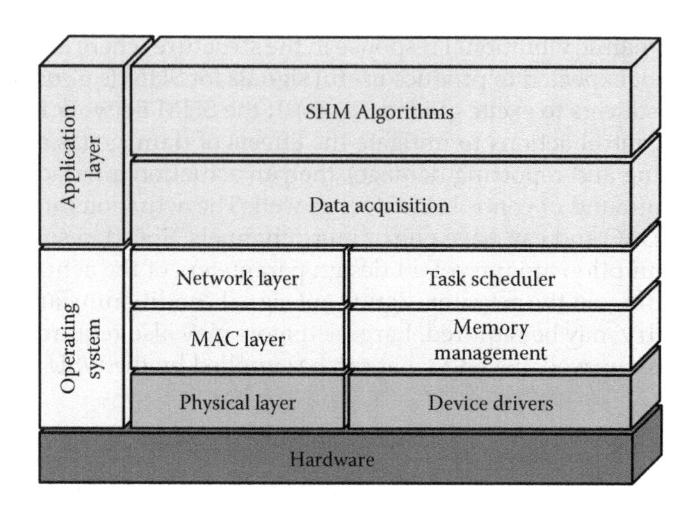

FIGURE 8.3
Organization of firmware components in a WSU.

8.3 Communication Strategies in Wireless Sensor Networks

Depending on the application, many communication topologies may be appropriate for a network of wireless sensors. Unlike a traditional cable-based sensor network, wireless sensors rely on the wireless RF medium to transmit data. Use of wireless communication limits not just the amount of data that can be transmitted in a wireless network during a given period, but also the distance that the sensors can communicate. For this reason, communication network topology is a major issue for wireless sensors. A number of communication topologies have been developed for wireless sensor networks, depending largely on the application. Some of the most common topologies will be presented in this section along with their strengths and weaknesses. In addition, when multiple wireless devices share a communication channel, it is important to establish rules that define how and when individual devices are allowed to access the communication medium (the RF channel) and how to handle contention between devices. These rules are referred to as MAC, defined in numerous wireless communication standards, and they are also a critical design feature for wireless sensor networks along with network topology (Wu and Pan 2008).

8.3.1 Star Networks

One of the most basic and useful network topologies is the star network topology. This topology utilizes a central WSU or even a PC server that communicates directly with some limited number of wireless sensors (Figure 8.4). These networks can have very simple communication rules and are therefore relatively easy to design and debug. This network topology requires that all wireless sensors in the network be able to communicate directly with the server in order to operate. When this condition is met, the MAC protocol can be significantly simplified because it is possible to have a completely server-initiated MAC strategy in which sensors within the network wait to communicate until prompted by the server (or central node). In addition, the presence of a central node that can be "heard" by all units also simplifies sensor synchronization considerably, which can be a major challenge in wireless sensor networks.

The disadvantages of this topology relate to range, robustness, and data throughput. Because all sensors must be able to communicate with the central server, the effective range of the network is limited to the distance that a single WSU can communicate. In addition, the central server will act as a single point-of-failure in the sensor network in a similar way that the central DAQ system is a nonredundant necessary component of a wired sensor network. That is, if this one component fails, the entire network will become unusable until repairs can be made. Finally, relying

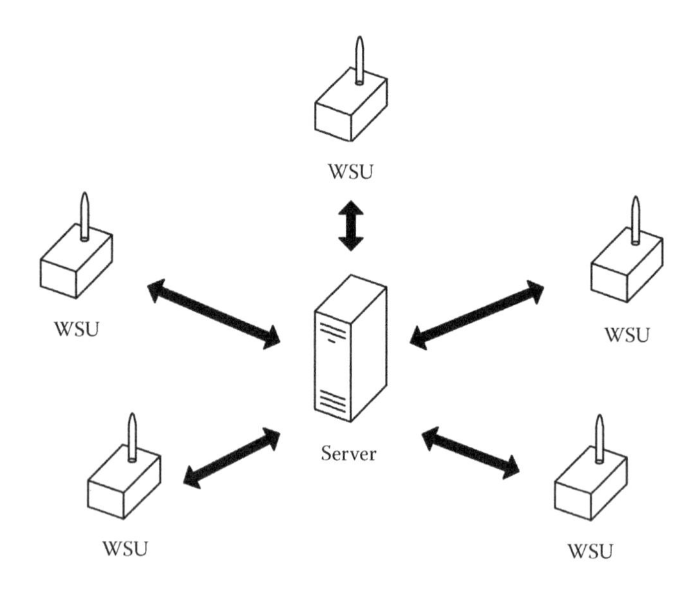

FIGURE 8.4
Star network wireless sensor network topology.

on a single node or server to aggregate data from the entire network may limit the amount of data that the network can handle. If the server is limited to a single modem and communication channel, then the total data throughput of the network is equal to just the throughput of a single device. However, practically speaking, this last limitation can be significantly offset by efficiency gains realized using simplified MAC protocols previously mentioned.

8.3.2 Multi-Hop Networks

To communicate over distances that are longer than the range of a single wireless sensor unit, a multi-hop network will be required (Figure 8.5). In multi-hop networks, data may originate from some physically distant wireless sensor and be transmitted to neighboring units until it can be routed to the central server. The principal advantage of this kind of network is the gain in communication range that can be realized when no longer limited to the communication range of a single device. MAC protocols may become more complicated though, and data transmission rates may suffer as well. With some units out of range of the central unit, a strictly server-initiated MAC strategy becomes impossible, although server commands can be relayed through appropriate nodes. Also, data throughput is limited compared with a single-hop network because data from outer nodes must be transmitted additional times in order to reach the server.

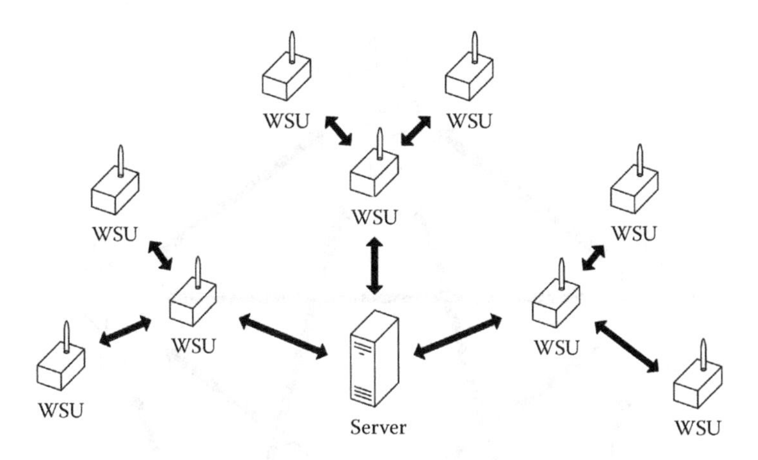

FIGURE 8.5
Multi-hop wireless sensor network topology.

8.3.3 Peer-to-Peer Networks

Star networks, whether single-hop or multi-hop, are limited by the central node or server. If the server fails, then the entire network fails. Furthermore, failure of noncentral nodes may cripple portions of a network if inflexible network configurations are used. Peer-to-peer networks (Figure 8.6) can be considerably more robust in this regard as there is no designated central node that can bring down the entire network. Also, network configurations are fluid, allowing WSUs to enter and exit the network in an ad hoc manner. The primary disadvantage of this kind of network is the additional overhead required to continually update the network, track network members, and maintain collision-free communication. Without a central server to coordinate network connections or medium access, sensors must establish their communication network paths and coordinate their own access to the wireless communication channel themselves. Such a practice, while robust, incurs overhead, thus reducing efficiency.

8.3.4 Multi-Tier and Hybrid Networks

Wireless communication networks excel at communicating data from spatially distributed or inaccessible locations in an economically efficient manner. However, they are inherently slower than wired networks and suffer from reliability issues in noisy RF environments, making them relatively poor substitutes for cable-based data networks where large amounts of consolidated data must be rapidly communicated. Multi-tier networks (Figure 8.7), consisting of both low-power wireless sensor nodes and highly capable master nodes, can help alleviate the challenge of communicating large quantities of consolidated data. Low-power wireless sensor nodes can collect data from far-flung locations and consolidate it at nearby master nodes.

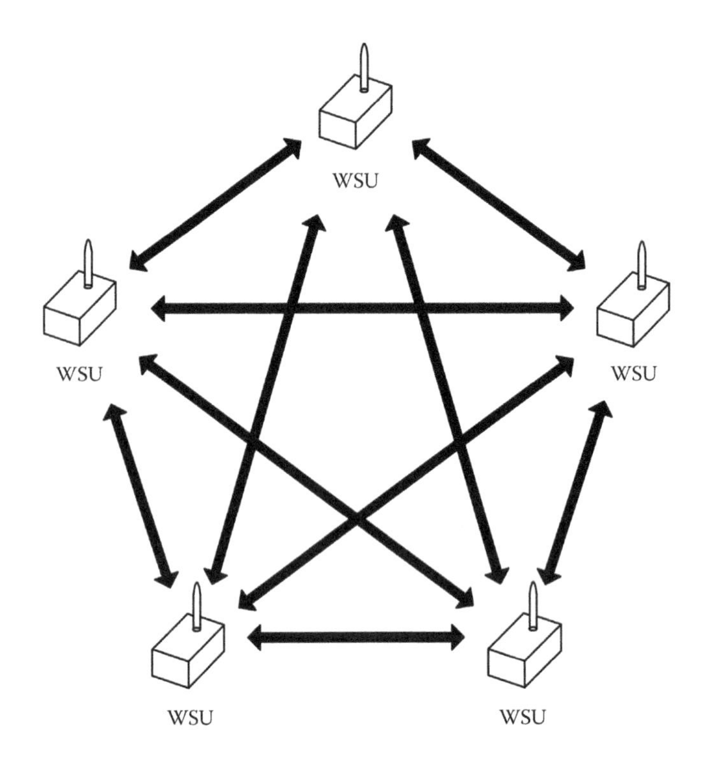

FIGURE 8.6
Peer-to-peer wireless sensor network topology.

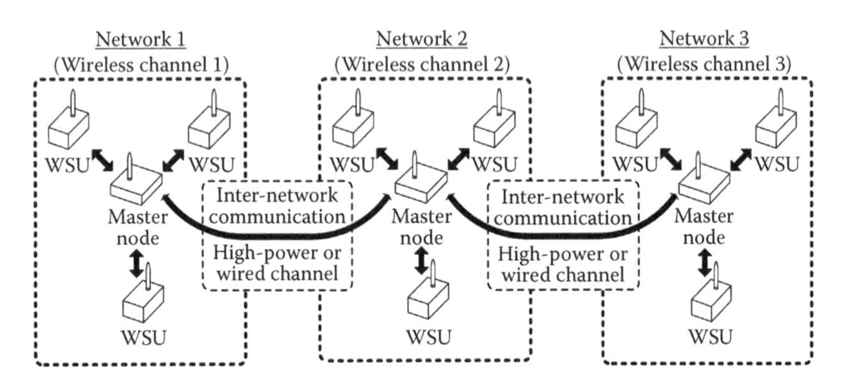

FIGURE 8.7
Multi-tier/hybrid wireless sensor network topology.

Master nodes are provided with additional computational power and memory resources to buffer, process, and transmit large quantities of data. They may have an additional communication link on a separate wireless channel to communicate with other master nodes or with a central server. Hybrid wired/wireless networks have wired communication links between master nodes to rapidly and reliably transmit data sent to them from the sensor nodes. Hybrid networks combine many of the best features of both technologies: economical data consolidation from distributed sensor locations and rapid transmission of data to a central location from each subnet. Whether multi-tier or hybrid, master nodes may be provided with a stable source of power (rather than batteries) in order to support their enhanced duties.

8.3.5 Medium Access Control

Medium access control strategies strongly affect the performance of wireless sensor networks and are tied to network topology. When multiple devices, such as WSUs, operate using separate controllers, it is possible that they will attempt to access the wireless communication channel simultaneously. If they do, their attempted transmissions will collide and be lost. MAC strategies are employed to prevent or mitigate this issue and affect both wired and wireless communications. Depending on the network topology and the level of flexibility needed in the network, various MAC approaches may be considered.

When simplicity and low overhead are important and flexibility is less of an issue, deterministic schedule-based strategies such as time division multiple access (TDMA) are often advantageous. Under TDMA, communication bandwidth use is scheduled for each device in the network. Such a scheme requires good synchronization between devices, making it especially useful in star networks. Owing to its deterministic nature, TDMA can be easier to implement and debug than a stochastic approach, but lacks flexibility. Devices that do not necessarily require use of the communication bandwidth still receive it on a regular basis. When communication needs are irregular, or when network configurations change, TDMA can be inefficient. Effective use of TDMA requires excellent *a priori* knowledge of network needs and relatively static demand; however, in real-time sensing and control applications, deterministic MAC strategies provide significant reliability benefits.

On the other hand, stochastic MAC approaches, such as carrier sense multiple access (CSMA), are more flexible, allowing devices on the network to attempt to access the communication channel as needed. Under CSMA, transmission packet collisions are managed rather than avoided. Devices check for network activity before initiating a transmission. If the channel is free, they will attempt to broadcast. CSMA approaches define the behavior of multiple devices when they initiate transmissions at the same instant, creating a collision. In such cases, the simultaneous packets will be corrupted and rendered unusable. Under CSMA with collision detection (CSMA/CD), both devices will cease transmitting when a collision is detected. Under CSMA with collision avoidance (CSMA/CA), devices will also pause some random amount of time before attempting a retransmission in order to avoid additional collisions. Because there is no fixed schedule, stochastic approaches such as CSMA offer flexibility that can be useful in peer-to-peer and ad hoc networks or in applications where communication demands are irregular; however, collision avoidance algorithms do add overhead to the communication activities within the sensor network that can be costly in terms of both energy and time.

8.4 Wireless Sensor Networks: Overcoming Challenges

This chapter began with a discussion of the relative advantages of wireless sensor technology as compared with traditional cable-based approaches, showing why an SHM system might benefit from going wireless. However, there are significant limitations to this technology that must also be carefully considered before its adoption. In some applications, these limitations will outweigh the advantages, indicating that a wired approach would be preferable. This section attempts to lay out these disadvantages in a way that allows the reader to decide how meaningful the disadvantages are for a given application and to aid in the selection of the best technology. Also presented are some of the approaches that may be of use in mitigating these drawbacks through proper selection of technology, hardware, firmware, network topologies, and embedded algorithms. These limitations are organized from general problems that will likely affect most systems, to more specific issues that only affect certain applications. Generally speaking, these issues revolve around five broad characteristics of wireless sensors. Those are

1. Limited energy resources
2. Limited wireless communication bandwidth
3. Distributed data and computing resources
4. Limited on-board memory and computational power
5. Reliability and security

Recognizing that the state of the art is especially fluid with respect to computers and computing technology, this chapter will attempt to differentiate between those issues that are inherent to wireless devices and those issues that may be expected to become less critical (or even be resolved) as incremental technological improvements are realized. For example, it is expected that faster, smaller, and more energy-efficient electronics will enter the market on a regular basis, allowing for the execution of more complex engineering algorithms within wireless sensors. However, wireless sensors will always collect data in a distributed fashion, requiring energy and bandwidth to centralize. Therefore, issues relating to memory and computing capacity will likely become less severe in the coming years, whereas issues relating to distributed data are not likely to go away regardless of future technological advancements.

8.4.1 Limited Energy Resources

Because they are not connected to a central DAQ system, wireless sensors do not have a dedicated cable supplying power. In fact, many wireless sensor systems rely on battery packs. Even utilizing electrical components designed for extremely low-power operation, sensor systems designed for SHM are expected to last many years without maintenance, making battery life a major concern. Battery power is necessary for all wireless sensor functions, including the transducer, sensing interface, computational core, and wireless communication interface. Battery power is a very critical concern in wireless SHM sensor networks because cost savings realized from the elimination of signal cables during the deployment of the wireless network can easily be eroded over the lifetime of the system by battery maintenance costs. While many electronic components used in wireless sensing do show incremental improvements in power consumption from generation to generation, there certainly exists a temptation to view increasingly efficient electronic components as an opportunity to adopt more computationally capable components rather than reduce power consumption. Whatever the approach taken with regard to technological improvement, power limitations are often the primary impediment to adoption of wireless technology in SHM applications. Because of its central importance, many strategies have been developed to mitigate the problem of limited energy resources, the first of which can even help to transform this weakness into a strength.

8.4.1.1 Embedded Data Interrogation

One important approach for mitigation of energy limitations in wireless sensor networks is the use of embedded data interrogation. Identified by Straser and Kiremidjian (1998) as an energy-savings strategy, embedded processing exploits the fact that, in many wireless sensing platforms, embedded computation consumes less energy than data transmission. Therefore, if data can be interrogated within the WSU as it is collected, and bulk transmission of raw sensor data can be avoided, significant energy savings may be realized. Because the WSU must contain a microprocessor to digitize and buffer data, as well as to perform communication activities, it is also available to execute engineering data interrogation algorithms useful for SHM. Just how much energy can be saved using this practice will depend on the hardware characteristics of the WSU unit, the network topology, the complexity of the algorithm, and the size of the algorithm's output (Lynch and Loh 2006). However, if it is possible to interrogate data locally and transmit only a small amount of processed information (rather than all of the collected raw data), energy savings will be possible (Lynch et al. 2004).

Embedded data interrogation is an important strategy in wireless sensing networks and will be mentioned repeatedly in this section. In addition to energy conservation, embedded data interrogation can help alleviate the bandwidth limitation inherent in the wireless communication channel by reducing the volume of data that must be transmitted through the network. Reductions in data volume will lead to increases in quality of service as well as preserve system scalability (the ability to add more sensor channels without saturating the wireless communication channels). Most important, SHM data is not particularly useful when it is not interrogated. Dense sensor networks (wired or wireless) routinely generate copious amounts of data. A continuously active SHM network may easily generate enough data to overwhelm available resources (technician or engineer time) for manual processing. Embedded data processing within wireless sensing nodes is autonomous and requires no additional resources beyond the sensor network itself. By utilizing on-board autonomous data processing to reduce energy consumption within the wireless network, data interrogation is also assured.

However, there are challenges associated with autonomous data interrogation within wireless sensor networks. As will be explored later in this section, components of WSUs are designed to use low-power components with relatively small amounts of memory and computational power. While holistically, a wireless sensor network may contain significant memory and processing power, individually, each node is relatively limited. Because of these limitations, not all SHM algorithms will work particularly well in the resource-constrained wireless sensor network environment. Similarly, data collected by the wireless sensors is also distributed throughout the network. Therefore, decentralized data processing strategies are required in order to realize the benefits of embedded data processing for energy conservation.

8.4.1.2 Sleep Mode

Depending on its operational environment, threats to the integrity of a structure may not happen continuously. SHM sensing networks may not need to operate when the likelihood of damage is low; at these times, sleep mode may be employed. In sleep mode, nonessential portions of the wireless sensing node are shut down to preserve power and may be brought back online by a trigger (e.g., some external stimulus or resumed use of the structure) or periodically to perform schedule-based SHM. The conditions to wake the network will necessarily reflect the usage of the structure as well as owner priorities. Depending on the duty cycle required for the SHM network, sleep mode can drastically increase the lifespan of a battery supply and is an important feature of battery-powered sensor networks.

8.4.1.3 Power from the Structure and Environment

Many structures are equipped with their own power supply network for lighting, equipment, and service activities. When a native power network is present, wireless sensors located near power connections can be powered from the network itself. In this case, battery power supplies would function purely as a backup for emergency situations and energy concerns can be significantly alleviated. This approach for powering wireless sensors makes sense where power supplies are distributed throughout the structure, but data network cables are not. Many existing structures that are equipped with SHM networks as a retrofit may fall into this category. New structures equipped with powered components but where extensive data cables may be too costly or heavy to be practical are also good candidates for this approach.

Even unpowered structures may have some form of energy that can be exploited by wireless sensing nodes. Power-harvesting and energy-scavenging techniques are becoming increasingly popular as a means of extending battery life in wireless sensors (Mateu and Moll 2005, Bogue 2009, Anton and Sodano 2007). Depending on the structure and the location of the sensor, energy may be harvested from solar power cells (Kurata et al. 2011) or wind (Weimer et al. 2006) to operate and to recharge batteries when external exposures are available. Structures prone to vibrations provide an opportunity to convert kinetic energy into electrical energy that can be used for sensors (Roundy et al. 2004, Park et al. 2008). In addition, noncontact battery recharging strategies have proven to be successful for difficult-to-reach sensing units using directed RF energy (Mascarenas et al. 2009, Farinholt et al. 2009), laser light (Park et al. 2012), and electrical induction coupling (Yao et al. 2006). In such cases, some effort is required to recharge the batteries, but may be less than that required for battery replacement.

8.4.1.4 Passive Wireless Sensors

One strategy that is gaining in popularity for overcoming the inherent power limitations in wireless sensors is to use passive and unpowered sensors. This approach eschews the traditional WSU architecture (i.e., ADC, microcontroller, wireless modem) due to its reliance on electronic components that require power, can be costly, and may be prone to failure, and instead leverages embeddable transducers that can operate without a continuous power supply and that can be interrogated wirelessly or remotely (Lynch and Loh 2006). The ability to operate without a constant power supply is a severe limitation; however, the benefits of eliminating battery packs and fragile electronics are tremendous. Many such sensors use relatively simple transducers that measure instantaneous or peak strain conditions, force, pressure, cracking, corrosion, or impedance (Mita and Takahira 2004, Mascarenas et al. 2007). Readings can be made when the sensor is queried, for example, using an inductively coupled antenna as part of a radiofrequency identification (RFID) system (Loh et al. 2008). The ability to deduce damage conditions from instantaneous measurements can be a challenge for this approach. As additional research is performed to demonstrate damage detection from limited measurements, it is reasonable to expect that passive wireless sensors will become more prevalent in wireless SHM applications.

8.4.2 Communication Bandwidth Limitations

Compared with wired computer networks, wireless networks require more time to transmit any given amount of data. RF transmission are regulated by governmental agencies; in the United States, it is the FCC that dictates which RF frequencies may be used for unlicensed ISM devices such as wireless sensors, and what power transmission levels are acceptable (FCC 2004). These rules place limits on the speed at which data can be transmitted (e.g., the frequency of the carrier wave) and also the reliability of the communication link. RF interference is a common phenomenon that leads to stochastic losses of wireless communication packets. Wireless sensors share the unlicensed ISM frequency bands with many other devices, including consumer electronics that operate on the WiFi and Bluetooth standards, cordless telephones, wireless security cameras, baby monitors, and many wireless toys. A profusion of such devices can create a noisy environment degrading channel quality and effective data transmission rates. Other environmental factors such as solar storms, electrical power transmission equipment, construction activities, and microwave ovens produce negative interference that can hamper wireless data transmission. One way to overcome RF interference is to boost transmission power, increasing the signal-to-noise ratio of the transmission. However, FCC limitations prevent any device from employing excessive transmission power in order to preserve fair usage of these unlicensed frequency bands, thus limiting the effectiveness of power boosting. However, even without the FCC limitation, battery life concerns will also limit the effectiveness of boosting transmission power as a means of improving channel quality. Limitations in data transmission rates in wireless sensor networks (relative to wired networks) require special attention to data processing and transmission strategies when wireless sensors are employed.

Communication bandwidth challenges can become a bottleneck on sensor network scalability. Large sensor networks collecting data at high sampling rates will generate copious amounts of data that will overwhelm the wireless communication channel if appropriate mitigation strategies are not used. Since many SHM applications rely on dense sensor installations, such strategies become vital. In large structures where single-hop communications are not effective, communication bandwidth issues become especially critical as multihop data communication is inherently slower than single-hop. This section will focus on three strategies for alleviating communication bandwidth issues:

1. Buffering and decreased duty cycle
2. Use of multiple wireless channels and hybrid networks
3. Local, embedded data interrogation

The simplest strategy to prevent saturation of the wireless communication channel is to collect less data. Recognizing that many structures do not require continuous monitoring, engineers can program the sensor network to spend significant portions of time idle or in sleep mode. Reducing the duty cycle of the sensor network is a good strategy to preserve battery energy, but it has the ancillary benefit of alleviating bandwidth concerns as well. Naturally, when data must be continuously collected and transmitted within the network, the maximum size of the network is dictated by the communication channel. For SHM applications where data is collected only periodically, it becomes less critical that all data transmissions be completed in real time. That is, collected data can be buffered locally and transmitted as the wireless channel becomes available.

Use of multiple wireless channels provides an additional means of overcoming bandwidth limitations. Breaking sensor networks into multiple channels increases the bandwidth available to the network considerably. For example, devices operating under the IEEE 802.11 communication standard (WiFi) can be assigned to one of four nonoverlapping channels within the 2.4-GHz frequency band (Gast 2005, IEEE 2007). IEEE 802.15.4 devices (Zigbee) may be assigned to any one of 16 nonoverlapping channels within the same range (IEEE 2006). Sensors can be assigned to subnets on different channels or can be programmed to hop from channel to channel to avoid crowded channels and RF interference.

Multi-tier sensor networks are particularly well suited to take advantage of multiple communication channels. Basic WSUs can be assigned to subnets on the basis of their proximity to local master nodes. Master nodes collect data and issue commands to local WSUs on their assigned channels and communicate with other master nodes on different frequencies, resulting in a significant increase in wireless bandwidth. This practice can also be used to eliminate inefficient multi-hopping within a single channel. Master nodes may even be equipped with multiple radios that allow them to monitor multiple channels at once. Hybrid networks that use wireless communication between sensors and master nodes, and wired communication between master nodes and other master nodes, are also effective remedies for channel saturation. In hybrid networks, subnets of WSUs can reside on different frequencies and consolidate their data to master nodes wirelessly. Then the master nodes take the consolidated data and exchange it (or communicate it to a data server) over a wired digital backbone, significantly reducing the burden on the wireless communication channels. In addition, since the wired data transmissions are made in the digital domain, the danger of analog signal corruption in the signal cables is eliminated and considerably less bulky cables may be employed.

Finally, embedded data interrogation was discussed as an important strategy for energy conservation; it is also highly useful in preserving wireless bandwidth. When sensor densities grow large enough, it becomes impractical to transmit, centralize, and archive all of this data wirelessly. In such cases, embedded data interrogation becomes necessary to process the data before it is transmitted. Utilizing the computational capabilities of the embedded microcontroller, engineering algorithms can be used to interrogate raw sensor data and extract useful information. That information should be of lower order than the full data set. Communicating, centralizing, and archiving the results from the engineering algorithms presents a lower burden to the communications channels, preserving channel quality and reducing energy usage as well. However, to take advantage of this strategy, the distributed nature of the wireless sensor network must be taken into account.

8.4.3 Distributed Computational Resources and Data

The decentralized nature of wireless sensor networks presents significant challenges in their implementation. As discussed in previous sections, the cost required to consolidate data collected from dense networks of wireless sensors can be significant in terms of energy and bandwidth. Embedded data processing within the wireless sensor node is an attractive approach to mitigate these problems as the need to transmit copious quantities of raw sensor data can be abated or even eliminated. However, wireless sensor networks should not be viewed as equivalent to their centralized wired counterparts. Memory and data processing resources are scattered throughout the network. Large, computationally intensive, algorithms for data interrogation become impractical when each networked device is relatively low powered. Therefore, embedded processing in low-power distributed networks favors SHM algorithms that rely on low-order models and those that necessitate less computing power and memory. Larger models require parallel processing techniques with especially heavy penalties given to data transmissions.

The fact that data is initially distributed is another major challenge for many potential embedded algorithms. SHM methods that rely on spatial models (e.g., those based on mode shapes) do not yield much information from a single sensor and exchange of some data is required. Therefore, there exists a tradeoff between energy/bandwidth consumption and embedded algorithm execution. Careful design of the SHM algorithm can preserve energy and bandwidth gains made by using embedded computation while still producing usable results. Innovative data flow and parallel processing techniques are a necessary consideration for these kinds of wireless sensor networks. These considerations depend on the algorithm, the hardware capabilities of the wireless sensor, and the network topology.

Data synchronization is often an additional challenge in wireless sensor network applications. Input–output models, multiple-output models, modal-based approaches, finite element method (FEM) model updating algorithms, and others require synchronized data from all sensors to be effective. In a traditional wired network, all sensors are connected to a DAQ system that operates using a single clock and are inherently synchronized. In wireless sensing networks, each WSU has its own independent clock. Synchronization of these clocks can be a challenge. In single-hop star networks, synchronization can be simplified using the central or server node as a beacon for the rest of the network, aligning all of the WSU clocks with the server. When sampling frequencies are many orders of magnitude slower than the WSU system clocks, beaconing can be a particularly effective means to synchronize the network (Kim et al. 2010). In wireless networks that are spread over large areas, where single-hop transmission are not possible, more sophisticated synchronization approaches are needed (Elson and Römer 2003, Sundararaman et al. 2005). In addition, the clocks in the WSU have limited precision, requiring occasional resynchronization when the network must operate over an extended period. Such resynchronizations incur energy and bandwidth overhead and may lead to lost data in heavily trafficked networks. Continuous synchronization of wireless sensor networks can help preserve data in these cases (Sundararaman et al. 2005).

While a challenge, distributed data processing can also be a strength of the wireless sensor framework. Traditional centralized systems use a single computer to interrogate data and lose robustness due to the potential for failure of the only computer in the system. Parallel, distributed data processing using sensors with some degree of redundancy can enhance the robustness of the distributed system. The need to generate statistically significant damage information necessitates that sensor networks are designed with some degree of redundancy anyway; however, it is important to recognize that any complex system made up of many components must be designed with robustness in mind as well. Failures of some portion of the network should not lead to systemic failure. Properly designed and implemented, wireless sensors with their inherent on-board computational capabilities offer an ideal platform to perform robust SHM activities that will yield useful information even when some sensors fail or malfunction.

8.4.4 Low-Power Components

Energy preservation is usually the overriding concern in the design of a wireless sensing node. To achieve low-power operation, electrical components used in the sensor must necessarily have fewer capabilities than their high-power counterparts. That is, the processor used in a wireless sensing device will not be the same one sold for use in a high-end PC, for instance. In composite structures, size can also be a critical concern, limiting performance even further. As such, the microcontroller used in a wireless sensing device is likely to be orders of magnitude slower than that used in the state-of-the-art PC. SHM algorithms that are to be embedded in wireless sensors must be selected and designed with this reality in mind. Low-order models and simple algorithms should be favored over massive, data-intensive models and computationally complex algorithms. For example, even though impressive results in SHM have been achieved using model updating methods for high-fidelity FEM models, large FEM models are impractical to execute embedded within a low-power wireless sensor. The updating process for any large model may take hours or even days on a wireless sensor, rendering that information useless by the time it becomes available. Low-order damage-sensitive features such as time-series models, modal parameters, outputs from neural networks and genetic algorithms, or data-driven damage indices are usually better candidates for use in wireless sensor networks.

This limitation of wireless sensor nodes is one that is expected to change as time progresses. New generations of computing platforms provide more computing power and more memory for less cost and less energy. Therefore, the model that may be impractical today may be reasonable tomorrow. However, it is likely that the capabilities of low-power wireless devices will lag behind those of large wired systems for quite some time, if not indefinitely. Therefore, hybrid two-tiered SHM sensor networks composed of a low-power wireless sensor network with access to a high-powered, wired application server (Figure 8.8) may be used to leverage state-of-the-art modeling and optimization tools. Under such a framework, wireless sensors collect data and provide a first layer of data interrogation, looking for irregularities that might be indicative of damage. When suspicious data patterns are identified, raw structural response data can be uploaded from the wireless sensor network to an application server equipped with a wireless interface. The application server has a computationally powerful processor that can execute large and complex model updating algorithms to try to verify the presence of damage, identify its source, and understand its extent.

Besides memory and computing power, current low-power data acquisition hardware (suitable for wireless applications) is relatively limited as compared with its wired equivalent. As of the writing of this chapter, wireless data acquisition hardware is largely limited to low resolution (12–16 bits) and low sampling speeds (low to sub kHz) (Lynch and Loh 2006), although recent industrial and academic devices have been released to address these shortcomings. As technology improves, it is expected that new wireless sensing

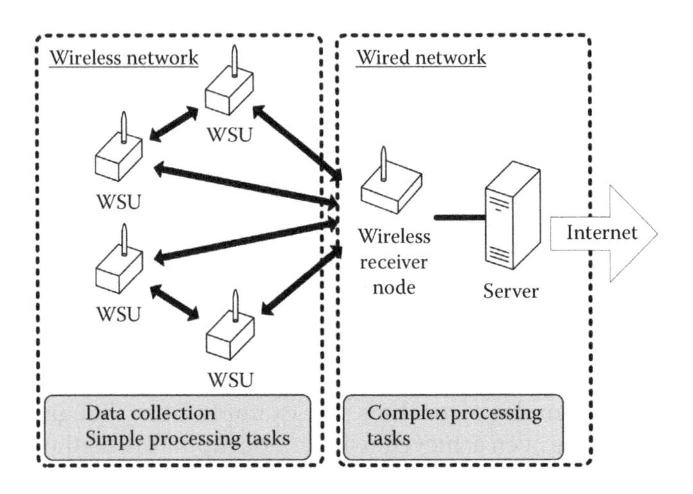

FIGURE 8.8
Hybrid wireless/wired in-network data interrogation.

nodes capable of high sampling rates and resolution will become more commonplace improving the capabilities of wireless devices and increasing the number of SHM applications that will be possible to use with wireless sensors.

8.4.5 Reliability and Security

Inexpensive wireless sensors can be expected to experience reliability issues during the lifetime of a structure. Therefore, it is necessary to build provisions into the design of a wireless sensor network for redundancy and maintenance in the event of sensor failure. Installing dense sensor networks helps increase redundancy and robustness in the face of uncertainty and sensor failure. Redundant sensor nodes are only part of the solution though. Network topologies should be flexible as well, allowing for failure of one or more nodes without compromising the entire network. SHM algorithms should be fault tolerant, not collapsing when input from a single sensor is lost. Such provisions are wise for any kind of sensor network, wired or wireless.

Wireless data communications invite security concerns as well (Xiao et al. 2007). Unlike wired networks, wireless communications are easy to monitor and easy to interrupt. Because wireless communication takes place in an open medium, accidental or malicious interference is always a possibility. As stated in prior sections, interference on the wireless data channel can easily lead to delayed or even lost information. For critical monitoring applications where data delays or network outages are not tolerable, wireless communication technologies may not be the best fit. Since wireless data is broadcast in all directions, anyone within the range of the transmitter will be able to intercept the transmission and may even try to insert their own malicious data packets. Encryption algorithms can prevent such an occurrence; however, they come with additional overhead as sensors must spend both time and energy encrypting and then decrypting every transmission. Also, disrupting a wireless sensor network by flooding the communication channel with high-power RF noise is illegal, but not difficult to accomplish. Critical networks that may be targeted by such an attack should not be considered to be good candidates for wireless sensor technology.

8.5 Conclusions

Wireless sensing technology is an important tool in many SHM applications. It offers advantages in terms of cost, weight, ease of installation, and certain aspects of robustness. However, wireless sensors should not be considered as a one-to-one replacement for cabled sensors. Reliance on batteries, low communication bandwidth, limited memory and computing power, distributed data and computing resources, synchronization, and security issues are all important aspects to consider when deciding whether to adopt this technology. Although these challenges can be formidable, many strategies are being advanced to alleviate these concerns and may tip the scales in favor of wireless sensing for a broad range of applications. Embedded data processing helps overcome energy and bandwidth limitations, and provides a guarantee of some minimum level of data utilization because it occurs autonomously at the sensor level. Energy-scavenging technologies can also mitigate energy concerns. In addition, novel passive and externally powered sensors can eliminate energy concerns entirely, as well as eliminate some of the more costly and vulnerable electronics from the sensor design. Careful consideration is required before adopting wireless technology to determine which approach, if any, is best suited to a given SHM application. Engineers must consider capabilities of both hardware and firmware components of wireless sensor nodes and take advantage of embedded processing as much as possible. Properly utilized, wireless sensors can help reduce economical and technical barriers to SHM implementation in composite structures. Improperly utilized, they can undermine the SHM application as well as the bottom line.

References

Anton, S. R. & Sodano, H. A. 2007. A review of power harvesting using piezoelectric materials (2003–2006). *Smart Materials and Structures*, 16, R1–R21.

Bogue, R. 2009. Energy harvesting and wireless sensors: A review of recent developments. *Sensor Review*, 29, 194–199.

Elson, J. & Römer, K. 2003. Wireless sensor networks: A new regime for time synchronization. *SIGCOMM Computer Communication Review*, 33, 149–154.

Farinholt, K. M., Gyuhae, P. & Farrar, C. R. 2009. RF energy transmission for a low-power wireless impedance sensor node. *Sensors Journal, IEEE*, 9, 793–800.

FCC 2004. *Title 47 of the Code of Federal Regulations—Chapter 1, Part 15, Radio Frequency Devices*. Washington DC, Federal Communications Commission.

Gast, M. 2005. *802.11 Wireless Networks: The Definitive Guide*. Sebastopol, CA, O'Reilly Media, Inc.

IEEE. 2006. *802.15.4: Standard for Information Technology—Telecommunications and Information Exchange between Systems—Local and Metropolitan Area Networks—Specific Requirements Part 15.4: Wireless Medium Access Control (MAC) and Physical Layer (PHY) Specifications for Low Rate Wireless Personal Area Networks (LR-WPANs)*. New York, IEEE Standards Association.

IEEE. 2007. *IEEE 802.11: Wireless LAN Medium Access Control (MAC) and Physical Layer (PHY) Specifications*. New York, IEEE Standards Association.

ISO/IEC. 1996. *Information Technology—Open Systems Interconnection—Basic Reference Model: The Basic Model*. Geneva, Switzerland, ISO/IEC Copyright Office.

Kim, J., Swartz, R. A., Lynch, J. P., Lee, J. J. & Lee, C. G. 2010. Rapid-to-deploy reconfigurable wireless structural monitoring systems using extended-range wireless sensors. *Smart Structures and Systems, TechnoPress*, 6, 505–524.

Kurata, M., Lynch, J. P., Linden, G. V. D., Jacob, V., Zhang, Y., Thometz, E., Hipley, P. & Sheng, L.-H. 2011. Long-term assessment of autonomous wireless structural health monitoring system at the new Carquinez Suspension Bridge. In: *SPIE Smart Structures/NDE*, 2011. San Diego, CA.

Loh, K. J., Lynch, J. P. & Kotov, N. A. 2008. Inductively coupled nanocomposite wireless strain and pH sensors. *Smart Structures and Systems*, 4, 531–548.

Lynch, J. P. & Loh, K. J. 2006. A summary review of wireless sensors and sensor networks for structural health monitoring. *The Shock and Vibration Digest*, 38, 91–128.

Lynch, J. P., Sundararajan, A., Law, K. H., Kiremidjian, A. S. & Carryer, E. 2004. Embedding damage detection algorithms in a wireless sensing unit for attainment of operational power efficiency. *Smart Materials and Structures, IOP*, 13, 800–810.

Mascarenas, D., Flynn, E., Farrar, C., Park, G. & Todd, M. 2009. A mobile host approach for wireless powering and interrogation of structural health monitoring sensor networks. *Sensors Journal, IEEE*, 9, 1719–1726.

Mascarenas, D. L., Todd, M. D., Park, G. & Farrar, C. R. 2007. Development of an impedance-based wireless sensing node for structural health monitoring. *Smart Materials and Structures*, 16, 2137–2145.

Mateu, L. & Moll, F. 2005. Review of energy harvesting techniques and applications for microelectronics. In: *Proc. SPIE*, 2005, 359–373.

Mita, A. & Takahira, S. 2004. Damage index sensor for smart structures. *Structural Engineering and Mechanics*, 17, 331–346.

Park, G., Rosing, T., Todd, M. D., Farrar, C. R. & Hodgkiss, W. 2008. Energy harvesting for structural health monitoring sensor networks. *Journal of Infrastructure Systems*, 14, 64–79.

Park, H.-J., Sohn, H., Yun, C.-B., Chung, J. & Lee, M. M. S. 2012. Wireless guided wave and impedance measurement using laser and piezoelectric transducers *Smart Materials and Structures*, 21, 035029.

Roundy, S., Wright, P. K. & Rabaey, J. M. 2004. *Energy Scavenging for Wireless Sensor Networks: With Special Focus on Vibrations*. Norwell, MA, Kluwer Academic Publishers.

Straser, E. & Kiremidjian, A. S. 1998. *Modular, Wireless Damage Monitoring System for Structures*. Stanford, CA, John A. Blume Earthquake Engineering Center.

Sundararaman, B., Buy, U. & Kshemkalyani, A. D. 2005. Clock synchronization for wireless sensor networks: A survey. *Ad Hoc Networks*, 3, 281–323.

Swartz, R. A., Jung, D., Lynch, J. P., Wang, Y., Shi, D. & Flynn, M. P. 2005. Design of a wireless sensor for scalable distributed in-network computation in a structural health monitoring system. In: *5th International Workshop on Structural Health Monitoring*, 2005. Stanford, CA.

Weimer, M. A., Paing, T. S. & Zane, R. A. Year. Remote area wind energy harvesting for low-power autonomous sensors. In: *Power Electronics Specialists Conference*, 2006. PESC '06. 37th IEEE, June 18–22 2006, 1–5.

Wu, H. & Pan, Y. 2008. *Medium Access Control in Wireless Networks*. New York, Nova Science Publishers, Inc.

Xiao, Y., Shen, X. & Du, D. 2007. *Wireless Network Security*. New York, Springer Science+Business Media, LLC.

Yao, W., Li, M. & Wu, M.-Y. 2006. Inductive charging with multiple charger nodes in wireless sensor networks. *Lecture Notes in Computer Science*, 3842/2006, 262–270.

9

Acoustic Emission of Composites: A Compilation of Different Techniques and Analyses

Piervincenzo Rizzo

CONTENTS

9.1 Introduction

The applications of composite materials are on the rise in many engineering fields, including civil and aerospace structures. In the aerospace industry, for instance, manufacturers have increasingly moved toward replacing lightweight metals with composites for large areas of primary structure. The motivation for this migration stems from the industry's need to improve the performance and the fuel efficiency of aircrafts that can be attained by reducing structural weight. These achievements are possible owing to the unique properties of composites that have been extensively discussed in the first part of this book. However, composite materials may be prone to different types of defects, including but not limited to fiber breaks, delaminations, and voids. Owing to the critical role that composites have in modern engineering, a great effort has been devoted by academia and industry to the development of reliable and robust tools for the nondestructive evaluation (NDE) and structural health monitoring (SHM) of such structural elements. In aeronautics, for instance, the increasing demand in lowering maintenance costs contributes to the growing development of SHM systems to ensure continuous knowledge of the structural state of the monitored aircraft components.

This chapter delves with the acoustic emission (AE) technique applied to fiber-reinforced composites. The technique can be used as a nondestructive tool to monitor structures in real time in order to prevent the onset of a structural failure. One of the advantages of AE is that it may identify the different types of failure mechanisms that typically occur at the microscopic level, namely matrix cracking, fiber–matrix debonding, fiber pullout, fiber bridging, inter-ply failure, and fiber breakage. A review of the capabilities of AE applied to fiber-reinforced polymer (FRP) testing was given in the 1980s by Hamstad (1986). By the end of the 1990s, several studies had been conducted in laboratory settings for the monitoring of fracture mechanisms and damage progression in laminated FRPs subject to fatigue loading (Awenbruch and Ghaffari 1988, Eckles and Awenbruch 1988, Tsamtsakis et al. 1998, Komai et al. 1991, Tsamtsakis and Wevers 1999), tensile loading (Komai et al. 1991, De Groot et al. 1995, Sato and Kurauchi 1995, Mizutani et al. 2000, Bohse 2000, Berthelot and Rhazi 1990, Favre and Laizet 1989, Luo et al. 1995), compressive loading (Byrne and Green 1994), or drilling (Ravishankar and Murthy 2000a,b).

By that time, most of the work available in the literature focused on small laboratory test coupons. However, there are differences between AE testing of small-scale and large-scale specimens. The latter suffer from wave attenuation and dispersion (Prosser 1996) that are caused by four factors: geometric spreading, internal friction, dissipation into adjacent media, and losses related to dispersion (Pollock 1996, Prosser 1996). Moreover, some acoustic waveform features such as frequency content are strongly influenced by the size of the test piece as shown by Hamstad and Downs (1995), which tested carbon fiber-reinforced polymer (CFRP) pressure vessels. Driven by this consideration, some of the works conducted in the last two decades focused on the AE of large-scale structures.

This chapter is organized as follows. For the sake of completeness, Section 9.2 introduces the general concepts of NDE and SHM, as well as the general principles, advantages, and limitations of the AE technique. Section 9.3 describes the overall setup and the common terminology used in AE testing, as well the parameter analysis, which is the simplest and more immediate approach to examine AE data. Section 9.4 delves with time-based and amplitude-based source localization methods aimed at identifying the position of AE sources. Section 9.5 reviews some case studies that span from large-scale engineering applications to impact-related damage detection. Finally, Section 9.6 illustrates some of the latest advancements in the field of pattern recognition and signal processing applied to AE raw data.

The scope of this chapter is twofold. First, it approaches senior undergraduates and graduate students to provide a first insight on the principles of AE method and its potential in the area of damage detection and structural monitoring. The intent is to make the chapter accessible to those readers that may not yet have a deep knowledge of the technique and desire to learn the necessary fundamentals to prepare an experimental setup and conduct data analysis effectively. Second, this chapter supports those researchers that are already working on composites and wish to acquire more knowledge with the state of the art of AE applied to composites. With the inclusion of Sections 9.5 and 9.6 and the citation of appropriate bibliography, those researchers may gain some insights on the application of AE to large-scale structures and on the use of advanced signal processing tools.

9.2 Acoustic Emission: Background

The American Society for Nondestructive Testing defines NDE as "the examination of an object with technology that does not affect the object's future usefulness." The basic principle of any NDE method is simple: use a physical phenomenon (the interrogating parameter) that interacts with and be influenced by the test specimen or the structure (the interrogated parameter) without altering or compromising the structure's operability. As defined by Adams (2007), SHM "is the scientific process of nondestructively identifying four characteristics related to the fitness of an engineered component (or system) as it operates

1. The operational and environmental loads that act on the component (or system)
2. The mechanical damage that is caused by that loading
3. The growth of damage as the component (or system) operates
4. The future performance of the component (or system) as damage accumulates"

Health-monitoring technologies must be nondestructive and are implemented with hardware/software to provide a diagnostics or the system in real time.

While NDE testing is performed periodically, e.g., every 2 years for bridges in the United States, health monitoring is performed continuously and it involves the observation of a system over time, the extraction of damage-sensitive features, and the analysis of these features by means of signal processing and pattern recognition algorithms.

Among the many NDE methods, AE is one of the more common and oldest. It is a passive method for examining the behavior of materials deforming under stress. A formal definition establishes that AE is "the release of transient elastic waves produced by a rapid redistribution of stress in a material." The elastic energy propagates as a stress wave (AE event) in the structure and is detected by one or more sensors. In composites, AE events may be generated by moving dislocations, crack onset and growth, fiber breaks, disbonds, plastic deformations, etc. (Degala et al. 2009, Spada et al. 2011). It is similar to seismology where elastic waves originate from a source inside the Earth and travel through the material so that they can be recorded by a network of seismometers distributed on the surface. As in seismology, the features of the "input" or emitted signal are relatively unknown, and the detected signal is dependent on the characteristics of the source but may also be affected by the traveling path from the source to the detector. As pointed out by Prosser (2002), AE monitoring is not a truly nondestructive method, as some damage or change in the material must occur to make the AE testing effective. As the method is based on the propagation of elastic waves at the ultrasonic range, the full understanding of this method cannot ignore the knowledge of the general principles of ultrasonic wave propagation, which for this reason is provided below.

In unbounded media, two types (modes) of ultrasonic waves can propagate: the longitudinal bulk mode, also referred as P-wave (pressure wave), and the shear bulk mode, sometimes referred as S-wave or T-wave. In the longitudinal mode, the particle displacement is parallel to the direction of the wave propagation; conversely, in the shear waves, the particle displacement is orthogonal to the propagation direction. The latter mode causes shearing stresses that cannot be supported in fluids such as water, and in gases. Bulk waves are widely used in ultrasonic testing because only two velocities need to be measured, the longitudinal C_L and the shear C_T wave velocities. When the bulk waves propagate in a material without boundaries and a planar wave front in an isotropic, homogeneous material can be assumed, these speeds are given by

$$C_L = \sqrt{\frac{E}{\rho}\frac{1-\nu}{(1+\nu)(1-2\nu)}} \quad (9.1a)$$

$$C_T = \sqrt{\frac{\mu}{\rho}} \quad (9.1b)$$

where E is the medium's Young's modulus; ρ is the medium's density; μ is one of the medium's second-order Lamé constants, namely the shear modulus; and ν is the Poisson's ratio. The longitudinal mode propagates with the fastest velocity, and therefore, it should be the first measured AE signal arrival if the source produced a sufficiently large longitudinal component.

Whenever an ultrasound propagates into a bounded medium, it has the characteristics of a guided ultrasonic wave (GUW). The wave is termed "guided" because it travels along the medium guided by the medium's geometric boundaries. GUWs propagate along, rather than across, the waveguide. Different types of GUWs exist and the main ones are Rayleigh waves, Lamb waves, and cylindrical waves (Kolsky 1963, Achenbach 1984, Auld 1990, Krautkrämer and Krautkrämer 1990, Graff 1991, Rose 1999). Rayleigh waves propagate along the surface of a semi-infinite space, and their particle motion is elliptical. Lamb waves exist in plate-like structures and occur in two different basic modes, the symmetrical or dilatational mode, and the asymmetrical or bending mode. The particles of the neutral fiber (the middle zone in an unstressed plate) perform pure longitudinal displacements in the case of symmetric modes and pure shear oscillations in the case of antisymmetric modes. In the partial wave representation, a guided wave can be thought of as a superposition of partial plane waves, which are reflected inside the bounded structure. In a free waveguide, i.e., surrounded by vacuum, the guided wave is attenuated by material damping only; when the waveguide is surrounded by another material, the attenuation is also due to the leakage of the waves into the surrounding medium. In a slender, isotropic, traction-free cylindrical waveguide, three types of vibrational modes can exist: longitudinal, flexural, and torsional waves. The first two waves are analogous to symmetric and antisymmetric Lamb waves, respectively. However, the family of the torsional mode results when only the angular displacement exists along the cylinder cross-sectional surface.

In NDE and SHM, GUWs are becoming very popular owing to the capability of inspecting moderately large areas using a single probe attached or embedded in the structure while maintaining high sensitivity to small flaws. GUWs can travel at relatively large distances with little attenuation and offer the advantage of exploiting one or more of the phenomena associated with transmission, reflection, scattering, mode conversion, and absorption of acoustic energy. Unfortunately, GUWs are multimode (many vibrating modes can propagate simultaneously) and dispersive (the propagation velocity and the attenuation depend on the wave frequency). The dispersive behavior is represented by the dispersion curves that describe the relation between the wave velocity and the frequency. Sometimes the frequency is replaced by the wavenumber k, which is related to wavelength λ by the equation $k = 2\pi/\lambda$. Each mode is characterized by two main speeds: the *phase velocity* and the *group velocity*. The phase velocity C_p is the speed at which an individual crest of a wave moves, and it is related to the wavelength and the circular frequency $\omega = 2\pi f$ by the equation $c_p = \omega\lambda/2\pi = \omega/k$. The group velocity c_g describes the speed at which a guided packet travels in the element, i.e., how quickly the ultrasonic energy propagates. The group velocity is related to the circular frequency and to the phase velocity by the following equations (Rose 1999):

$$c_g = \frac{d\omega}{dk} = c_p - \lambda \frac{dc_p}{d\lambda} \tag{9.2}$$

For illustrative purposes, the dispersion curves of a steel rod in vacuum are presented in Figure 9.1a,b. The following material parameters were considered $C_L = 5.890$ km/s, $C_T = 3.233$ km/s, and $\rho = 7843$ kg/m^3. The notation of Meitzler (1961) is employed for mode identification. The letters L, F, and T

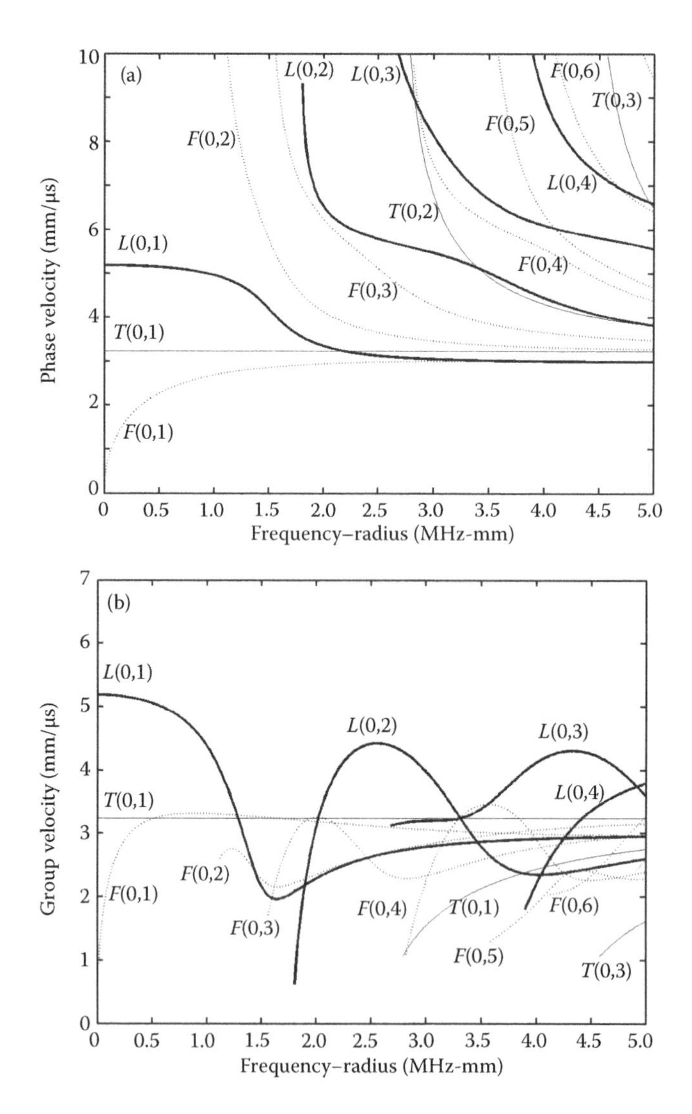

FIGURE 9.1
Dispersion curves for a steel bar in vacuum. (a) Phase velocity. (b) Group velocity.

indicate the longitudinal, flexural, and torsional waves, respectively. The letter is then followed by two integer numbers in parentheses. The first number refers to the variation in displacement around the circumference of the bar and it is zero, by definition, for the longitudinal modes. For the F modes, the first counter is determined by the number of wavelengths around the circumference of the solid cylinder (Pavlakovic 1998, Beard 2002, Rizzo 2004). The second number is a counter variable that represents the order modes of the branches. The numbering starts with 1 for those modes (the fundamental modes) that can propagate at zero frequency.

The AE technique can be interpreted as the passive detection and subsequent analysis of ultrasonic waves generated by material cracks, dislocation, or other kinds of structural anomalies. Thus, the characteristics of the detected wave in an AE setting obey the features associated with the wave propagation in bulk materials, semi-infinite solids, and waveguides. AE differs from other nondestructive methods in two ways. First, the signal has its origin in the material itself and is not introduced from an external source, i.e., it is a passive method. Second, AE detects movement or strain, whereas most other methods detect existing geometric discontinuities or breaks (Prosser 2002). The use of an AE testing setup for NDE and SHM has one or more of the following objectives:

1. Localize the source where the transient wave are released
2. Discriminate among different sources of damage, thereby attributing each emission to a particular source type or failure mode
3. Monitor the onset and progression of damage
4. Predict the incumbency of major failure to enable the adoption of proper measures to prevent catastrophic failures

The principles, advantages, and disadvantages of AE can be summarized as follows.

Principles. The method is based on the detection of bursts (emissions/events) that are originating from the material being monitored. Special transducers, conventionally denominated AE sensors, are used to detect these emissions. These sensors convert the displacement or acceleration generated by the propagation of the elastic wave into an analog signal that is digitized by a dedicated electronic apparatus. The signal is then processed in real time or post processed to attain one or more of the four objectives described earlier.

Advantages. AE can be used for global monitoring; an array of sensors opportunely spaced in the structure or subsystem to be monitored can provide an efficient way to determine the onset of new damage or the growth of existing damage. This arrangement avoids time-consuming and expensive point-by-point scanning as is, for instance, necessary in ultrasonic testing by means of bulk waves. The sensor system can be remotely located in the structure and can communicate with the processing unit either via a tethered system or wirelessly. This implies that an AE unit can be used *in situ*, for instance in bridges, while the structure remains in service. If proper processing is developed, the method can, in principle, predict failure and identify the origin and the location of the AE events.

Limitations. Like any other NDE technique, AE also has some limitations. As it is a method based on wave propagation, AE suffers from signal attenuation and multimodal dispersive behavior. This applies to structures made of composites that are anisotropic and in most cases have the shape of a waveguide (plates, shells, hollow cylinders, etc.). The presence of existing defects along the path between the source and the detector causes the AE event to be reflected, refracted, and mode converted. Even simple signals, such as those generated by the pencil-lead break, known as a Hsu–Nielsen source—a standard method for simulating AE signals—that consists of a fast rise time or step function, can result in very complicated detected acoustic waveform. Thus, care must be paid to the number and location of the sensors to be deployed on the structure to be monitored. The data may be affected by extraneous noise, which, if not identified, can lead to the generation of false positives. Sensors must be attached to the material and therefore contact must be ensured throughout the monitoring period. Contact can be guaranteed using adhesive bond or mechanical holders. Both must be robust against hostile environmental conditions such as temperature gradients, wind, rain, or moisture that may alter the composition of the bond or displace the position of the holders. When aimed at providing permanent structural monitoring, both the sensors and the hardware must be protected against vandalism. To localize the flaw, multiple sensors are required. Owing to its principles, AE may not be suitable to recognize the presence of existing damage; therefore, it may not be able to map the health of the structure at the moment of the AE installment. Finally, this technique cannot capture those damaging phenomena, such as for instance corrosion, that evolves without significant emission of bursts or with the emission of bursts that are below the ground noise.

In the AE of composites, there are two predominant kinds of AE testing: fundamental and applied. At the fundamental level, dominant in the academic research, AE is carried out experimentally; the setup is straightforward and applied to both simple and complex geometries. A quasi-static tensile test is accompanied with AE registration and full-field strain mapping. The test is stopped anytime, and visual inspection of the sample is carried out using x-ray, scanning electron microscopy, or other forms of inspection. The inspection serves to associate the characteristics of the AE features recorded during the experiment with the type of cracks and damages that occurred in the sample. The tensile tests are usually performed according to the American Society for Testing and Materials (ASTM) D3039 standard (ASTM D3039/D3039M—08 Standard Test Method for Tensile Properties of Polymer Matrix Composite Materials). At the applied level conducted by both industry and academia, the AE monitoring is performed in very large structures or in the field. In the latter case, several factors must be taken into account: sensor ruggedness against environmental conditions such as temperature, wind, and moisture; false positives triggered by operating conditions such as the passage of heavy trucks or by weather-related phenomena such as rain; and vandalism.

9.3 Fundamentals

The overall scheme of an AE testing is presented in the chart of Figure 9.2. The wave triggered by an event reaches an AE sensor. The sensor is usually a piezoelectric transducer that converts the displacement or the acceleration proportional to the strength of the wave into voltage. On the basis of the material and the geometry of the structure, the transducer can be wideband or narrowband. The transducer is connected to a dedicated signal conditioner and event detector. A preamplifier and a filter may be interposed between the transducer and the conditioner. The preamplifier increases the amplitude of the signal usually in the order of +20, +40, or +60 dB, whereas the filter contributes to suppress AE signatures that are outside the frequency spectrum of interest. The first step in the AE analysis is the acquisition of the elastic wave (Figure 9.2b). The most common method used in commercial systems to discriminate the signal from the background noise is by comparing the signal against a certain threshold. If the threshold, which can be fixed or floating, is exceeded, then a hit is said to be detected. Once the signal amplitude exceeds the threshold, the time waveform from the transducer is stored by the event detector. Here a preliminary analysis can be carried out by determining certain parameters such as rise time, duration, peak amplitude, AE energy, and number of counts (Figure 9.2b). This approach is conventionally called parameter analysis, and these features are used to assess the mechanical/structural event that generated the release of transient waves and to ascertain the presence of the Kaiser effect or the Felicity effect. The Kaiser effect is the absence of detectable AE at a fixed sensitivity level, until previously applied stress levels are exceeded and it is used as an indicator of the stress history of a specimen. This effect states that if a sample is loaded, unloaded, and then reloaded under the same conditions, AE events should not be detected until the previous load peak is achieved or passed (Gostautas et al. 2005). As such, the Kaiser effect is often used to assess the existence of permanent damage, because this effect is satisfied when a test object has not experienced permanent damage. When there is an occurrence of emissions at loads below the previous maximum load, the Kaiser effect is replaced by the Felicity effect, whose significance and related damage is defined by a Felicity ratio. This is the ratio between the load at which AE events are first generated on reloading and the previously applied maximum load (Gostautas et al. 2005). A Felicity ratio equal to 1 or greater is

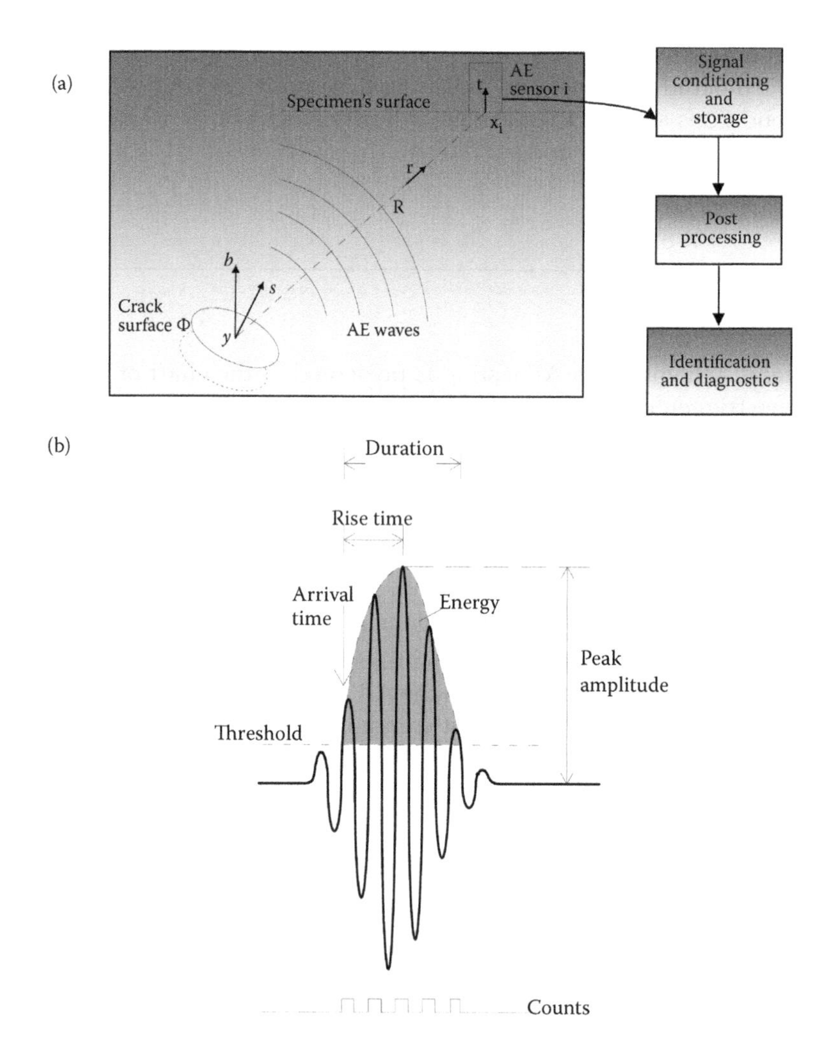

FIGURE 9.2
Fundamentals of AE. (a) Formation, propagation, and acquisition of an AE wave and subsequent digitization, storage, and postprocessing. (b) Main characteristics of an AE wave. (From Spada, A., Rizzo, P. and Giambanco, G., *ASCE J. Eng. Mech.* 137, 854, 2011.)

representative of a structure that has not been subjected to damage since the last AE inspection. A decreasing ratio may be symptomatic of cumulative, permanent damage.

Closely related to the parameter analysis, the intensity analysis (IA) evaluates the structural significance of an AE event as well as the level of deterioration of a structure by calculating two values called the historic index (HI) and severity (S_r) (Goustautas et al. 2005, Degala et al. 2009). The HI compares the signal strength of the most recent emissions to the signal strength of all emissions.

This requires estimating the slope changes of the cumulative signal strength (CSS) plotted as a function of time. The presence of one or more peaks may reveal the occurrence of new damage or the propagation of damage, respectively. The severity is the average of the J largest signal strength emissions received at a sensor. As the severity is a measure of structural damage, an increase in severity often corresponds to new structural damage. Analytically, the HI and the S_r are defined as (Goustautas et al. 2005, Degala et al. 2009)

$$\mathrm{HI} = \frac{N}{N-K}\left(\frac{\sum_{i=K+1}^{N} S_{oi}}{\sum_{i=1}^{N} S_{oi}}\right) \tag{9.3}$$

$$S_r = \frac{1}{J}\left(\sum_{m=1}^{J} S_{om}\right) \tag{9.4}$$

where N is number of hits up to time t; S_{oi} is the signal strength of the i-th event; and K and J are empirical constants based on the material under investigation. Common values found in the literature are as follows: for $N \le 100$, K is not applicable; for $101 < N < 500$, $K = 0.8N$; and for $N > 500$, $K = N - 100$; for $N \le 20$, J is not applicable, whereas for $N > 20$, $J = 20$ (Gostautas et al. 2005, Degala et al. 2009, Rizzo et al. 2010).

When the parameter analysis is not sufficient in providing insight on the health of the structure, the time waveforms are postprocessed using a wide variety of algorithms that overarch across disciplines, such as digital signal processing and statistical pattern recognition. The use of "b-value analysis" (Colombo et al. 2003, Schumacher et al. 2011), improved b-value (Shiotani et al. 1994, Aggelis et al. 2010), or wavelet transforms (Mizutani et al. 2000, Qi 2000) belongs to the category of advanced processing.

9.4 Source Localization

There are several different methods to localize AE sources. Those employing the triangulation technique are suitable for locating AE source in a planar structure without structural discontinuities (Banerjee et al. 2005, Mal et al. 2005, Prosser et al. 1995). Similarly, in bulk geometries made of isotropic materials, the differences among the time of arrival of the elastic waves at multiple sensors are considered by exploiting the propagation of p-waves. Let us assume that a bulk isotropic structure is monitored by using n AE transducers. The time of arrival $t_{s,i}$ of the p-wave at sensor (s) can be written as

$$t_{s,i} = \frac{|\mathbf{r}_{s,i} - \mathbf{r}|}{v_p} + T = \frac{\sqrt{(x_{s,i} - x)^2 + (y_{s,i} - y)^2 + (z_{s,i} - z)^2}}{C_L} + T \quad (9.5)$$

where $\mathbf{r}_{s,i} = (x_{s,i}, y_{s,i}, z_{s,i})$ is the position vectors of sensor ($i = 1,...,n$), $\mathbf{r} = (x, y, z)$ is the position vector of the AE source, and C_L is the velocity of the longitudinal bulk wave. Clearly, Equation 9.5 assumes that the wave velocity is independent of the direction of propagation and that the wave is not dispersive. The parameter T identifies the instant at which the release of transient energy occurred (Degala et al. 2009). When extended to complex geometries, to waveguides, or to anisotropic materials, the multimode nature of the waves makes the localization of the AE source approximate. In plate-like structures, one problem with time-of-flight triangulation is the dispersive nature of the flexural mode, which makes it difficult to identify an accurate arrival time from, for example, the conventional first-threshold crossing (Matt and Lanza di Scalea 2007).

To avoid the approximation on the localization of AE events in composite materials, Gangadharan et al. (2009) proposed a geodesic approach using Voronoi construction for AE source localization in a composite structure. The approach assumes that the wave takes a minimum energy path to travel from the source to any other point. With imposed material limitations, this path is generated by geodesics. Once the geodesic paths are extracted in a given geometry, the defect location is reached by back-propagating the waves along those paths from the sensor locations. This geodesic approach was validated on composite plate specimens of simple and complex geometry.

As the estimation of the time-of-flight and triangulation at multiple receiving points can be challenging, methods based on the signal amplitudes rather than time of arrival can be very attractive, as they avoid the detrimental limitations associated with the multimodal behavior of guided waves. The use of piezoelectric transducer rosettes goes toward this direction. Matt and Lanza di Scalea (2007) proposed the use of rosettes each composed of rectangular macrofiber composite (MFC) transducers to develop a method based on signal amplitudes rather than time of arrival. These transducers exhibit a highly directional response to stress and strain, and therefore exhibit high directivity to ultrasonic-guided wave propagation. The MFC response to the propagation of antisymmetric modes in plate-like structures was decomposed into axial and transverse sensitivity factors, which allow extraction of the direction of an incoming wave. Therefore, the wave source location in a plane can be determined by intersecting the wave directions detected by two rosettes. The source localization procedure proposed by Matt and Lanza di Scalea is based on the computation of the wave strain principal angle, and it does not require knowledge of the wave speed in the medium. The performance of the rosettes for source location was validated through pencil-lead breaks performed on an aluminum plate, an anisotropic CFRP laminate, and a complex CFRP–honeycomb sandwich panel (Figure 9.3). In plate-like structures, the rosette concept can also, in principle, be applied to locating sources of symmetric modes by looking at the symmetric mode response spectra derived for monolithic zirconate titanate sensors (Lanza di Scalea et al. 2007), where the desired directivity behavior is apparent.

While triangulation can be used to localize the position of the source, the simplified Green's function for moment tensor analysis (SiGMA), proposed by Ohtsu (1995), can be applied to determine the orientation, direction, and volume of cracks (Ohtsu 1995, Ohno and Ohtsu 2010). The method, based on the processing of bulk waves, applies to volumetric structures and serves to identify the type of failure that has determined the AE event. The following formulation is extracted from Degala et al. (2009). Consider a crack motion vector $\mathbf{b}(\mathbf{y}, t)$, at a point $\mathbf{y}$ and instant t, of a fracture surface Φ with normal vector s as depicted in Figure 9.2a. When $\mathbf{b}$ is parallel to $\mathbf{s}$, a tensile crack is propagating. Conversely, when $\mathbf{b}$ is orthogonal to $\mathbf{s}$, a shear crack is

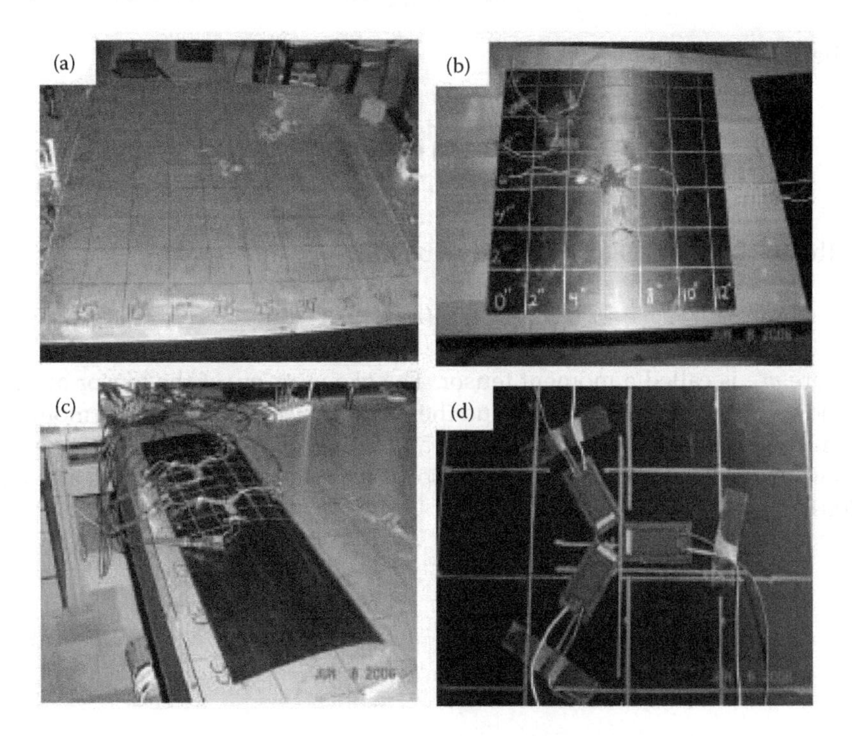

FIGURE 9.3
Specimens tested at the University of California, San Diego, to prove the feasibility of transducers' rosette for the localization of AE sources. (From Lanza di Scalea F. et al., *J. Acoust. Soc. Am.*, 121: 175–187, 2007. Reprinted with permission from IOP.)

generated. Because vector **b** is time dependent, its variation with time generates the elastic wave $\mathbf{u}_i(\mathbf{x}_i, t)$ detected by the AE transducers (Ohtsu 1995). If $\mathbf{b}(\mathbf{y}, t)$ is decomposed as

$$\mathbf{b}(\mathbf{y}, t) = b(\mathbf{y})\, \mathbf{l}\, S(t) \tag{9.6}$$

with **l** the unit vector of the crack motion, $b(\mathbf{y})$ the magnitude of the crack displacement at point **y**, and $S(t)$ the source time function of crack motion, the theoretical waveform at sensor *i* can be written as

$$\mathbf{u}_i(\mathbf{x}_i, t) = \int_{\Phi} G_{ip,q}(\mathbf{x}_i, \mathbf{y}, t) C_{pqkl} b(\mathbf{y}) l_k n_l * S(t) \mathrm{d}\Phi \tag{9.7}$$

where $G_{ip,q}$ is the spatial derivative of Green's function, C_{pqkl} are the elastic constants of the material, and the symbol * represents the convolution integral. Assuming

$$m_{pq} = C_{pqkl} l_k n_l \Delta V \tag{9.8}$$

where

$$\Delta V = \int_{\Phi} b(\mathbf{y}) \mathrm{d}\Phi \tag{9.9}$$

is the crack volume, Equation 9.7 can be rewritten as

$$\mathbf{u}_i(\mathbf{x}_i, t) = G_{ip,q}(\mathbf{x}_i, \mathbf{y}, t) m_{pq} * S(t) \tag{9.10}$$

where m_{pq} is called a moment tensor. The elements m_{pq} of the tensor are the product of a volume by a stress, and therefore, they have the unit of a moment.

In the SiGMA approach, only the amplitude of the first cycle of the AE waveform is considered. Thus, Equation 9.10 is simplified as (Chang and Lee 2004)

$$A(\mathbf{x}) = \frac{C_s \mathrm{Ref}(\mathbf{t},\mathbf{r})}{\mathrm{R}} \begin{pmatrix} r_1 & r_2 & r_3 \end{pmatrix} \begin{pmatrix} m_{11} & m_{12} & m_{13} \\ m_{21} & m_{22} & m_{23} \\ m_{31} & m_{32} & m_{33} \end{pmatrix} \begin{pmatrix} r_1 \\ r_2 \\ r_3 \end{pmatrix} \tag{9.11}$$

where $A(\mathbf{x})$ is the amplitude of the first motion, C_s is the calibration coefficient of the sensor, and Ref(**t**, **r**) is the reflection coefficient between vectors **t** and **r**. As shown in Figure 9.2a, vector **t** is the direction of AE transducer sensitivity, whereas vector **r** represents the unit vector that identifies the wave propagation direction from the source to the sensor, separated by the distance R. Vector **t** and the waveform amplitudes recorded at each sensor are known. If the AE transducers used for the localization are identical, i.e., they possess the same sensitivity, then the coefficient C_s can be neglected. The moment tensor m_{pq} is symmetric, and only six elements are independent. The distance R and the vector **r** are determined by localizing the position of the source. Therefore, Equation 9.11 has six unknowns and at least six sensors are needed to calculate the elements m_{pq}. Once the elements m_{pq} are determined, the eigenvalue analysis of the moment tensor is performed to determine the crack type, orientation, and direction of the crack. From the tensor eigenvalues λ_1, λ_2, and λ_3, with $\lambda_1 > \lambda_2 > \lambda_3$, the values of the shear ratio X, deviatoric tensile ratio Y, and isotropic tensile ratio Z can be calculated by solving the following system:

$$\begin{cases} \dfrac{\lambda_1}{\lambda_1} = 1 = X + Y + Z \\ \dfrac{\lambda_2}{\lambda_1} = 0 - 0.5Y + Z \\ \dfrac{\lambda_3}{\lambda_1} = -X - 0.5Y + Z \end{cases} \tag{9.12}$$

When $X > 60\%$, the crack is referred to as a shear crack; when $X < 40\%$ and contemporary $Y + Z > 60\%$, the sources are referred to as tensile cracks; finally, if $40\% < X < 60\%$, the source is considered a mixed crack.

The computation of the eigenvectors $\mathbf{e}_1$, $\mathbf{e}_2$, and $\mathbf{e}_3$ is obtained by means of the following system:

$$\begin{cases} \mathbf{e}_1 = \mathbf{l} + \mathbf{s} \\ \mathbf{e}_2 = \mathbf{l} \times \mathbf{s} \\ \mathbf{e}_3 = \mathbf{l} - \mathbf{s} \end{cases} \tag{9.13}$$

and it yields to the determination of the unit crack motion vector and the unit crack normal vector.

9.5 Acoustic Emission for Composites: Case Studies

The use of composite materials in civil structures is relatively recent, and the application of AE testing methods on large structures began in the late 1990s. The use of AE on large carbon fiber–reinforced polymer (CFRP) stay cables in the form of strands and parallel wires was demonstrated for the first time by Lanza di Scalea et al. (2000) and Rizzo and Lanza di Scalea (2001) using parameter analysis. The cables, ranging from 5500 to 5870 mm, were part of the design of the I-5 Gilman Drive Bridge, a composite bridge planned to be built in San Diego, California. The study included the characterization of acoustic attenuation and dispersion phenomena that are relevant to AE testing of large-scale CFRP cables. Despite their large size, CFRP cables are excellent acoustic waveguides exhibiting very low attenuation and therefore suitable to be monitored *in situ* via AE. Three kinds of cables were tested. The fibers were all oriented along the cable axial direction. The first type was a PAN/modified epoxy twisted cable with a 0.64 fiber volume fraction. It consisted of seven, 15.2-mm-diameter, strands each made of seven twisted wires with a 5-mm diameter. The second type was a PAN/epoxy cable with a 0.60 fiber volume fraction and it consisted of 31, 5-mm-diameter, individual wires and was tested at a length of 5500 mm. Finally, a PAN/epoxy with a 0.60 fiber volume fraction made of 28, 5-mm-diameter, individual wires was tested, at a length of 5870 mm.

Shorter glass fiber–reinforced polymer (GFRP) and CFRP strands were monitored by Li et al. (2011) using AE signal features and Higuchi's fractal dimension (FD). Each test cable consisted of seven GFRP or CFRP strands; the diameter of each FRP strand was 7 mm and the length between the two anchorages was 800 mm. The cables were covered with polyethylene sheaths. The tensile strength, elastic modulus, and ultimate strain of the GFRP strands were 1000 MPa, 125 GPa, and 2.31%, respectively, and for the CFRP strands, they were 2200 MPa, 200 GPa, and 2.42%, respectively. The analytical results indicated that the FD was associated with the frequency response. The technique was proven by analyzing the time history responses and frequency responses of the AE signals and the corresponding damage modes through fatigue testing. The FD value depends on the frequency distribution and is larger if the signal contains more high-frequency components. As the FD-based damage index obtained from AE signals increased with the number of loading cycles, it could be used to provide advanced warning incipient failures (Li et al. 2011).

Gostautas et al. (2005) tested six full-scale glass FRP bridge deck panels with nominal cross-sectional depths varying from 152 to 800 mm and tested under static loading to failure. The panels were manufactured using a hand wet lay-up process, and each panel consisted of two outer-face panels and a honeycomb core. The constituent materials of the panels were E-glass fiber in the form of mats and a polyester/vinyl resin. The E-glass fiber mats were fabricated using four types of architecture. The face panels were a combination of bidirectional (0°/90°) stitched fabric in an orthogonal direction with a balanced number of CM3205 fibers and nine layers of 0° unidirectional UM1810 fiberglass designation. In the experimental testing, the panels were subsequently repaired to evaluate repair technique performance and to allow a direct AE comparison of both the original and repaired systems. The objective of the work was to characterize damage, e.g., fiber breakage, matrix cracking, and delamination. The conventional parameter analysis together with certain AE signature characteristics, such as the Kaiser effect, Felicity ratio, and the IA, were considered.

Another application of AE is the evaluation of the effects of mechanical impacts. Wang et al. (2011) studied experimentally low-velocity impact on plates made of a carbon fabric–reinforced silicon carbide (C/SiC) ceramic matrix composites. The C/SiC composite specimens were affected at different energy levels ranging from 1 to 9 J, and the AE technique was used to detect the damage process. They used a parameter analysis based on the computation of hits and the emissions' amplitude. The study determined that when the impact load reaches its peak, delamination and fiber fracture start to occur. In the unloading stage, the damage modes are mainly matrix cracking and delamination. By measuring the number of AE events, it was found that at the lower impact energy (<3 J) matrix cracking is the main damage mode. Between 3 J and 6 J impact energy, delamination can also occur and become the main damage modes. At higher impact energy, fiber fracture is the main damage mode.

In recent years, carbon nanotubes (CNT) have been used to improve the toughness of polymers. The presence of 0.25 wt.% of CNTs added to the epoxy matrix can reduce the crack growth rate under fatigue loading with a factor of 20 (Zhang et al. 2009). The effect of CNTs on the damage development in woven carbon fiber/epoxy composites can be evaluated by using AE in conventional quasi-static tension tests. It was found that when compared with non-nano-modified composites, the addition of CNT increases the amount of AE activities, both in terms of the number of events and energy level, but lowers crack density (De Greef et al. 2011).

9.6 Acoustic Emission and Advanced Signal Processing

The diffusion of advanced signal processing methods and the availability of inexpensive computation tools have sparked interest on coupling raw AE data, i.e., the time waveforms detected by the AE sensors, to signal processing methods.

Lu et al. (2011) applied the Hilbert–Huang transform to the AE signals recorded during conventional tensile testing of fiber-reinforced composites. The results confirmed what has been known since the 1980s, i.e., that the frequency range of matrix crack AE signals is below 200 kHz, the frequency range of fiber breakage is above 200 kHz, and the frequency associated with delaminations phenomena is distributed over a wide range.

One way to correlate the mechanical behavior of a given structure and the AE events generated during its operation is by means of the sentry function.

The sentry function $f(x)$ is defined as the logarithm of the ratio between the strain energy and the cumulative acoustic energy as

$$f(x) = \ln \frac{E_s(x)}{E_a(x)} \tag{9.14}$$

where $E_s(x)$ is the strain energy, $E_a(x)$ is the AE event energy, and x is the displacement. The strain energy is determined by using a conventional load–displacement diagram, and the cumulative AE energy is considered as the summation of the AE event energy (Fotouhi et al. 2011). The function $f(x)$ is defined over displacement domain in which the acoustic energy is nonzero. This function identifies five different behaviors that are schematized in Figure 9.4:

- Bottom-up (BU) trend that is indicative of an instantaneous energy-storing capability in the material induced by strengthening event
- Decreasing trend P1 related to the fact that the material strain energy-storing capability is lower than the AE activity
- Constant behavior P2 associated with the progressive strain energy-storing phases due to material damage progression
- Increasing trend P3 corresponding to the strain energy-storing phases
- Sudden drop due to immediate release of stored energy caused by internal material failure (Oskouei and Ahmadi 2010, Oskouei et al. 2011, Fotouhi et al. 2011)

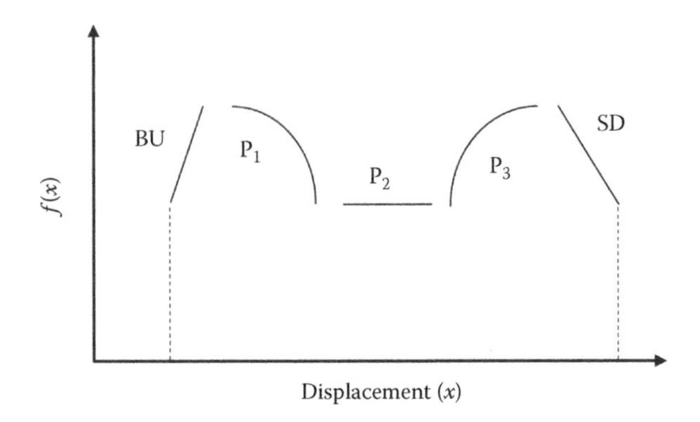

FIGURE 9.4
Behaviors identified by the sentry function. (Adapted from Fotouhi, M., Pashmforoush, F., Ahmadi, M. and Refahi Oskouei, A., *J. Reinf. Plast. Comp.* 30, 1481, 2011.)

Fotouhi et al. (2011) used the sentry function to determine the critical force at which the onset of delamination occurred when drilling glass/epoxy composite material. The function was used to study the initiation and growth of the delamination process on two different layups, woven $[0, 90]_s$ and unidirectional $[0]_s$, subjected to a three-point bending test.

To identify the different damage mechanisms in composites that may occur at the microscopic level, Sause et al. (2012) proposed a clustering technique to discriminate matrix cracking, interface failure, and fiber breakage. The technique is based on frequency-based features, including but not limited to average, reverberation, and initiation frequencies. The approach considered all possible combinations of these features and investigated the classification performance of the *k*-means algorithm. In general, the *k*-means algorithm is a clustering method aimed at partitioning *n* input vectors into *k* clusters. Each input vector is allocated to the cluster with the nearest mean (Gutkin et al. 2011). In Sause et al. (2012), this unsupervised learning algorithm was applied to the experimental datasets obtained by testing specimens made of T800/913 prepreg system with a $[0°/90°/90°/90°/90°]_{sym}$ stacking sequence. The specimens with dimensions of 100 × 15 × 1.4 mm were loaded in a four-point bending setup according to DIN-EN-ISO 14125. Figure 9.5 shows some of the results of the pattern recognition algorithm. The figure shows the partial power, defined as the power of the signal in the 150–300 kHz bandwidth, as a function of the weighted peak frequency $\langle f_{peak} \rangle$ defined as

$$\langle f_{peak} \rangle = \sqrt{f_{peak} \cdot f_{centroid}} \tag{9.15}$$

where f_{peak} and $f_{centroid}$ are known AE parameters. The three clusters discriminate the matrix cracking, fiber breakage, and the interface failure created by

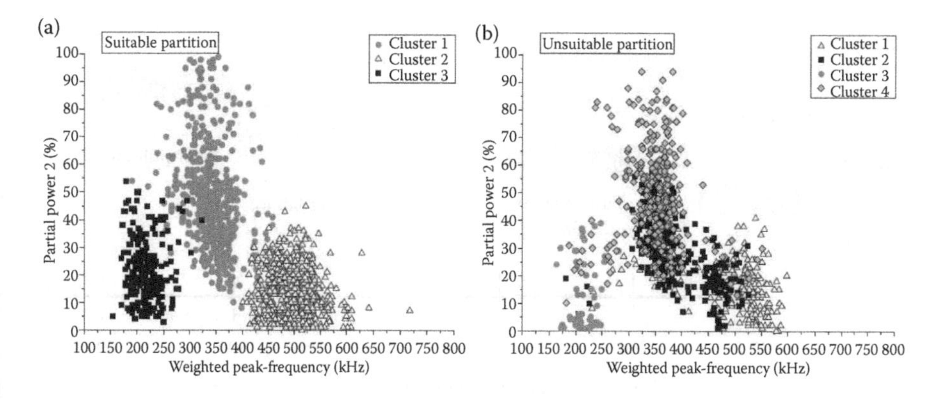

FIGURE 9.5
Clustering of acoustic emission data to discriminate among different damage mechanisms. (Reprinted from *Pattern Recogn. Lett.*, 33, Sause, M.G.R., Gribov, A., Unwin, A.R. and Horn, S., Pattern recognition approach to identify natural clusters of acoustic emission signals, 17–23, Copyright 2012, with permission from Elsevier.)

a four-point bending testing. The performance of three pattern algorithms, namely *k*-means, self-organizing map (SOM), and competitive neural network (CNN) showed that the SOM combined with *k*-means appeared as the most effective of the three algorithms (Gutkin et al. 2011). The SOM is an artificial neural network where neighboring neurons compete and develop through mutual interaction to recognize patterns in a given set of data. A SOM consists of neurons organized on a regular low-dimensional grid (Kohonen 1990, Suresh et al. 2004, Gutkin et al. 2011). The most similar neuron to an input vector is modified so that it becomes even more similar. However, not only the most similar neuron is modified, but also its neighbors on the map are moved toward the input vector (Gutkin et al. 2011). CNNs are unsupervised artificial neural networks similar to the SOM, with the exceptions that only the winning neuron is updated, and updating only takes place in the input space as there is no projection on low-dimensional space (Rumelhart 1985). In the study conducted by Gutkin et al. (2011), the AE signals were collected from various test configurations: tension, compact tension, compact compression, double cantilever, and four-point bend end-notched flexure. The specimens consisted of a 0.25 mm thick high-performance unidirectional carbon/epoxy prepreg (IM7/8552). The outcome of the study is summarized in Figure 9.6 where the frequency regions associated with matrix cracking, delamination, debonding, fiber break, and fiber pullout are shown. The results agreed with past works (de Groot et al. 1995, Ramirez-Jimenez et al. 2004) and are similar to the findings of Fotouhi et al. (2012), which applied wavelet packet transform and fuzzy C-means clustering to detect the different fracture mechanisms during delamination in quasi-static three-point bending of woven $[0, 90]_s$ and unidirectional $[0]_s$ glass/epoxy composites.

Fuzzy clustering methods allow the objects to belong to some other clusters concurrently, with different degrees of membership between 0 and 1

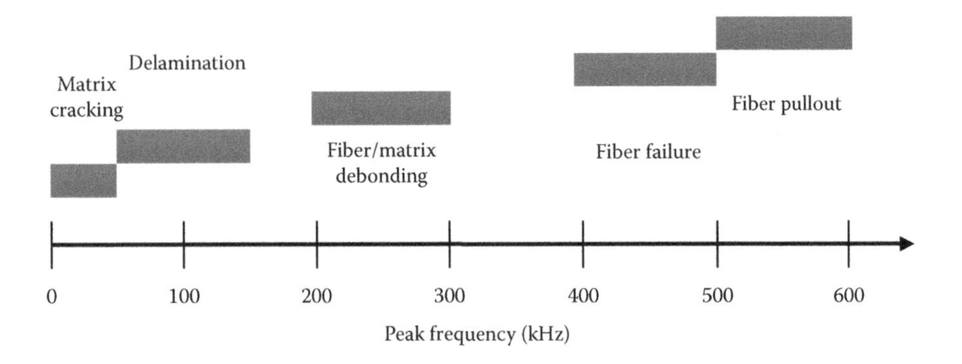

FIGURE 9.6
Frequency bandwidths of different types of damage in composites. (Adapted from Gutkin, R., Green, C.J., Vangrattanachai, S., Pinho, S.T., Robinson, P. and Curtis P.T., *Mech. Syst. Signal Process* 25, 1393, 2011.)

(Fotouhi et al. 2012, Theodorids and Koutroumbas 1999, Marec et al. 2008, Yuan and Li 1998). The data are obtained through the observations of various physical phenomena. Each observation is composed of n measured variables, separated into an n-dimensional column vector. The findings of Fotouhi et al. (2012) were corroborated using scanning electron microscopy to determine the different fracture mechanisms.

Other clustering techniques used to classify AE signals include artificial neural networks (Philippidis et al. 1998, Yan et al. 1999, Godin et al. 2004, de Oliveira and Marques 2008, Huguet et al. 2002) and genetic programs (Suresh et al. 2004).

Fatigue testing is often an integral part of the certification program for composites, and it may consist of subjecting the structure to compression, tension, shear, and torsion stresses simultaneously. The fatigue failure of composite structures is more complicated than that of metallic structures. The cyclical S–N curves (where S is the cyclical stress and N is the cycles to failure) are only applicable to specific laminate sequences, shapes, and loading conditions. Small changes in these or other properties, such as the ply thickness and fiber orientation, will result in different S–N curves (Knops and Bogle 2006, Unnthorsson et al. 2008).

In fatigue testing, AE signals can be roughly divided into bursts, continuous, and mixed. Bursts are transient signals generated by the formation of damage, e.g., fiber breaking and delamination. Continuous signals can be generated by several sources, including noise and rubbing, and consist of many overlapping transients that cannot be separated. Finally, mixed-type signals contain both bursts and continuous signals, and it is the type that is normally encountered (Unnthorsson et al. 2008). The concurrent application of different loading types makes the AE monitoring of composites subjected to fatigue testing complicated. AE signals may contain a high number of overlapping AE transients that can be emitted from many AE sources in the material. If the sources are emitted simultaneously, the dispersive nature of the guided waves, which result in an anisotropic speed of propagation and attenuation, will delay the time of arrival of the corresponding signals. Vice versa, if the sources are emitted separately, they may arrive simultaneously. As such, important signals can be buried in secondary AE signals such as those generated by the friction and rubbing of crack surfaces (Mouritz 2003), for instance. During cyclic testing of composites, the AE signal is emitted from both actual damage progression and cumulated damage, i.e., friction. The majority of the emissions come from cumulated damage. The values of the AE signal features from the cumulated damage usually fall in the same range as the ones from damage growth (Unnthorsson et al. 2008).

Unnthorsson et al. (2008) presented a study aimed at determining an AE-based failure criterion for determining the failure of complex-shaped composites under multiaxial loading conditions. The AE-based failure criterion is equivalent to the stiffness-based criterion and it determines a failure when the value of a parameter being monitored exceeds a specified threshold, equivalent to a 10% displacement failure, i.e., a 10% change in displacement, with respect to initial value. The criterion was tested on 75 CFRP prosthetic feet called VariFlex© made using the same laminate sequences. They used parameter-based AE with adaptive threshold and short-time Fourier transform. The AE signals were split into four subsignals. It was found that shortly before or at the time when the 10% displacement criterion was met, the value of the AE features increased significantly in two to four of the subsignals.

References

Achenbach, J.D. 1984. *Wave Propagation in Elastic Solids*. New York: North-Holland.

Adams, D. 2007. *Health Monitoring of Structural Materials and Components: Methods with Applications*. Chichester, England: John Wiley & Sons, Ltd.

Aggelis, D.G., Shiotani, T. and Terazawa, M. 2010. Assessment of construction joint effect in full-scale concrete beams by acoustic emission activity. *J. Eng. Mech.* 136(7): 906–912.

Auld, B.A. 1990. *Acoustic Fields and Waves in Solids*, 2nd ed. Malabar, FL: R.E. Krieger Publ.

Awenbruch, J. and Ghaffari, S. 1988. Monitoring progression of matrix splitting during fatigue loading through acoustic emission in notched unidirectional graphite/epoxy composite. *J. Reinf. Plast. Comp.* 7: 245–264.

Banerjee S., Prosser, W. and Mal, A. 2005. Calculation of the response of a composite plate to localized dynamic surface loads using a new wave number integral method. *J. Appl. Mech.* 72: 18–24.

Beard, M.D. 2002. Guided wave inspection of embedded cylindrical structures. PhD Dissertation, Imperial College, London.

Berthelot, J.M. and Rhazi, J. 1990. Acoustic emission in carbon fiber composites. *Compos. Sci. Technol.* 37: 411–428.

Bohse, J. 2000. Acoustic emission characteristics of micro-failure processes in polymer lends and composites. *Compos. Sci. Technol.* 60: 1213–1226.

Byrne, C. and Green, R.E. Jr. 1994. Acoustic emission monitoring of thick composite laminates under compressive loads. *Nondestr. Charact. Mater.* 6: 191–198.

Chang, S.H. and Lee, C.I. 2004. Estimation of cracking and damage mechanisms in rock under triaxial compression by moment tensor analysis of acoustic emission. *Int. J. Rock Mech. Mining Sci.* 41: 1069–1086.

Colombo, S., Main, I.G. and Forde, M.C. 2003. Assessing damage of reinforced concrete beam using b-value analysis of acoustic emission signals. *ASCE J. Mat. Civ. Eng.* 15(3): 280–286.

Degala, S., Rizzo, P., Ramanathan, K. and Harries, K.A. 2009. Acoustic emission monitoring of CFRP reinforced concrete slabs. *Constr. Build. Mat.* 23(5): 2016–2026.

De Greef, N., Gorbatikh, L., Lomov, S.V. and Verpoest, I. 2011. Damage development in woven carbon fiber/epoxy composites modified with carbon nanotubes under tension in the bias direction. *Composites A* 42: 1635–1644.

de Groot, P.J., Wijnen, P.A.M. and Janssen, R.B.F. 1995. Real-time frequency determination of acoustic emission for different fracture mechanisms in carbon/epoxy composites. *Compos. Sci. Technol.* 55: 405–412.

de Oliveira, R. and Marques, A.T. 2008. Health monitoring of FRP using acoustic emission and artificial neural networks. *Comput. Struct.* 86: 367–373.

Eckles, W. and Awenbruch, J. 1988. Monitoring acoustic emission in cross-ply graphite/epoxy laminates during fatigue loading. *J. Reinf. Plast. Comp.* 7: 265–283.

Favre, J.P. and Laizet, J.C. 1989. Amplitude and counts per event analysis of the acoustic emission generated by the transverse cracking of cross-ply CFRP. *Compos. Sci. Technol.* 36: 27–43.

Fotouhi, M., Pashmforoush, F., Ahmadi, M. and Refahi Oskouei, A. 2011. Monitoring the initiation and growth of delamination in composite materials using acoustic emission under quasi-static three-point bending test. *J. Reinf. Plast. Comp.* 30: 1481–1493.

Fotouhi, M., Heidary, H., Ahmadi, M. and Pashmforoush, F. 2012. Characterization of composite materials damage under quasi-static three-point bending test using wavelet and fuzzy C-means clustering. *J. Compos. Mat.*, published online February 1 2012 doi: 10.1177/0021998311425968.

Gangadharan, R., Prasanna, G., Bhat, M.R., Murthy, C.R.L. and Gopalakrishnan S. 2009. Acoustic emission source location in composite structure by Voronoi construction using geodesic curve evolution. *J. Acoust. Soc. Am.* 126(5): 2324–2330.

Godin, N., Huguet, S., Gaertner, R. and Salmon, L. 2004. Clustering of acoustic emission signals collected during tensile tests on unidirectional glass/polyester composite using supervised and unsupervised classifiers. *NDT&E Int.* 37: 253–264.

Gostautas, R.S., Ramirez, G., Peterman, R.J. and Meggers, D. 2005. Acoustic emission monitoring and analysis of glass fiber-reinforced composites bridge decks. *ASCE J. Bridge Eng.* 10(6): 713–721.

Graff, K.F. 1991. *Wave Motion in Elastic Solids*. New York: Dover Publications, Inc.

Gutkin, R., Green, C.J., Vangrattanachai, S., Pinho, S.T., Robinson, P. and Curtis P.T. 2011. On acoustic emission for failure investigation in CFRP: Pattern recognition and peak frequency analyses. *Mech. Syst. Signal Process* 25: 1393–1407.

Hamstad, M.A. 1986. A review: Acoustic emission, a tool for composite-materials studies. *Exp. Mech.* 26: 7–13.

Hamstad, M.A. and Downs, K.S. 1995. On characterization and location of acoustic emission sources in real size composite structures—A waveform study. *J. Acoust. Emiss.* 13(1): 31–41.

Huguet, S., Godin, N., Gaertner, R., Salmon, L. and Villard, D. 2002. Use of acoustic emission to identify damage modes in glass fiber reinforced polyester. *Compos. Sci. Technol.* 62(10): 1433–1444.

Knops, M. and Bogle, C. 2006. Gradual failure in fibre/polymer laminates. *Compos. Sci. Technol.* 66(5): 616–625.

Kohonen, T. 1990. The self-organizing map. *Proc. IEEE* 78: 1464–1480.

Kolsky, H. 1963. *Stress Waves in Solids*. New York: Dover Publications Inc.

Komai, K., Minoshima, K. and Shibutani, T. 1991. Investigations of the fracture mechanism of carbon/epoxy composites by AE signal analyses. *JSME Int. J.* 34-I(3): 381–388.

Krautkrämer, J. and Krautkrämer, H. 1990. *Ultrasonic Testing of Materials*, 4th ed. Berlin: Springer-Verlag.

Lanza di Scalea, F., Rizzo, P., Karbhari, V.M. and Seible, F. 2000. Health monitoring of UCSD's I-5/Gilman advanced technology bridge. *Smart Mater. Bull.* 1(2): 6–10.

Lanza di Scalea, F., Matt, H.M. and Bartoli, I. 2007. Response of rectangular piezoelectric sensors to Rayleigh and Lamb ultrasonic waves. *J. Acoust. Soc. Am.* 121: 175–187.

Li, H., Huang, Y., Chen, W.L., Ma, M.L., Tao, D.W. and Ou, J.P. 2011. Estimation and warning of fatigue damage of FRP stay cables based on acoustic emission techniques and fractal theory. *Comput.-Aided Civ. Inf.* 26: 500–512.

Lu, C., Ding, P. and Chen, Z. 2011. AE signal analysis of carbon fiber reinforced composites based on Hilbert–Huang transform. *Adv. Mat. Res.* 301–303: 447–451.

Luo, J.-J., Wooh, S.-C. and Daniel, I.M. 1995. Acoustic emission study of failure mechanisms in ceramic matrix composite under longitudinal tensile loading. *J. Comp. Mat.* 29(15): 1946–1961.

Mal, A.K., Ricci, F., Banerjee, S. and Shih, F. 2005. A conceptual structural health monitoring system based on vibration and wave propagation. *Struct. Health Monit.* 4: 283–293.

Marec, A., Thomas, J.H. and Guerjouma, E. 2008. Damage characterization of polymer-based composite materials: Multivariable analysis and wavelet transform for clustering acoustic emission data. *Mech. Syst. Signal Process* 22: 1441–1464.

Matt, H.M. and Lanza di Scalea, F. 2007. Macro-fiber composite piezoelectric rosettes for acoustic source location in complex structures. *Smart Mater. Struct.* 16: 1489–1499.

Meitzler, A.H. 1961. Mode coupling occuring in the propagation of elastic pulses in wires. *J. Acoust. Soc. Am.* 33(4): 435–445.

Mizutani, Y., Nagashima, K., Takemoto, M. and Ono, K. 2000. Fracture mechanism characterization of cross-ply carbon fiber composites using acoustic emission analysis. *NDT&E Int.* 33: 101–110.

Mouritz, A.P. 2003. Non-destructive evaluation of damage accumulation. In: Harris, B., editor. *Fatigue in Composites*. Cambridge: Woodhead Publishing Ltd., 242–266.

Ohno, K. and Ohtsu, M. 2010. Crack classification in concrete based on acoustic emission. *Constr. Build. Mater.*—Special Issue on Fracture, Acoustic Emission and NDE in Concrete (KIFA-5), 24(12): 2339–2346.

Ohtsu, M. 1995. Acoustic emission theory for moment tensor analysis. *Res. Nondestr. Eval.* 6(1): 169–184.

Oskouei, A.R. and Ahmadi, M. 2010. Acoustic emission characteristics of mode I delamination in glass/polyester composites. *J. Compos. Mat.* 44: 793.

Oskouei, A.R., Zucchelli, A., Ahmadi, M. and Minak, G. 2011. An integrated approach based on acoustic emission and mechanical information to evaluate the delamination fracture toughness at mode I in composite laminate. *Mat. Design* 32: 1444–1455.

Pavlakovic, B.N. 1998. Leaky guided ultrasonic waves in NDT. PhD Dissertation, Imperial College, London.

Philippidis, T.P., Nikolaidis, V.N. and Anastassopoulos, A.A. 1998. Damage characterization of carbon/carbon laminates using neural network techniques on AE signals. *NDT&E Int.* 31: 329–340.

Pollock, A. 1996. Classical wave theory in practical AE testing. Progr. in Acoust. Emiss. 3, *Proc. of the 8th Int. Symp. of the Japanese Society for Non-Destructive Testing*, Tokyo, 708–721.

Prosser, W.H. 1996. Advanced AE techniques in composite materials research. *J. Acoust. Emiss.* 14(3): s1–s11.

Prosser, W.H. 2002. Acoustic emission, Chapter 6. In: Shull, P.J., editor. *Non-Destructive Evaluation. Theory, Techniques, and Applications*. New York: Marcel Dekker.

Prosser, W., Jackson, K.E., Kellas, S., Smith, B.T., McKeon, J. and Friedman, A. 1995. Advanced waveform based acoustic emission detection of matrix cracking in composites. *Mater. Eval.* 53: 1052–1058.

Qi, G. 2000. Wavelet-based AE characterization of composite materials. *NDT&E Int.* 33(3): 133–144.

Ramirez-Jimenez, C.R., Papadakis, N., Reynolds, N., Gan, T.H., Purnell, P. and Pharaoh, M. 2004. Identification of failure modes in glass/polypropylene composites by means of the primary frequency content of the acoustic emission event. *Compos. Sci. Technol.* 64: 1819–1827.

Ravishankar, S.R. and Murthy, C.R.L. 2000a. Characteristic of AE during drilling composites laminates. *NDT&E Int.* 33: 341–348.

Ravishankar, S.R. and Murthy, C.R.L. 2000b. Application of AE in drilling of composites laminates. *NDT&E Int.* 33: 429–435.

Rizzo, P. 2004. Health monitoring of tendons and stay cables for civil structures. PhD Dissertation, University of California, San Diego.

Rizzo, P. and Lanza di Scalea, F. 2001. Acoustic emission monitoring of carbon-fiber-reinforced-polymer bridge stay cables in large-scale testing. *Exp. Mech.* 41(3): 282–290.

Rizzo, P., Spada, A., Degala, S. and Giambanco, G. 2010. Acoustic emission monitoring of chemically bonded anchors. *J. Nondestruct. Eval.* 29: 49–61.

Rose, J.L. 1999. *Ultrasonic Waves in Solid Media.* Cambridge, UK: Cambridge University Press.

Rumelhart, D.E. and Zipser, D. 1985. Feature discovery by competitive learning. *Cogn. Sci.* 9: 75–112.

Sato, N. and Kurauchi, T. 1995. Interpretation of acoustic emission signal from composite materials and its application to design of automotive composite components. *Res. Nondestr. Eval.* 9: 119–136.

Sause, M.G.R., Gribov, A., Unwin, A.R. and Horn, S. 2012. Pattern recognition approach to identify natural clusters of acoustic emission signals. *Pattern Recogn. Lett.* 33: 17–23.

Schumacher, T., Higgins, C.C. and Lovejoy, S.C. 2011. Estimating operating load conditions on reinforced concrete highway bridges with b-value analysis from acoustic emission monitoring. *Int. J. Struct. Health Monit.* 10: 17–32.

Shiotani, T., Fujii, K., Aoki, T. and Amou, K. 1994. Evaluation of progressive failure using AE sources and improved b-value on slope model tests. *J. Acoust. Emiss.* VII(7): 529–534.

Spada, A., Rizzo, P. and Giambanco, G. 2011. Elastoplastic damaging model for adhesive anchor systems. Part II: Numerical and experimental validation. *ASCE J. Eng. Mech.* 137(12): 854–861.

Suresh, S., Omkar, S.N., Mani, V. and Menaka, C. 2004. Classification of acoustic emission signal using genetic programming. *Int. J. Aerospace Sci. Technol.* 56: 26–40.

Theodorids, S. and Koutroumbas, K. 1999. *Pattern Recognition.* Boston: Academic Press.

Tsamtsakis, D. and Wevers, M. 1999. Acoustic emission to model the fatigue behavior of quasi-isotropic carbon-epoxy laminate composites. *Insight* 41: 513–516.

Tsamtsakis, D., Wevers, M. and De Meester, P. 1998. Acoustic emission from CFRP laminates during fatigue loading. *J. Reinf. Plast. Comp.* 17: 1185–1201.

Unnthorsson, R., Runarsson, T.P. and Jonsson, M.T. 2008. Acoustic emission based fatigue failure criterion for CFRP. *Int. J. Fatigue* 30: 11–20.

Wang, J.S., Yao, L.J., Li, Z., Bin, L., Chen, L.D. and Tong, X.Y. 2011. Acoustic emission analysis on low velocity impact damage of 2D C/SiC composites. *Adv. Mat. Res.* 301–303: 1367–1371.

Yan, T., Holford, K., Carter, D. and Brandon, J. 1999. Classification of acoustic emission signatures using a self-organization neural network. *J. Acoust. Emiss.* 17: 49–59.

Yuan, X. and Li, Z. 1998. Tool wear monitoring with wavelet packet transform—Fuzzy clustering method. *Wear* 219: 145–154.

Zhang, W., Srivastava, I., Zhu, Y.F., Picu, C.R. and Koratkar, N. 2009. Heterogeneity in epoxy nanocomposites initiates crazing: Significant improvements in fatigue resistance and toughening. *Small* 5: 1403–1407.

10

Neural Network Nondestructive Evaluation of Composite Structures from Acoustic Emission Data

Eric v. K. Hill, Michele D. Dorfman, and John A. Capriolo II

CONTENTS

10.1 Background

Acoustic emission (AE) is a nondestructive testing method based on the emission of acoustic waves by materials undergoing deformation. AE is defined as the elastic waves generated by the rapid release of energy from sources such as failure mechanisms within a material. This test method is employed in various industries because of its ability to provide valuable insight into how a structure would behave under loading.

An AE testing instrumentation system consists of one or more transducers, preamplifiers, filters, a data acquisition system, and a display, as depicted in Figure 10.1. The mechanical stresses from the AE stress waves are converted into electrical voltage signals by the piezoelectric transducer. These voltage signals are then sent to the data acquisition system to obtain graphical outputs of the various signal parameters for analysis.

Piezoelectric transducers are the most commonly used transducers in AE testing. They operate through the principles of piezoelectricity in which a small crystal inside the transducer is deformed. This deformation causes the crystal to generate a small voltage that is passed through the filters and amplifiers then measured by the data acquisition system. The transducers are attached to the test specimen using hot melt glue or a coupling gel. These transducers are oftentimes equipped with built-in preamplifiers to reduce noise interferences, as shown in Figure 10.2. Typical preamplifiers have a gain of either 40 or 60 dB. The 150-kHz resonant frequency sensor is the most popular transducer used in AE testing because of its high sensitivity. Here the most common frequency range for conducting tests in field applications is 100–300 kHz. However, broadband sensors are oftentimes used in laboratories for a wider range of study, their drawback being that they are less sensitive, which is typically not a problem in composite structures because of the large energies associated with their failure mechanisms. During a test, a specific frequency range is usually chosen to eliminate low-frequency mechanical noises and high-frequency disturbances such as electromagnetic interference.

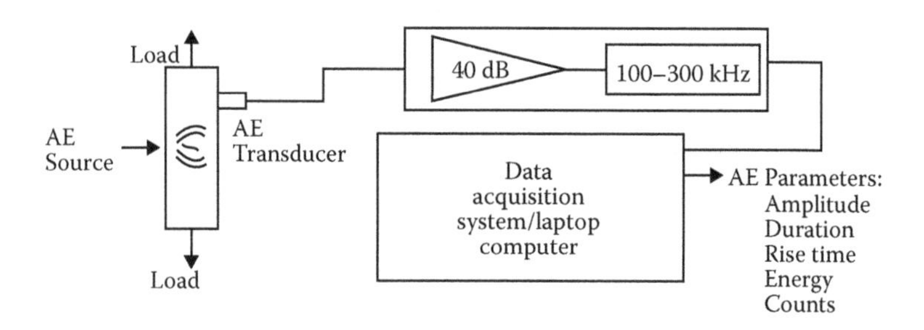

FIGURE 10.1
Basic acoustic emission test system setup.

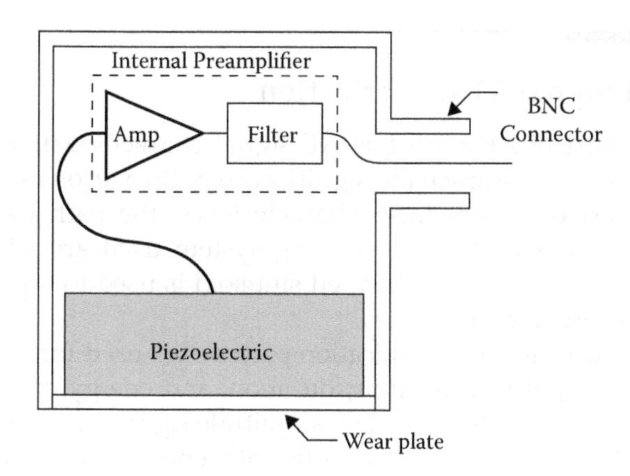

FIGURE 10.2
Piezoelectric transducer with integral preamplifier to eliminate noise.

The AE created by a source is an expanding spherical wave packet with an intensity diminishing in the near field at a rate proportional to the inverse of the radius of distance travelled squared. Some common sources of AE are categorized into four parts and are listed in Table 10.1.

From Table 10.1, there appear to be many pseudo sources of AE. These are created from the high sensitivity of the transducers, which is one of the greatest advantages, but also one of the greatest disadvantages, of AE testing. Because of the high sensitivity, the transducers are able to pick up elastic waves generated by movements on the order of a billionth of an inch (25.4×10^{-12} m). This allows for analysis of very small discontinuities in a structure. However, also because of this high sensitivity, there is a need for filtering systems to eliminate unwanted signals. Often, data produced from an AE test will contain noise from background sources, such as rubbing of grips, electromagnetic interference, pump noise in hydraulic equipment, etc.; thus, it is important to be able to set a threshold and filters to eliminate or at least minimize these noise sources in the data set.

TABLE 10.1
Different Sources of Acoustic Waves

AE Sources	Examples
Unflawed metals	Dislocation movement, dislocation multiplication, slip, twinning, inclusion fracture, inclusion debonding
Pseudo sources	Crack closure, frictional rubbing, liquid and gas leakage, loose particles and loose parts, cavitation, phase transformations, boiling, freezing, melting, electric discharge
Subcritical crack growth	Crack front movement, plastic zone growth
Composite materials	Transverse matrix cracking, matrix splitting (fiber–matrix debonding), delamination, fiber pullout, fiber fracture

10.2 Digital Signal Characterization

An important part of AE testing is AE signal characterization. The objective is to identify and estimate the significance of the various sources of AE. Source characteristics, transducer characteristics, the path travelled from transducer to source, and the measuring system used are all factors that affect the output waveform. A damped sinusoid is used to represent an AE burst for characterization purposes.

The main AE waveform quantification parameters used for signal analysis include counts, amplitude, rise time, duration, and energy counts. To begin with, a threshold is set to filter out low amplitude signals associated with system noise. For those signals that are sufficiently energetic to cross the threshold, the number of times the signal crosses the detection threshold is known as the counts. Amplitude is the largest voltage peak in the signal waveform measured in either volts (V) or decibels (dB). A logarithmic amplifier is used when measuring the peak amplitude since AE signal sources vary between 1 μV for transverse matrix cracking signals to signals in excess of 10 V for fiber bundle breaks that occur during final failure of the structure. Using a logarithmic scale allows the amplitudes of the large dynamic range of AE signals to all be plotted on the same scale. Here the amplifier gain is given by the equation

$$\Delta\text{dB} = 20\log\left(\frac{V_{\text{out}}}{V_{\text{ref}}}\right)$$

where the gain is measured in decibels (dB), V_{out} is the output voltage, and V_{ref} is the reference voltage. The latter is commonly set at 1 μV at the transducer input, this representing the smallest signal that can be detected by the AE analyzer above the system noise.

AE amplitudes are typically plotted between 0 and 100 dB, while common threshold settings for composite materials are between 50 and 60 dB since their failure mechanisms are relatively loud. An amplitude distribution histogram can be graphed by plotting AE hits versus amplitude. This differential AE amplitude distribution data, which includes all the failure mechanisms leading up to failure, has proven to be an extremely useful input when using neural networks to predict structural integrity measures such as ultimate load.

The signal rise time is the time in microseconds it takes from the first threshold crossing to when the peak amplitude is reached. The duration is the time in microseconds measured from the signal's first crossing of the threshold to when the signal last crosses the threshold. Energy, also known as the measured area under the rectified signal envelope (MARSE), is quantitatively represented by the area under the signal curve and is expressed in energy counts. Figure 10.3 provides visual definitions of these five parameters.

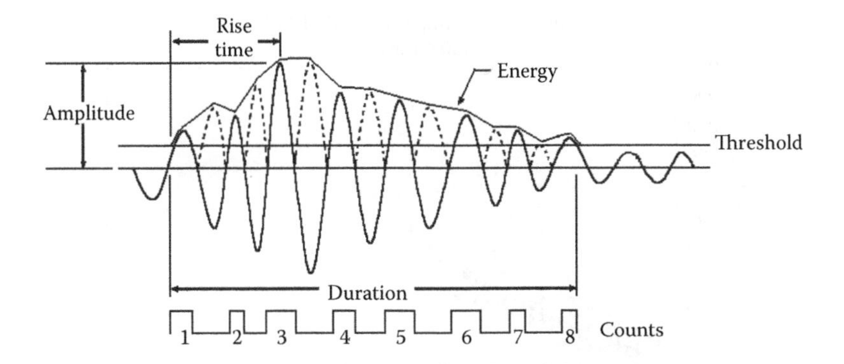

FIGURE 10.3
AE waveform quantification parameters.

Generally, AE data are captured by the processing system, and subsequently, various waveform parameters are plotted to make qualitative assessments concerning the structural integrity. Figures 10.4 and 10.5 are two examples of such plots.

Analysis of these data plots allows for the identification of the failure mechanisms involved from Figure 10.5, as well as some quantitative predictions to be made concerning their structural integrity from Figure 10.4. An example of ultimate strength prediction from low proof load AE data will be given later in the chapter.

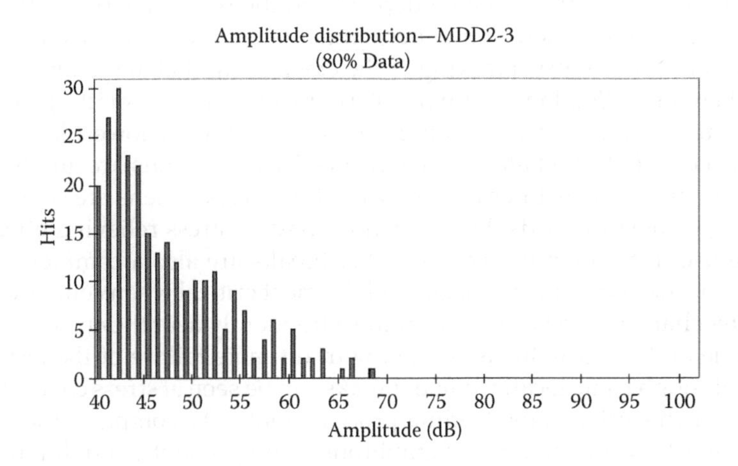

FIGURE 10.4
Amplitude (hits vs. amplitude) distribution or histogram.

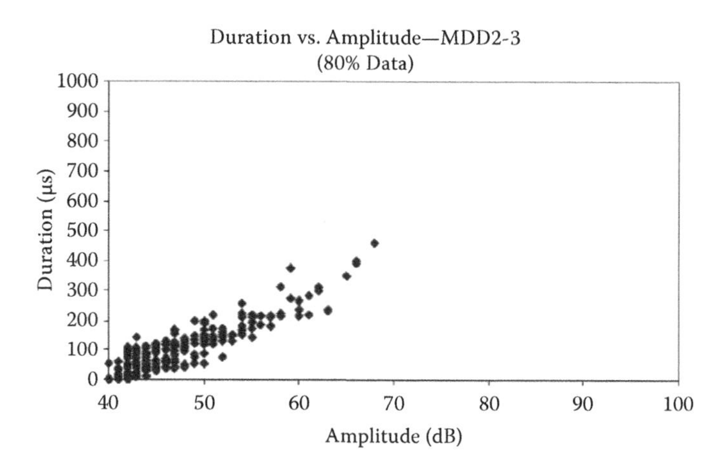

FIGURE 10.5
Duration versus amplitude plot.

10.3 Failure Mechanisms in Composites

Composite materials encounter three major failure mechanisms. These are matrix cracking, delaminations, and fiber breaks. To characterize the type of failure mechanism, the magnitude of the amplitude, duration, counts, rise time, and energy of each AE signal (hit) have to be analyzed.

Matrix cracking can be observed in two types: transverse and longitudinal. The type of matrix cracking depends on the relation between the crack and the fiber orientation. If it is perpendicular to the fiber orientation, it is called transverse matrix cracking. If the crack is parallel to the fiber orientation, then it is called longitudinal matrix cracking matrix splitting, or fiber/matrix debonding. Matrix cracking usually occurs at the lowest loads and is the weakest of all the failure mechanisms. Delaminations are another type of failure mechanism and can be observed when specimens are subjected to bending or flexural loads. Delaminations have a stress relieving effect that can be useful for pressure vessels. Fiber breaks are also a common type of failure mechanism and are considered the most critical when compared with other mechanism types. They occur most frequently near the end of the loading cycle and are usually associated with ultimate failure of the test specimen, although some localized fiber breaks can be seen at stress concentration points much earlier in the loading process. Table 10.2 compares these three failure mechanisms in terms of amplitude, energy, counts, and duration.

To characterize the failure mechanism data obtained during AE testing and to predict failure loads or burst pressures, neural networks are usually necessary to analyze the typically noisy data sets involved.

TABLE 10.2

AE Parameters and Associated Failure Mechanisms in Fiberglass/Epoxy Pressure Vessels

AE Parameter	Transverse Matrix Cracking	Longitudinal Matrix Cracking	Delaminations	Fiber Breaks
Amplitude	Low	Medium	High	Low–medium
Energy	Low	Medium	High	Very high
Counts	Low	High	High	Medium–high
Duration	Short	Long	Long	Short–medium

Source: E.v.K. Hill, *Materials Evaluation*, 50, 1439–1445, 1992.

10.4 Neural Networks

An artificial neural network is a mathematical modeling and information processing tool with performance characteristics similar to those of a biological neural network. A neural network consists of a network of massively parallel interconnected processing elements (PEs) or neurons. A typical PE is shown in Figure 10.6.

Each PE receives a number of input signals that may or may not generate an output signal based on the given inputs. Each input has a relative weight associated with it such that the effective input to the PE is a summation of the inputs multiplied by their associated weights. This value is then modified by a transfer or activation function (Figure 10.7) and passed directly to the output path of the PE. These outputs can either be excitatory or inhibitory. An excitatory output will cause the PE to fire; an inhibitory output will keep the PE from firing. This output signal can then be interconnected to the input paths of other PEs.

Processing elements are typically organized into groups called layers. In general, a network will consist of an input layer, one or more hidden layers, and an output layer. Data is presented to the network at the input layer,

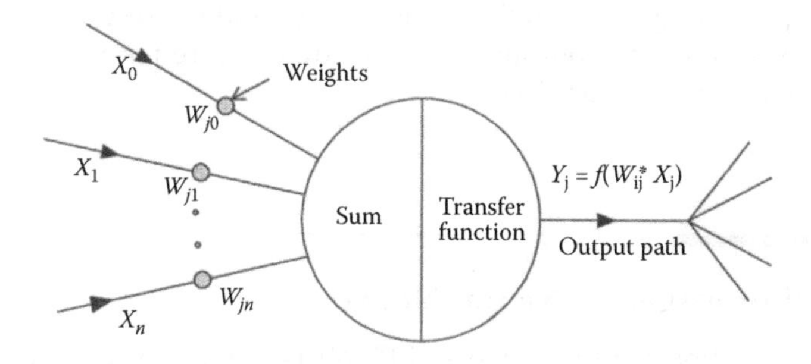

FIGURE 10.6
Processing element (neuron).

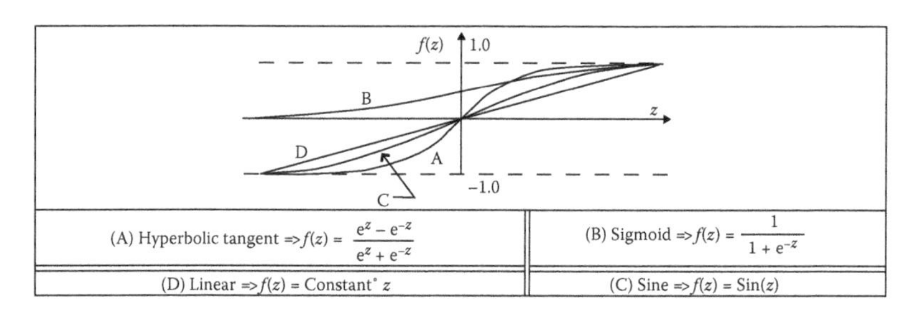

FIGURE 10.7
Transfer or activation functions. (From J.L. Walker II and E.v.K. Hill, "An introduction to neural networks: A tutorial," *First International Conference on Nonlinear Problems in Aviation & Aerospace*, Embry-Riddle Aeronautical University Press, Daytona Beach, FL, 1997, pp. 667–672.)

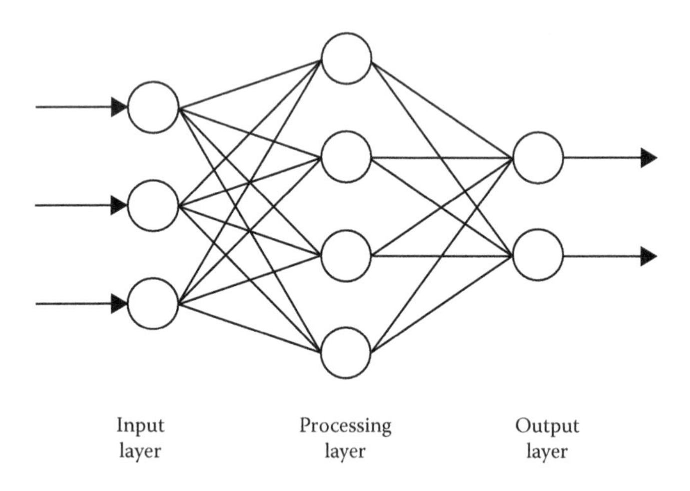

FIGURE 10.8
Generic neural network architecture.

processing is accomplished in the hidden layers, and the response of the network is presented at the output layer. The architecture for a generic neural network is shown in Figure 10.8.

10.5 Backpropagation Neural Networks

A backpropagation neural network (BPNN) is a multilayered, supervised, feedforward network, as shown in Figure 10.9. Supervised learning means that the network is given defined information such as input data and target

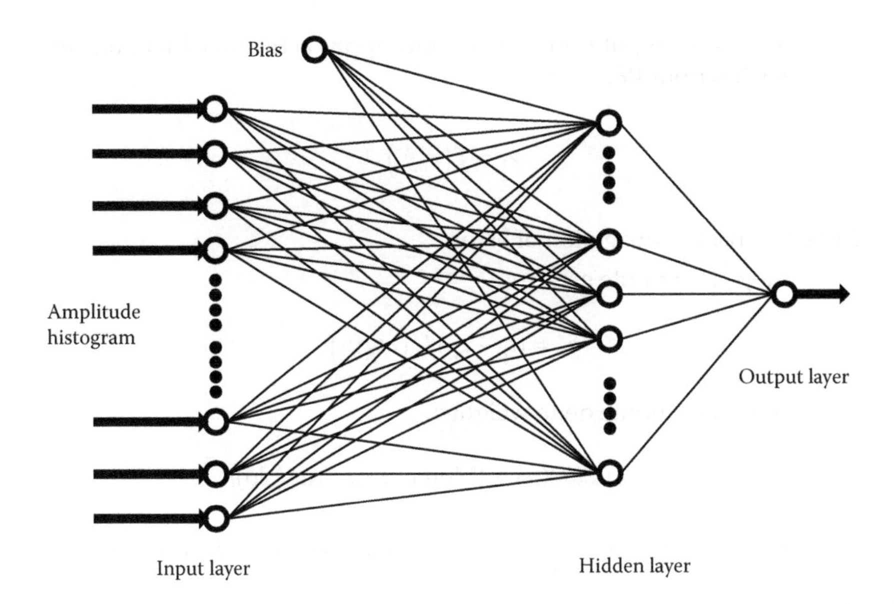

FIGURE 10.9
Backpropagation neural network. (From J.L. Walker II and E.v.K. Hill, "An introduction to neural networks: A tutorial," *First International Conference on Nonlinear Problems in Aviation & Aerospace*, Embry-Riddle Aeronautical University Press, Daytona Beach, FL, 1997, pp. 667–672.)

output data. A feedforward network means that the input follows through the network in a forward direction with no connections to previous layers.

A BPNN learns the relationship between the given input and the target output vector by minimizing the difference between the target and actual output vectors. The learning process consists of two stages. In the first stage, the input vectors are fed through the network to generate a response vector. In the second stage, the output error is computed for each input response based on the target output values. The overall network error is then reduced by back propagating the error through the network weights to make updates. This process is repeated through many iterations until the output error reaches a specified level.

The algorithm for a simple BPNN is given by Walker and Hill (1997):

STAGE 1: Forward propagation of input vector

Step 1: Initialize weights to small random values

Step 2: Do while stopping condition is false

Step 3: Compute input sum and apply activation function for each middle PE:

$$y_j = f\left(w_{ij}x_i\right)$$

Step 4: Compute input sum and apply activation function for each output PE:

$$z_k = f\left(v_{ij}y_i\right)$$

STAGE 2: Back propagation of error

Step 5: Compute error:

$$\delta_k = (t_k - z_k)f'\left(w_{jk}y_i\right)$$

Step 6: Compute delta weights:

$$\Delta v_{jk} = (\alpha)(\delta_k)(y_j) + \{\text{Momentum* } \Delta v_{ij}(\text{old})\}$$

Step 7: Compute error contribution for each middle layer PE:

$$\delta_j = \delta_k w_{jk} f'\left(w_{ij}x_i\right)$$

Step 8: Compute delta weights:

$$\Delta w_{ij} = (\alpha)(\delta_j)(x_i) + \{\text{Momentum } \Delta w_{ij}(\text{old})\}$$

Step 9: Update weights:

$$Q_{rs}(\text{new}) = Q_{rs}(\text{old}) + \Delta Q_{rs}$$

Step 10: Test stopping condition

Stopping conditions for a BPNN occur when the weight changes have reached some minimal value or when the average error across a series of input vectors is below some desired level.

EXAMPLE 10.5.1

Consider a backpropagation network with two inputs, two hidden or middle layer PEs, and a single output (Walker and Hill 1997). Find the new weights when the network is presented with an input vector x_i = [0.0, 1.0] and target vector $T_1 = 1.0$ using a learning coefficient of 0.25 and a sigmoid activation function (Figure 10.10).

The initial weights are given as

$$w_{ij} = \left|\begin{array}{cc|c} 0.7 & -0.4 & 0.4 \\ -0.2 & 0.3 & 0.6 \end{array}\right| \qquad v_k = \left|\begin{array}{cc|c} 0.5 & 0.1 & -0.3 \end{array}\right|$$

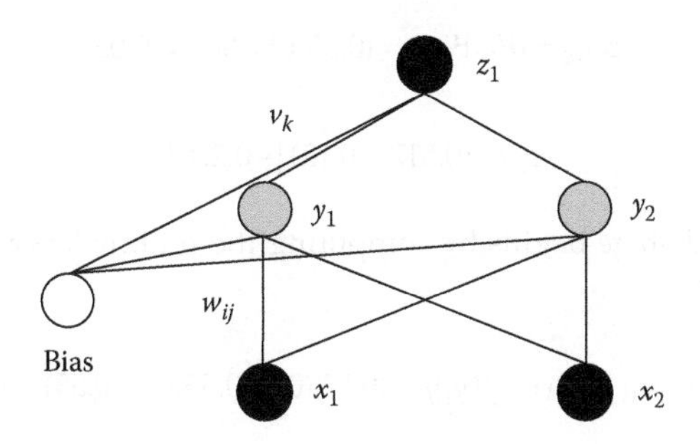

FIGURE 10.10
Example BPNN. (From J.L. Walker II and E.v.K. Hill, "An introduction to neural networks: A tutorial," *First International Conference on Nonlinear Problems in Aviation & Aerospace*, Embry-Riddle Aeronautical University Press, Daytona Beach, FL, 1997, pp. 667–672.)

First, compute the middle layer output using the relationship $y_j = w_{ij}x_i$

$$y_1 = w_{11}x_1 + w_{21}x_2 + w_{1B} = (0.7)(0) + (-0.2)(1.0) + 0.4 = 0.2$$

$$y_2 = w_{12}x_1 + w_{22}x_2 + w_{2B} = (-0.4)(0) + (0.3)(1.0) + 0.6 = 0.9$$

$$y_1(\text{out}) = f(y_1) = \frac{1}{1+e^{-y_1}} = 0.55$$

$$y_2(\text{out}) = f(y_2) = \frac{1}{1+e^{-y_2}} = 0.71$$

Next, compute the network output and associated error using the relationship $z_k = v_{ij}y_i$

$$z_1 = v_{11}y_1 + v_{12}y_2 + v_{1B} = (0.5)(0.55) + (0.1)(0.71) - 0.3 = 0.046$$

$$z_1 = f(z_1) = \frac{1}{1+e^{-z_1}} = 0.51$$

$$\delta_k = (T_k - z_k(\text{out}))\, f'(z_k(\text{out}))$$

$$\delta_{z1} = (T_1 - z_1(\text{out}))\, f(z_1)(1 - f(z_1)) = (1.0 - 0.51)(0.51)(1 - 0.51) = 0.12$$

The middle to output layer weights can now be updated using $\Delta v_{jk} = \alpha\delta_k y_j(\text{out})$

$$\Delta v_{11} = \alpha\delta_{z1}y_1(\text{out}) = (0.25)(0.12)(0.55) = 0.017$$

$$\Delta v_{12} = \alpha\delta_{z1}y_2(\text{out}) = (0.25)(0.12)(0.71) = 0.021$$

$$\Delta v_{1B} = \alpha\delta_{z1}\text{Bias} = (0.25)(0.12)(1) = 0.030$$

$$v_k = |0.517 \quad 0.121|-0.270|$$

The second stage begins by computing the middle layer error as $\delta_j = \delta_k v_{kj} f'(y_j(\text{out}))$

$$\delta_{y1} = \delta_{z1} v_{11} f(y_1)(1 - f(y_1)) = (0.12)(0.5)(0.55)(1 - 0.55) = 0.015$$

$$\delta_{y2} = \delta_{z1} v_{12} f(y_2)(1 - f(y_2)) = (0.12)(0.1)(0.71)(1 - 0.71) = 0.0025$$

The input to middle layer weights are then updated using $\Delta w_{ij} = \alpha\delta_i x_j$

$$\Delta w_{11} = \alpha\delta_{y1} x_1 = (0.25)(0.015)(0) = 0$$

$$\Delta w_{12} = \alpha\delta_{y1} x_2 = (0.25)(0.015)(1.0) = 0.0038$$

$$\Delta w_{21} = \alpha\delta_{y2} x_1 = (0.25)(0.025)(0) = 0$$

$$\Delta w_{22} = \alpha\delta_{y2} x_2 = (0.25)(0.025)(1.0) = 0.0006$$

$$\Delta w_{1B} = \alpha\delta_{y1}\text{Bias} = (0.25)(0.015)(1.0) = 0.0038$$

$$\Delta w_{2B} = \alpha\delta_{y2}\text{Bias} = (0.25)(0.025)(1.0) = 0.0006$$

Finally, the new updated weights are given as

$$w_{ij}(\text{new}) = \begin{vmatrix} 0.70000 & -0.3962 & | & 0.4038 \\ -0.20000 & 0.3006 & | & 0.6006 \end{vmatrix}$$

10.6 Kohonen Self-Organizing Maps

A Kohonen self-organizing map (SOM) is a single-layered, unsupervised, competitive neural network, as shown in Figure 10.11.

A SOM is a neural network that sorts data into different categories, or creates a two-dimensional map from multidimensional inputs. When trained properly, a SOM can take data that is difficult to separate accurately, and divide it into different groups or clusters with common characteristics.

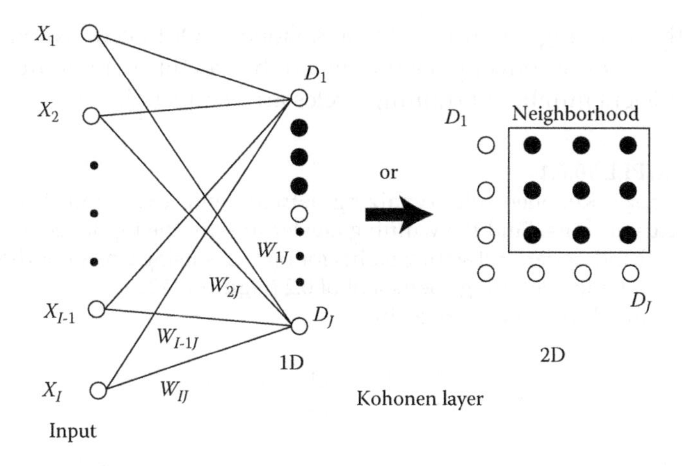

FIGURE 10.11
Kohonen self-organizing map. (From J.L. Walker II and E.v.K. Hill, "An introduction to neural networks: A tutorial," *First International Conference on Nonlinear Problems in Aviation & Aerospace*, Embry-Riddle Aeronautical University Press, Daytona Beach, FL, 1997, pp. 667–672.)

A SOM has an architecture that usually consists of an input layer and a two-dimensional Kohonen layer. The PEs in the input layer are not connected to each other; however, each PE in the input layer is connected to all the PEs in the Kohonen layer. Furthermore, the PEs in the Kohonen layer are connected to each other. All of these connections have an associated weight for calculations.

A SOM learns by minimizing the Euclidean distance between the weights and the input vectors. The network attempts to cluster the input vectors on a mapping layer. The network not only clusters the input vectors but also locates groups with like behaviors close to each other. The algorithm for a simple Kohonen SOM is given as follows (Walker and Hill 1997):

Step 1: Initialize weights, and set neighborhood and learning rate parameters

Step 2: Do while stopping condition is false

Step 3: For each input vector, x_i

Step 4: Compute for each PE: $D_j = \sqrt{\sum (w_{ij} - x_i)^2}$

Step 5: Find index j for D_j minimum

Step 6: Update all weights in neighborhood of j

$$w_{ij}(\text{new}) = w_{ij}(\text{old}) + \alpha(x_i - w_{ij}(\text{old}))$$

Step 7: Update learning rate and neighborhood parameters

Step 8: Test stopping condition

Typically, stopping conditions for a Kohonen SOM occur when the network is said to have converged either when the weight changes are small, or after a sufficient number of training cycles are completed.

EXAMPLE 10.6.1

Consider a Kohonen self-organizing network with two input PEs and five cluster units. Find the winning cluster unit for the input vector $x_i =$ [0.5, 0.2] and update network weights for one pass using a neighborhood factor of 1 and a learning coefficient of 0.2 (Figure 10.12).

The initial weights are given by

$$w_{ij} = \begin{vmatrix} 0.3 & 0.6 & 0.1 & 0.4 & 0.8 \\ 0.7 & 0.9 & 0.5 & 0.3 & 0.2 \end{vmatrix}$$

First, the Euclidean distances are computed using $D_j = \sqrt{\sum (w_{ij} - x_i)^2}$

$$D_1 = \sqrt{(w_{11} - x_1)^2 + (w_{21} - x_2)^2} = \sqrt{(0.3 - 0.5)^2 + (0.7 - 0.2)^2} = 0.54$$

$$D_2 = \sqrt{(w_{12} - x_1)^2 + (w_{22} - x_2)^2} = \sqrt{(0.6 - 0.5)^2 + (0.9 - 0.2)^2} = 0.71$$

$$D_3 = \sqrt{(w_{13} - x_1)^2 + (w_{23} - x_2)^2} = \sqrt{(0.1 - 0.5)^2 + (0.5 - 0.2)^2} = 0.50$$

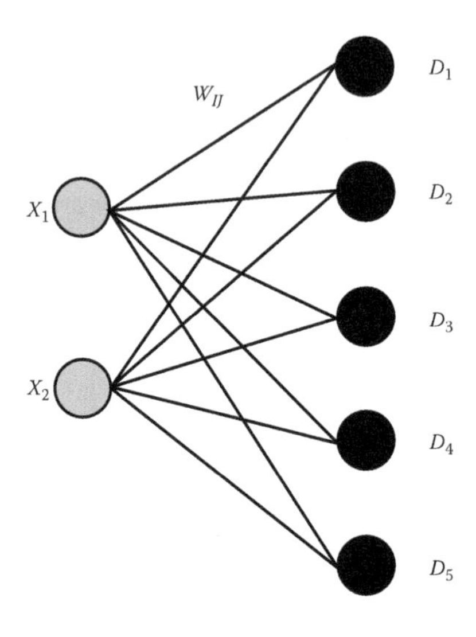

FIGURE 10.12
Example Kohonen SOM. (From J.L. Walker II and E.v.K. Hill, "An introduction to neural networks: A tutorial," *First International Conference on Nonlinear Problems in Aviation & Aerospace*, Embry-Riddle Aeronautical University Press, Daytona Beach, FL, 1997, pp. 667–672.)

$$D_4 = \sqrt{(w_{14} - x_1)^2 + (w_{24} - x_2)^2} = \sqrt{(0.4 - 0.5)^2 + (0.3 - 0.2)^2} = 0.14$$

$$D_5 = \sqrt{(w_{51} - x_1)^2 + (w_{25} - x_2)^2} = \sqrt{(0.8 - 0.5)^2 + (0.2 - 0.2)^2} = 0.30$$

Since D_4 is the closest to zero, it is deemed the winning PEs. With a neighborhood factor of 1, this implies that the weights for PEs j = 3, 4, and 5 will be updated using $w_{ij}(\text{new}) = w_{ij}(\text{old}) + \alpha(x_i - w_{ij}(\text{old}))$:

$$w_{13}(\text{new}) = w_{13}(\text{old}) + \alpha(x_1 - w_{13}(\text{old})) = 0.1 + 0.2\ (0.5 - 0.1) = 0.18$$

$$w_{14}(\text{new}) = w_{14}(\text{old}) + \alpha(x_1 - w_{14}(\text{old})) = 0.4 + 0.2\ (0.5 - 0.4) = 0.42$$

$$w_{24}(\text{new}) = w_{24}(\text{old}) + \alpha(x_2 - w_{24}(\text{old})) = 0.3 + 0.2\ (0.2 - 0.3) = 0.28$$

$$w_{15}(\text{new}) = w_{15}(\text{old}) + \alpha(x_1 - w_{15}(\text{old})) = 0.8 + 0.2\ (0.5 - 0.8) = 0.74$$

$$w_{25}(\text{new}) = w_{25}(\text{old}) + \alpha(x_2 - w_{25}(\text{old})) = 0.2 + 0.2\ (0.2 - 0.2) = 0.20$$

Finally, the new weight matrix is given as

$$w_{ij}(\text{new}) = \begin{vmatrix} 0.3 & 0.6 & 0.18 & 0.42 & 0.74 \\ 0.7 & 0.9 & 0.44 & 0.28 & 0.20 \end{vmatrix}$$

If this cycle is repeated, w_{14} approaches a value of 0.50, while w_{24} approaches a value of 0.20. When the fourth column weights almost exactly match the input vector [0.50, 0.2], the network is said to have learned the input vector, the network is considered to be trained, and the weights are fixed in preparation for the testing (or classification) phase.

Sometimes a fully connected, two PE (x,y) output layer will be added to the SOM of Figure 10.12 to graph the x-y points representing the various classification clusters onto a 2D plot. Here, each cluster maps to its own x-y point on the output plot. Hence, if there are three source mechanisms operative in a given material, there will be three output points on the 2D output plot for the SOM, and every point associated with each of the three source mechanisms will have the same (x,y) coordinate. ■

SOM neural networks are useful for classifying failure mechanisms in AE data. By analyzing data such as the hits versus amplitude histogram shown in Figure 10.4, the failure mechanism can be sorted. Generally, this is an iterative process, requiring large SOMs to be generated first and then slowly reducing the size of the grid until it appears that the correct number of failure mechanisms have been found (Hill et al. 2013). This requires a good understanding of the material and specimen under test and of AE testing practices.

The use of neural networks is a powerful tool, especially for AE testing. Using these computational algorithms, large amounts of data can be analyzed. As this is typically a highly iterative and computationally tedious operation, the process of using neural networks has been computerized.

10.7 Backpropagation Neural Network Analysis Tutorial

The purpose of this section is to provide a real-world example of using AE testing data to predict ultimate load. This involves the AE data from unidirectional fiberglass/epoxy beams in three-point bending as seen in Figure 10.13 (Hill et al. 2013).

In this tutorial, a BPNN will be generated using Microsoft Excel to format the input data and NeuralWorks Professional II/Plus to build the network. This tutorial will demonstrate the ability of a neural network to solve prediction-type problems, in this case, ultimate loads. The data sets used in this tutorial are AE amplitude frequency distribution histograms from the various fiberglass/epoxy beams. The amplitude histogram for one of the samples is shown in Figure 10.14.

FIGURE 10.13
Fiberglass/epoxy beam in three point bending.

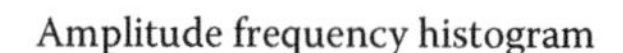

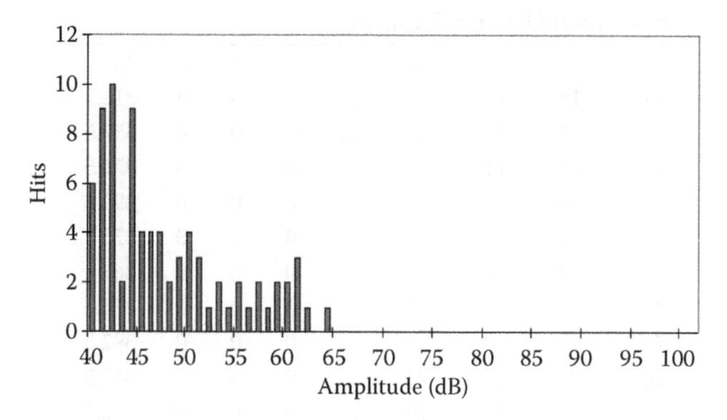

FIGURE 10.14
Hits versus amplitude histogram.

10.8 Experimental Data

The fiberglass/epoxy beams were loaded to failure in a three-point bending test fixture. AE data were recorded from the onset of loading until failure. The AE data were then filtered to include only the data acquired up to 80% of the average ultimate load with a total of 15 specimens being tested. The network was trained on the AE amplitude histograms from seven specimens and was tested on the data from the remaining eight for prediction of ultimate loads. This tutorial will show how a BPNN can be used to accurately predict the ultimate failure loads within ±5% error using the 80% of ultimate load AE data.

10.9 Creating Training and Testing Files and Results Spreadsheet

10.9.1 Required Materials

- Computer equipped with Microsoft Excel
- Experimental data (provided in Appendix A)

10.9.2 Steps for Creating Training File

1. Open Microsoft Excel.
2. Input each specimen's AE amplitude distribution data in the following format (see Appendix A for complete data set). The network will

TABLE 10.3
Specimen Data for Training

3	9	6	2	...	0	0	0	336
14	18	18	9	...	0	0	0	312.5
20	27	30	23	...	0	0	0	357.5
4	17	14	10	...	0	0	0	372.5
7	10	9	7	...	0	0	0	392.5
2	2	4	3	...	0	0	0	375
4	12	7	6	...	0	0	0	365
2	2	4	3	...	0	0	0	375
7	10	9	7	...	0	0	0	392.5
20	27	30	23	...	0	0	0	357.5
4	17	14	10	...	0	0	0	372.5
7	10	9	7	...	0	0	0	392.5
3	9	6	2	...	0	0	0	336
4	12	7	6	...	0	0	0	365
4	17	14	10	...	0	0	0	372.5
14	18	18	9	...	0	0	0	312.5
3	9	6	2	...	0	0	0	336
2	2	4	3	...	0	0	0	375
4	12	7	6	...	0	0	0	365
20	27	30	23	...	0	0	0	357.5
14	18	18	9	...	0	0	0	312.5

be trained on only seven specimens; however, to provide sufficient training data, the data should be duplicated for each specimen. Here, the data is repeated three times to help the software learn on a larger set of data, a technique called bootstrapping in statistical analyses. The specimens' amplitude histogram data are then entered in random order. There should be a total of 21 rows and 62 columns: three rows for each of the seven training specimens and 61 columns for each frequency distribution, with the last column being the actual ultimate load (Table 10.3).

3. Save the training file data as a text file. For example, the training file for this tutorial will be referred to as 80%MDDtraining2.txt. (Note: Here MDD represents the initials of Michele D. Dorfman, the coauthor who performed the original research.)

10.9.3 Steps for Creating Test Files

1. Open Microsoft Excel.
2. Input each specimen's AE amplitude distribution data in the following format (see Appendix A for complete data set). There should be a total of 15 rows and 62 columns: one row for each of the

15 specimens, 61 columns for each frequency distribution, and one column for the ultimate load. The first seven specimens entered in the test file are the specimens used to train the network, and the next eight specimens are used for testing (prediction of ultimate loads) (Table 10.4).

3. Save the testing file data as a text file. For example, the testing file for this tutorial will be referred to as 80%MDDtesting2.txt.

10.9.4 Steps for Creating Network Results Spreadsheet

1. Open Microsoft Excel.
2. Create a worksheet containing the following network parameters: number of inputs, number of neurons in the hidden layer, number of outputs, learning coefficient, momentum, transition point, learning coefficient ratio, F' offset, learning rule, transfer function, epoch size, and root mean square (RMS) error. This spreadsheet will be used to track the various network permutations throughout the optimization process (Table 10.5) (Dorfman 2004).
3. Create another worksheet that will track the percent error between the output from the neural network and the actual ultimate load. Note the worst-case error in each network run (Table 10.6).
4. Save the results spreadsheet as an Excel file. For example, the results file for this tutorial will be referred to as NetworkData.xls.

TABLE 10.4

Specimen Data for Testing

3	9	6	2	...	0	0	0	336
4	17	14	10	...	0	0	0	372.5
20	27	30	23	...	0	0	0	357.5
14	18	18	9	...	0	0	0	312.5
7	10	9	7	...	0	0	0	392.5
2	2	4	3	...	0	0	0	375
4	12	7	6	...	0	0	0	365
6	9	10	2	...	0	0	0	375
14	19	24	13	...	0	0	0	312.5
5	11	15	8	...	0	0	0	365
9	15	8	14	...	0	0	0	327.5
3	14	11	6	...	0	0	0	340
3	4	7	5	...	0	0	0	363
1	3	4	2	...	0	0	0	372.5
0	5	2	1	...	0	0	0	367.5

TABLE 10.5

Network Parameter Iteration

Network	1	2	3	4	5
Inputs	61	61	61	61	61
Hidden 1	2	3	4	5	2
Output	1	1	1	1	1
L. coef.	0.3	0.3	0.3	0.3	0.3
	0.15	0.15	0.15	0.15	0.15
Momentum	0.4	0.4	0.4	0.4	0.4
Trans. pt.	10,000	10,000	10,000	10,000	10,000
L. coef. ratio	0.5	0.5	0.5	0.5	0.5
F' offset	0.1	0.1	0.1	0.1	0.1
Learn rule	NCD	NCD	NCD	NCD	NCD
Transfer	tanH	tanH	tanH	tanH	tanH
Epoch	21	21	21	21	21
RMS error	0.03	0.03	0.03	0.03	0.03

Note: For a complete list of definitions of network parameters, see the NeuralWorks Professional II/PLUS help manual.

TABLE 10.6

Training and Test Data Errors

	Actual	Net 1	% Error	Net 2	% Error
Training data	336	333.72	0.679	334.36	0.487
	372.5	372.69	–0.051	372.50	0.000
	357.5	357.58	–0.023	357.75	–0.071
	312.5	312.26	0.077	312.13	0.117
	392.5	392.15	0.090	392.13	0.095
	375	378.19	–0.849	378.32	–0.884
	365	364.68	0.088	363.85	0.315
Test data	375	369.20	1.546	393.64	–4.971
	312.5	394.96	–26.387	354.24	–13.356
	365	395.42	–8.335	396.01	–8.496
	327.5	373.99	–14.196	389.70	–18.993
	340	349.48	–2.787	376.81	–10.825
	363	366.05	–0.839	375.45	–3.429
	372.5	395.91	–6.283	378.30	–1.556
	367.5	341.85	6.980	399.83	–8.798
		Worst	–26.387	Worst	–18.993

10.9.5 Expected Outcome

Upon completion of Sections 10.9.2 through 10.9.4, the training and testing files should have been created and saved in text format. An Excel spreadsheet to track the results and calculate the percent errors should have also been created.

10.10 Creating a Backpropagation Neural Network

10.10.1 Required Material

- Computer equipped with NeuralWorks Professional II/Plus and Microsoft Excel
- Training File: 80%MDDtraining2.txt
- Testing File: 80%MDDtesting2.txt
- Results File: NetworkData.xls

10.10.2 Creating the Network

1. Open the NeuralWorks Professional II/Plus application.
2. Click the **Back Propagation** command in the **InstaNet Menu** to open the Instanet/Back Propagation dialog box. Enter the parameters as shown in the dialog box and Click the **OK** command. There are 61 neurons in the input layer for the AE amplitude distribution frequencies and one neuron in the output layer for the ultimate load. The normalized-cumulative-delta rule is used as the learning rule, and the hyperbolic tangent is used as the transfer function. In this case, the epoch is set to be the number of specimens in the training file (three repeats of the seven specimens in the test set = 21 epochs). The remaining parameters are the software defaults and will be varied later in the tutorial (Figure 10.15).
3. When you have completed the dialog box and clicked **OK**, the Instrument/Create dialog box appears. Ensure that all instruments are highlighted and click **OK** (Figure 10.16).
4. Double click the **RMS Error** instrument to open the Instrument Parameters dialog box. Check the **Convergence Criterion** box and enter 0.03 as the **Threshold** as shown in the dialog box. Click the **OK** command. This step tells the network to train until the RMS error is 3%. In this tutorial, the network trains x number of cycles until the RMS error converges to 3%. Another method of training, not used here, is to tell the network the exact number of cycles to learn (Figure 10.17).

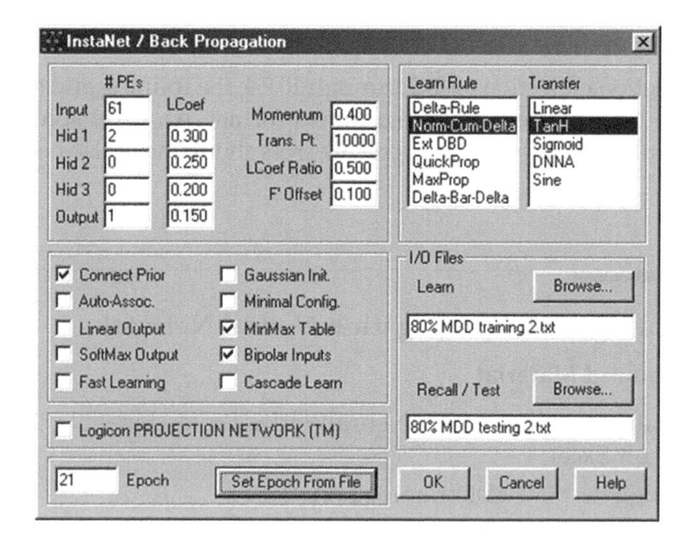

FIGURE 10.15
Backpropagation dialog box.

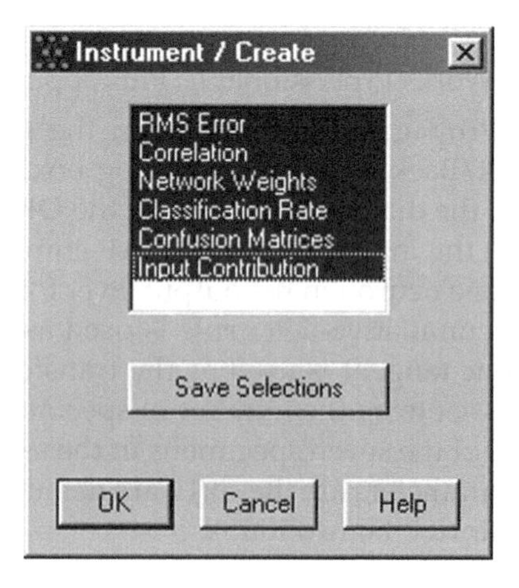

FIGURE 10.16
Instrument dialog box.

5. Click the **Learn** command in the **Run Menu** to open the Run/Learn dialog box. Enter the parameters as shown in the dialog box. Click the **OK** command (Figure 10.18).
6. Click the **Test** command in the **Run Menu** to open the Run/Test dialog box. Enter the parameters as shown in the dialog box. Click the **OK** command (Figure 10.19).

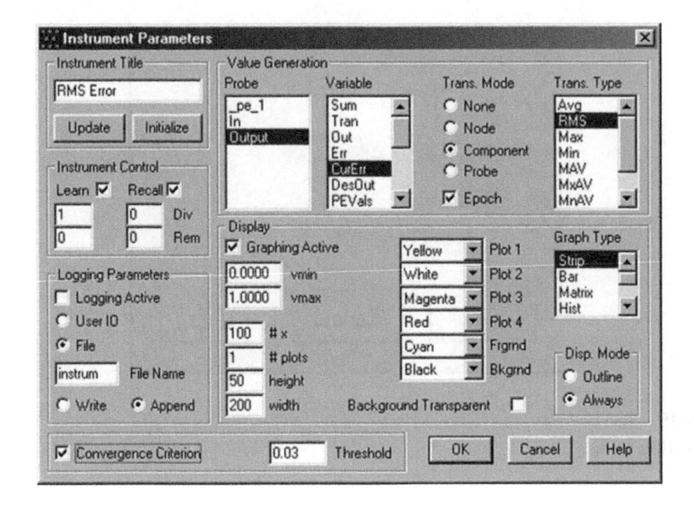

FIGURE 10.17
Instrument Parameters dialog box.

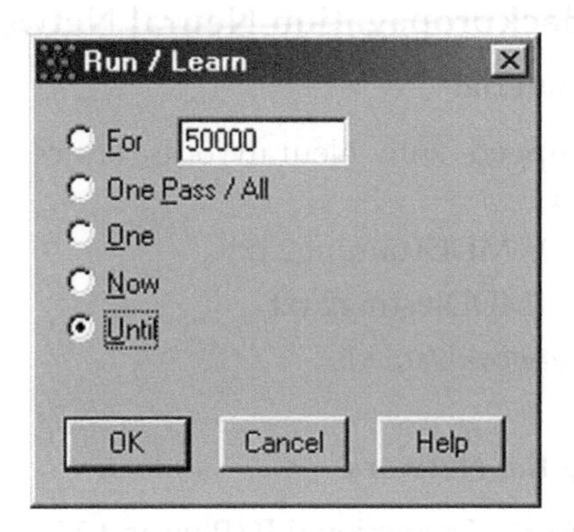

FIGURE 10.18
Run/learn dialog box.

7. NeuralWorks Professional II/Plus has now created an output file in the folder that your testing file is stored in. The file should be called 80%MDDtesting2_txt.nnr. Open this file and copy the output information into your results spreadsheet. The percent errors should automatically be calculated.

10.10.3 Expected Outcome

Upon completion of Section 10.10.2, the reader should be able to create a neural network with the desired architecture, can train the network to a desired RMS error, test the network, and extract the output data.

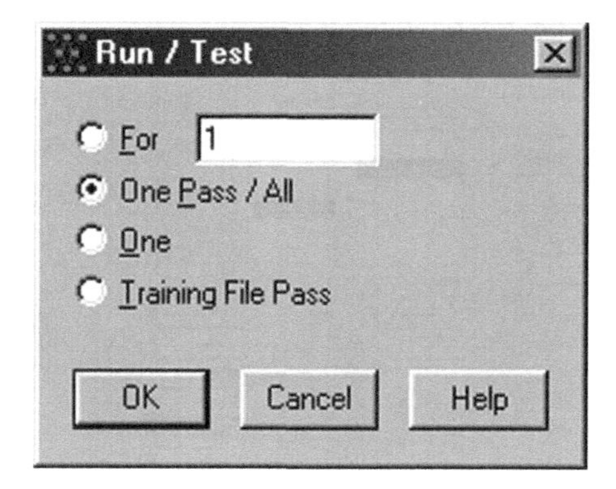

FIGURE 10.19
Run/test dialog box.

10.11 Testing a Backpropagation Neural Network

10.11.1 Required Material

- Computer equipped with NeuralWorks Professional II/Plus and Microsoft Excel
- Training File: 80%MDDtraining2.txt
- Testing File: 80%MDDtesting2.txt
- Results File: NetworkData.xls

10.11.2 Optimizing the Network

1. Open NeuralWorks Professional II/Plus and Microsoft Excel.
2. The first step is to determine the optimum number of neurons in the hidden layer. Theoretically, the number of neurons in the hidden layer is related to the number of failure mechanisms in the test specimen. If it is assumed that $n = 6$ failure mechanisms are present, $2n + 1 = 13$ neurons in the hidden layer will be expected to be the optimum number (Hill et al. 1993). Using the following network parameters, vary only the number of neurons in the hidden layer from 2 through 23. Record the results in the spreadsheet for ease of tracking the optimization (Table 10.7).
3. After determining the number of neurons in the hidden layer that produces the least error, fix that parameter and vary only the F' offset. The F' offset is a constant offset added to the derivative of the hyperbolic transfer function. Vary the F' offset from 0.01 through 0.15 in increments of 0.01. Record the results in the spreadsheet.

TABLE 10.7

Network Parameters

Network Number	1
Inputs	61
Hidden 1	?
Output	1
L. coef.	0.3
	0.15
Momentum	0.4
Trans. pt.	10,000
L. coef. ratio	0.5
F' offset	0.1
Learn rule	NCD
Transfer	tanH
Epoch	21
RMS error	0.03

4. After determining the best F' offset, fix that parameter and vary only the transition point. The transition point dictates when the learning coefficient is multiplied by the learning coefficient ratio, which decreases the learning step size as it homes in on the minimum error. Vary the transition point from 5000 through 15,000 in increments of 1000. Record the results in the spreadsheet.
5. After determining the best transition point, fix that parameter and vary only the momentum. The momentum prevents the network from getting stuck in localized minima during training. Vary the momentum from 0.2 through 0.7 in increments of 0.1. Record the results in the spreadsheet.
6. After determining the best momentum, fix that parameter and vary only the hidden layer learning coefficient. The learning coefficient works in conjunction with the transition point and the learning coefficient ratio values to configure a decaying learn schedule. Vary the hidden layer learning coefficient from 0.1 through 0.5 in increments of 0.05. Record the results in the spreadsheet.
7. After determining the best hidden layer learning coefficient, fix that parameter and vary only the output layer learning coefficient. Vary the output layer learning coefficient from 0.05 through 0.3 in increments of 0.05. Record the results in the spreadsheet.
8. After determining the best output layer learning coefficient, fix that parameter and vary only the learning coefficient ratio. Vary the learning coefficient ratio from 0.1 through 0.9 in increments of 0.1. Record the results in the spreadsheet.

TABLE 10.8

Optimized Network Parameters

Network	82
Inputs	61
Hidden 1	13
Output	1
L. coef.	0.3
	0.15
Momentum	0.4
Trans. pt.	7000
L. coef. ratio	0.35
F' offset	0.05
Learn rule	NCD
Transfer	tanH
Epoch	21
RMS error	0.03

9. After determining the best learning coefficient ratio, fix that parameter and vary only the RMS error. The RMS error is the root mean square of the errors from the 3 × 7 = 21 (epoch size) training specimens. Vary the RMS error from 0.01 through 0.05 in increments of 0.01. Record the results in the spreadsheet.

10.11.3 Expected Outcome

Upon completion of the above steps, the network should be optimized to predict the ultimate loads, hopefully within ±5% of the actual loads. Owing to the randomization of the initial connection weights, your final parameters may not exactly match the values shown in Table 10.8, but they should be close.

10.12 Tutorial Conclusions and Recommendations

Upon completion of this tutorial, the reader should be familiar with the backpropagation portion of the NeuralWorks Professional II/Plus software. The reader should also have gained the knowledge of how to use AE data to create training and testing files as inputs for the neural network and how to create a spreadsheet for tracking the network results during optimization of the network parameters. The reader at this point will understand that the optimization process takes time and patience. Although two similar networks can be created by using the same parameters, the reader should be aware that the results may vary slightly due to the randomization of initial or starting point weights.

It is also important to understand that there is no one-size-fits-all approach to neural networks. Various architectures are necessary for analysis of different problems. Different neural network architectures should be explored, and the most appropriate architecture should be selected for the given problem. Furthermore, networks can be created using more than one hidden layer in the architecture for solving more complex problems.

10.13 Chapter Summary

In progressing through this chapter, the reader should have become acquainted with the basics of AE nondestructive testing and evaluation. This includes an understanding of what AE is, how it is measured, and how the acquired data is analyzed to make predictions on test specimen performance. The reader should also understand some of the basics of neural networks and their importance in evaluating AE data. It should be noted that there are whole books written on AE testing, and the reader is encouraged to explore further. To this end, several technical publications have been listed in the Bibliography that may provide more engineering applications of interest.

Appendix 10.A

Training Set Data

Specimen ID	Ultimate Load (lb)	Amplitude Histogram Data
MDD3-1	336	3 9 6 2 4 2 0 0 1 0 2 0 0 0 1 0 0 1 0 1 0 0 0 0 0 0 0 0 0 0 0 0 0 0 0 0 0 0 0 0 0 0 0 0 0 0 0 0 0 0 0 0 0 0 0 0 0 0 0 0 0
MDD4-2	372.5	4 17 14 10 10 11 12 7 9 7 6 3 1 7 3 7 1 1 2 6 1 1 1 0 0 1 0 0 0 0 0 0 0 0 0 0 0 0 0 0 0 0 0 0 0 0 0 0 0 0 0 0 0 0 0 0 0 0 0 0 0
MDD2-3	357.5	20 27 30 23 22 15 13 14 12 9 10 10 11 5 9 7 2 4 6 2 5 2 2 3 0 1 2 0 1 0 0 0 0 0 0 0 0 0 0 0 0 0 0 0 0 0 0 0 0 0 0 0 0 0 0 0 0 0 0 0 0
MDD1-2	312.5	14 18 18 9 12 9 10 14 15 11 5 14 12 5 1 4 8 1 3 2 2 2 2 4 1 0 3 3 3 1 2 0 1 0 0 1 0 0 0 0 0 0 0 0 0 0 0 0 0 0 0 0 0 0 0 0 0 0 0 0 0
MDD4-3	392.5	7 10 9 7 8 2 5 3 1 4 2 1 2 2 2 3 4 4 3 4 1 3 4 5 0 2 0 0 2 0 1 2 1 1 1 0 0 0 0 0 0 0 0 0 0 0 0 0 0 0 0 0 0 0 0 0 0 0 0 0 0
MDD5-2	375	2 2 4 3 2 2 1 2 1 3 1 1 2 0 1 0 0 1 0 1 0 0 0 0 0 0 0 0 0 0 0 0 0 0 0 0 0 0 0 0 0 0 0 0 0 0 0 0 0 0 0 0 0 0 0 0 0 0 0 0 0
MDD5-3	365	4 12 7 6 5 3 1 2 3 1 1 0 0 1 1 0 1 1 0 1 3 0 0 0 0 1 0 0 0 0 0 0 0 0 0 0 0 0 0 0 0 0 0 0 0 0 0 0 0 0 0 0 0 0 0 0 0 0 0 0 0

Test Set Data

Specimen ID	Ultimate Load (lb)	Amplitude Histogram Data
MDD1-1	375	6 9 10 2 9 4 4 4 2 3 4 3 1 2 1 2 1 2 1 2 2 3 1 0 1 0 0 0 0 0 0 0 0 0 0 0 0 0 0 0 0 0 0 0 0 0 0 0 0 0 0 0 0 0 0 0 0 0 0 0 0
MDD3-2	312.5	14 19 24 13 16 12 9 5 4 6 5 2 4 9 3 5 4 7 4 2 3 2 3 2 3 0 0 0 0 1 0 0 0 0 1 0 1 2 0 0 0 0 0 0 0 0 0 0 0 0 0 0 0 0 0 0 0 0 0 0
MDD2-2	365	5 11 15 8 9 8 5 9 7 6 1 0 3 3 3 4 4 4 3 3 3 4 5 5 2 3 4 2 3 2 2 2 1 2 1 1 0 0 0 1 0 0 0 0 0 0 0 0 0 0 0 0 0 0 0 0 0 0 0 0 0
MDD1-3	327.5	9 15 8 14 10 12 9 3 5 5 3 5 2 1 1 1 2 1 1 0 1 1 0 0 0 0 0 0 0 0 0 0 0 0 0 0 0 0 0 0 0 0 0 0 0 0 0 0 0 0 0 0 0 0 0 0 0 0 0

Test Set Data

Specimen ID	Ultimate Load (lb)	Amplitude Histogram Data
MDD3-3	340	3 14 11 6 7 7 7 1 1 0 0 1 0 2 0 0 1 0 1 0 0 1 1 0 0 0 0 0 0 0 0 0 0 0 0 0 0 0 0 0 0 0 0 0 0 0 0 0 0 0 0 0 0 0 0 0 0 0 0 0 0
MDD4-1	363	3 4 7 5 3 5 3 1 2 0 1 0 0 0 0 1 0 0 0 0 0 0 1 0 0 1 2 0 0 0 0 0 0 0 0 0 0 0 0 0 0 0 0 0 0 0 0 0 0 0 0 0 0 0 0 0 0 0 0 0 0
MDD2-1	372.5	1 3 4 2 4 1 1 0 1 3 2 0 1 0 0 0 0 1 2 3 0 0 0 0 0 0 0 0 0 0 0 0 0 0 0 0 0 0 0 0 0 0 0 0 0 0 0 0 0 0 0 0 0 0 0 0 0 0 0 0
MDD5-1	367.5	0 5 2 1 2 1 1 0 1 1 0 1 2 1 0 1 1 2 1 0 1 2 0 0 0 1 1 0 0 0 0 0 0 0 0 0 0 0 0 0 0 0 0 0 0 0 0 0 0 0 0 0 0 0 0 0 0 0 0 0 0

References

1. E.v.K. Hill, "Predicting burst pressures in filament wound composite pressure vessels by using acoustic emission data," *Materials Evaluation*, Vol. 50, No. 12, 1992, pp. 1439–1445.
2. J.L. Walker II and E.v.K. Hill, "An introduction to neural networks: A tutorial," *First International Conference on Nonlinear Problems in Aviation & Aerospace*, Embry-Riddle Aeronautical University Press, Daytona Beach, FL, 1997, pp. 667–672.
3. E.v.K. Hill, M.D. Dorfman and Y. Zhao, "Ultimate load prediction in fiberglass/epoxy beams from acoustic emission data using neural networks and statistical analyses," *Materials Evaluation*, Vol. 71, No. 8, 2013, pp. 977–986.
4. M.D. Dorfman, Ultimate strength prediction in fiberglass/epoxy beams subjected to three-point bending using acoustic emission and neural networks, M.S.A.E Thesis, Embry-Riddle Aeronautical University, Daytona Beach, Florida, 2004.
5. E.v.K. Hill, P.E. Israel and G.L. Knotts, "Neural network prediction of aluminum-lithium weld strengths from acoustic emission amplitude data," *Materials Evaluation*, Vol. 51, No. 9, 1993, pp. 1040–1045 and 1051.

Bibliography

1. E.v.K. Hill, J. Iizuka, I.K. Kaba, H.L. Surber (now McCann) and Y.P. Poon, "Neural network burst pressure prediction in composite overwrapped pressure vessels using mathematically modeled acoustic emission failure mechanism data," *Research in Nondestructive Evaluation*, Vol. 23, 2012, pp. 89–103.
2. E.v.K. Hill, M.D. Scheppa, Z.D. Sager and I. Prata Thisted, "Neural network techniques for burst pressure prediction in Kevlar/epoxy pressure vessels using acoustic emission data," *Journal of Acoustic Emission*, Vol. 29, 2011, pp. 106–112.
3. E.v.K. Hill, S.-A.T. Dion, J.O. Karl, N.S. Spivey and J.L. Walker II, "Neural network burst pressure prediction in composite overwrapped pressure vessels," *Journal of Acoustic Emission*, Vol. 25, 2007, pp. 187–193.
4. E.v.K. Hill and R.J. Demeski, "Classification of failure mechanism data from fiberglass epoxy tensile specimens," *Nondestructive Testing Handbook*, Third Edition: Volume 6, Acoustic Emission Testing, American Society for Nondestructive Testing, Columbus, OH, 2005, pp. 167–170.
5. E.v.K. Hill and J.L. Walker II, "Neural network prediction of burst pressures in graphite epoxy pressure vessels," *Nondestructive Testing Handbook*, Third Edition: Volume 6, Acoustic Emission Testing, American Society for Nondestructive Testing, Columbus, OH, 2005, pp. 171–176.
6. E.v.K. Hill and T.J. Lewis, "Acoustic emission monitoring of a rocket motor case during hydrostatic testing," *Nondestructive Testing Handbook*, Third Edition: Volume 6, Acoustic Emission Testing, American Society for Nondestructive Testing, Columbus, OH, 2005, pp. 377–381.
7. E.v.K. Hill, "Acoustic emission prediction of burst pressures in fiberglass epoxy pressure vessels," *Nondestructive Testing Handbook*, Third Edition: Volume 6, Acoustic Emission Testing, American Society for Nondestructive Testing, Columbus, OH, 2005, pp. 382–387.
8. E.C. Fatzinger and E.v.K. Hill, "Low proof load prediction of ultimate loads of fiberglass/epoxy resin I-beams from acoustic emission," *Journal of Testing and Evaluation*, Vol. 33, No. 5, September 2005, 8 pages (online).
9. M.E. Fisher and E.v.K. Hill, "Burst pressure prediction in filament wound composite pressure vessels using acoustic emission," *Materials Evaluation*, Vol. 56, No. 12, 1998, pp. 1395–1401.
10. J.L. Walker II, S.S. Russell, G.L. Workman and E.v.K. Hill, "Neural network/acoustic emission burst pressure prediction for impact damaged composite pressure vessels," *Materials Evaluation*, Vol. 55, No. 8, 1997, pp. 903–907.
11. M. Kouvarakos and E.v.K. Hill, "Isolating tensile failure mechanisms in fiberglass/epoxy from acoustic emission signal parameters," *Materials Evaluation*, Vol. 54, No. 9, 1996, pp. 1025–1031.
12. E.v.K. Hill, J.L. Walker II and G.H. Rowell, "Burst pressure prediction in graphite/epoxy pressure vessels using neural networks and acoustic emission amplitude data," *Materials Evaluation*, Vol. 54, No. 6, 1996, pp. 744–748, 754.
13. E.v.K. Hill and J.L. Walker II, "Backpropagation neural networks for predicting ultimate strengths of unidirectional graphite/epoxy tensile specimens," *Advanced Performance Materials*, Vol. 3, No. 1, 1996, pp. 75–83.

14. T.M. Ely and E.v.K. Hill, "Longitudinal splitting and fiber breakage characterization in graphite/epoxy using acoustic emission data," *Materials Evaluation*, Vol. 53, No. 2, 1995, pp. 134–140.
15. E.v.K. Hill and T.J. Lewis, "Acoustic emission monitoring of a filament wound composite rocket motor case during hydroproof," *Materials Evaluation*, Vol. 43, No. 7, 1985, pp. 859–863.

11

Prediction of Ultimate Compression after Impact Loads in Graphite-Epoxy Coupons from Ultrasonic C-Scan Images Using Neural Networks

Eric v. K. Hill and Nikolas L. Geiselman

CONTENTS

11.1 Introduction

11.1.1 Overview

Over the past decade, composite materials have become increasingly prevalent in aerospace structures because they provide excellent stiffness properties and high strength-to-weight ratios. This major advantage of composites over metals has led to a huge increase in their use, especially as major structural elements in aerospace applications. Both the military and commercial aviation sectors have adopted composites heavily, as shown in Table 11.1.

In certain circumstances, the weight savings of composites dictates their use, such as during the development of the B-2 stealth bomber. The addition of the radar absorbing paint on the fuselage caused the aircraft to be overweight; to reduce this weight penalty, composites were employed over the majority of the aircraft [1]. Weight savings is also a crucial factor in the design of commercial aircraft because it controls the number of passengers and fuel economy; therefore, weight savings lowers operating costs and increases profits. Airbus utilized a fiber metal laminate in the design of the A380 called GLARE (glass laminate aluminum reinforced epoxy), which is composed of several very thin layers of aluminum interspersed with layers of unidirectional glass fiber prepreg bonded together with epoxy. This provides a weight savings of between 15% and 30% over standard aerospace aluminums [1].

Although they have high strength-to-weight ratios and excellent stiffness properties, polymer matrix composites are susceptible to barely visible impact damage (BVID). BVID can be caused by many different things that aircraft encounter on a daily basis, such as being struck with runway

TABLE 11.1

Composites by Weight of Various Aircraft

Aircraft Name	Composites by Weight
Boeing 787	50%
Boeing V-22 Osprey	50%
Eurofighter	40%
Airbus A320	28%
Dassault Rafale	26%
Lockheed F-22 Raptor	24%
Airbus A380	22%
Boeing 777	20%
FA-18 Hornet	19%

Source: Quilter, A., Composites in Aerospace Applications. [Online] [Cited: April 12, 2011.] http://cis.ihs.com/NR/rdonlyres/AEF9A38E-56C3-4264-980CD8D6980A4C84/0/444.pdf.

debris, tools being dropped by aircraft mechanics, or bird strikes. Although damage may be barely visible to the naked eye on the surface, significant damage may exist underneath the surface in the form of matrix cracking, delamination between plies, or even fiber breaks, substantially weakening the composite and causing the part to fail at a load up to 60% lower than it was designed to withstand. Because of the inherent danger of BVID existing in structures such as aircraft wings or other load-bearing parts, it becomes necessary to develop a method of quantitatively evaluating the severity of BVID without reliance on visual inspection.

11.1.2 Previous Research

Artificial neural networks have been used to predict ultimate compressive loads of impact-damaged composite laminates from ultrasonic C-scan image data. Hess [2] originated this research in 2003 when he obtained a worst-case error of 16.62% in predicting the compression after impact (CAI) loads of graphite-epoxy coupons. He used three sets of 16-ply graphite-epoxy coupons and damaged them with known impact energies ranging from 0 to 20 ft-lb$_f$. Each coupon was then ultrasonically C-scanned, and a 16-color image was generated of the impact damage. In the image file, each pixel was assigned a numerical value between 0 and 15, with 0 corresponding to black and 15 to white. The coupons were then compressed to failure in a Boeing CAI test fixture to determine their ultimate compressive load. The numerical value of each pixel was input into an artificial neural network in order to predict the ultimate CAI load of the coupons. In 2005, Nguyen [3] improved on Hess's results slightly with a worst-case error of 14.61% using fiberglass–epoxy coupons. Subsequently, Gunasekera [4] in 2009 obtained a worst-case error of –11.53% using the acoustic emission data taken during compression of the same graphite-epoxy coupons that were used in the current research.

11.1.3 Current Research

The current approach was to explore the ability for an artificial neural network to predict the ultimate compressive load of graphite-epoxy coupons that have BVID using improved ultrasonic C-scan images as inputs. Twenty-one test specimen coupons were impacted at known energies, scanned using an ultrasonic C-scan machine, and then compressed to failure with their ultimate CAI loads being recorded for use in an artificial neural network. After the C-scan images were recorded for each damaged coupon, the images were cropped into 100 × 100 pixel squares centered on the damaged area and input into a MATLAB code for image preprocessing. The MATLAB code quantified the damaged area of the image and created a 300 row × 1 column matrix of the image, which contained three colors: red, green, and blue. By inspection, the red and blue layers contained only noise; thus, only the green layer was used as the input to the neural network. Fifteen coupons were employed for training the neural network, with the remaining six being used to test the network. The target goal of the artificial neural network was twofold: (1) having the ability to predict the ultimate compressive load of each composite coupon within ±10% of the actual failure load, and hopefully, (2) predicting within the statistical B-basis allowables of the graphite-epoxy composite coupons as well.

11.2 Background

11.2.1 Ultrasonic C-Scan

Ultrasound is a volumetric method of nondestructive testing that uses high-frequency sound waves to analyze a part, point by point, without destroying it [5]. Ultrasonic waves are high-frequency sound waves that are outside the range of human hearing, normally well above 20 kHz. A typical ultrasonic system consists of a pulser/receiver transducer that contains a piezoelectric ceramic crystal that converts an electric signal into a sound wave; conversely, the piezoelectric element will produce an electrical signal in response to an incident sound wave. To scan a part, the ultrasonic transducer emits a sound wave then switches to listen mode to receive the echoes from the part. Whenever the wave encounters a change in density, it is both reflected and refracted; these changes in density can be caused by a defect under the surface.

As the ultrasonic sound wave enters the part from left to right in Figure 11.1, it reflects and refracts. The reflected wave returns to the transducer first as the "front surface echo" or "main bang." Then, the wave propagates through the material until it reaches the crack or discontinuity and is again reflected and refracted, whereupon the reflected wave returns to the transducer as the "crack echo." Note that there is also some noise between the main bang and the crack echo. The wave continues to propagate through the part and eventually reaches the back surface, and is again reflected and refracted. Here the reflected wave returns to the transducer as the "back surface echo." Thus, the three echoes in Figure 11.1 are formed.

The specific application of ultrasonic nondestructive testing used in this research was the ultrasonic C-scan, which is shown in Figure 11.2. A C-scan is a planar image generated by compiling all of the ultrasonic echoes for each point along the surface of the part. This allows the user to see the size and location of flaws both underneath and atop the surface. This image is generated by moving the ultrasonic transducer in a sweeping pattern over the part while recording the amplitude and time-of-flight

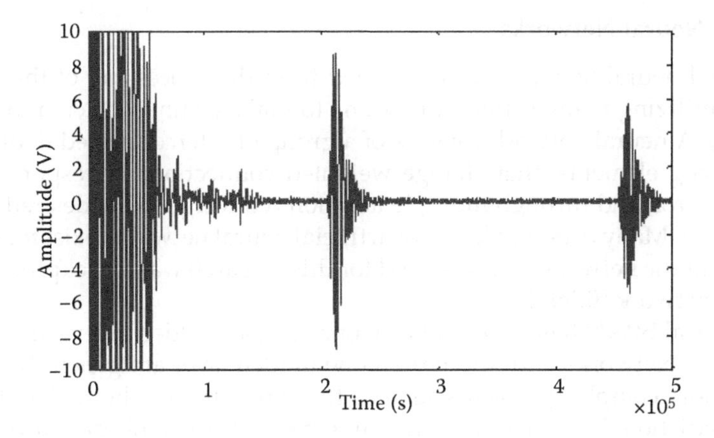

FIGURE 11.1
Ultrasonic waveform. (From Park, I.K., Park, U.S., Ahn, H.K., Kwun, S.I. and Byeon, J.W., "Experimental Wavelet Analysis and Applications to Ultrasonic Non-destructive Evaluation," 15th World Conference on Non-Destructive Testing, 15–21 October 2000, Rome. http://www.ndt.net/article/wcndt00/papers/idn347/idn347.htm.)

of the received pulses as it proceeds. These signals are displayed on a screen at each position of the transducer using either a color scale or a gray scale. The C-scan system used in this research was a water coupled immersion scanning machine, which means that both the transducer and the part being inspected were under water, allowing the water to transmit the sound waves across the distance from the transducer to the part and back again.

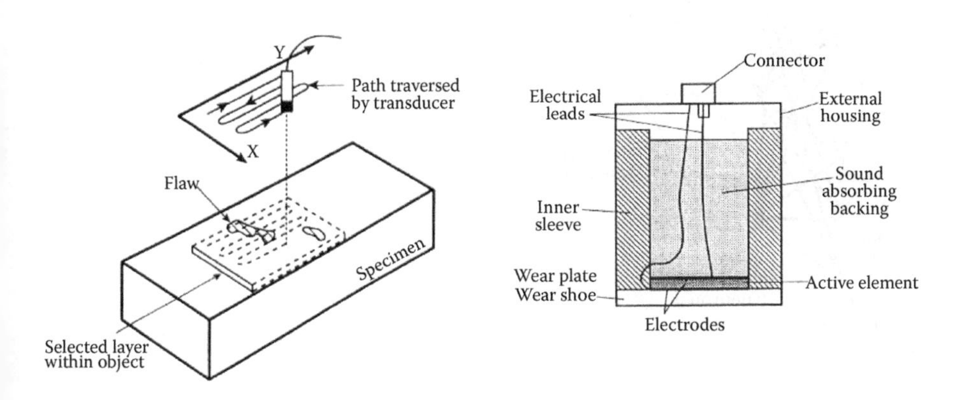

FIGURE 11.2
C-scan pattern (left); ultrasonic transducer cutaway (right). (From Cartz, L., *Nondestructive Testing: Radiography, Ultrasonics, Liquid Penetrant, Magnetic Particle, Eddy Current*. Materials Park, OH: ASM International, 1995.)

11.2.2 Neural Networks

Artificial neural networks were derived from the processing of the human brain, utilizing many different neurons to make complex calculations very quickly. A neural network consists of a group of interconnected neurons, or processing elements, that change weighted connections in response to an output error, and through multiple iterations converge to the desired output or answer. Many different kinds of artificial neural networks exist for special tasks, but the network that was used for this research was a backpropagation neural network (BPNN).

A typical BPNN has an input layer, one or more hidden layers, and an output layer. Networks with more than one hidden layer are generally used to solve more complex problems such as those that require both classification and prediction. Each layer is fully connected to the neighboring layers with information passing from the input layer through the hidden layers and on to the output layer. During the learning phase, the output error is propagated back through the network to update all the connection weights, which is where the BPNN gets its name.

The BPNN is a feedforward, multilayered, supervised learning system. Feedforward refers to the direction of movement of information in the network. In the BPNN of Figure 11.3, information enters the network through the input layer, is then passed forward (from left to right) through one or

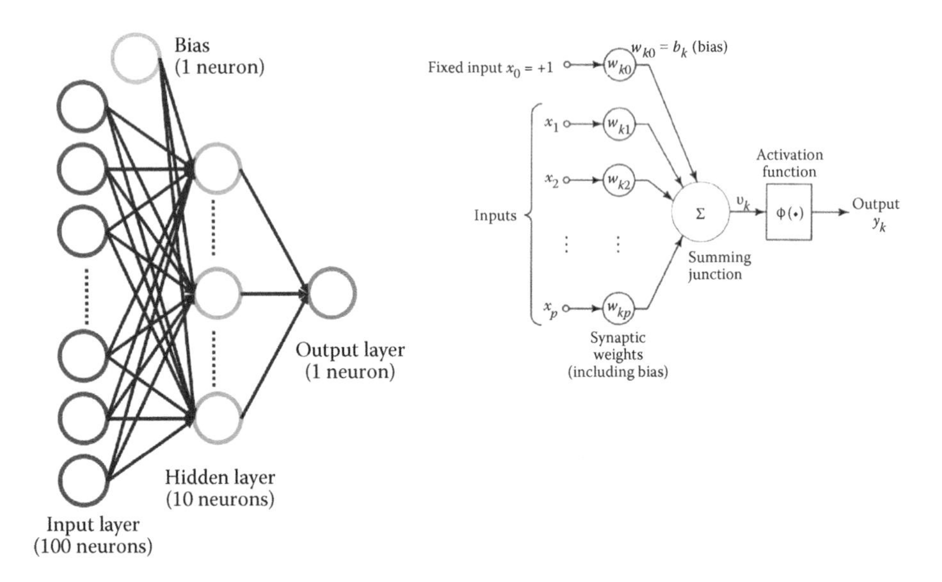

FIGURE 11.3
Backpropagation neural network architecture (left); neuron architecture (right). (From Haykin, S.S., *Neural Networks: A Comprehensive Foundation*. Hamilton, ON, Canada: Macmillan, 1994.)

more hidden layers, and finally exits the network through the output layer. Information is not passed backward through the network, and it is therefore designated as a feedforward network. Multilayered alludes to the number of layers of a neural network. The number of hidden layers can be varied as needed. A supervised network such as the BPNN requires the operator to give it a desired result or output, which the network trains toward during the training phase.

The BPNN initializes by assigning random weights between 0 and 1 to all the neuron connections. The inputs are then multiplied by these weights and passed through a transfer or squashing function, which normalizes the data before it is passed on to the next layer. After each training iteration, the resulting answer from the single output layer neuron is compared with the desired answer, and the error is determined. This error is used to calculate weight adjustments, which are then propagated back through all the network connections, after which the next training iteration begins. This process continues iteratively until the answer in the output layer approaches the desired answer, at which point training is considered to be complete. Once the network has been trained, the network weights are held fixed and are no longer changed. The iterative nature of BPNN training leads to an optimum solution that is relatively insensitive to moderate amounts of noise in the input data.

The testing phase begins when the trained network with its fixed weights is presented another input file that it has not seen before; the data from this file are then run through the network. The resulting answer given by the output layer is compared with the desired answer, and the prediction error is calculated. This final prediction error is what is being minimized for this research. The desired result is to optimize the BPNN such that it will predict the ultimate CAI loads of BVID graphite-epoxy coupons to within a ±10% worst-case error from the ultrasonic C-scan images of the coupons.

The learning algorithm used for the BPNN herein was the normalized cumulative delta rule. This rule adds up all the squared errors over the training set or epoch, then takes the square root of the sum and divides this value by the epoch (training data set) size to normalize it. As mentioned previously, this normalized root mean square (RMS) error is then used to update the weighted connections between the output and the hidden layer(s), and the hidden layer(s) and the input. Thus, all network weights are updated at the end of each pass through the training set or training epoch (which in this case comprised the data from 15 of the 21 total coupons repeated three times in random order). When this RMS output error reaches a user-defined value, typically 5% or less, or the network completes a specified number of training cycles, the training is considered complete, the network weights are fixed, and the testing or prediction phase can begin on data inputs that have not yet been considered.

The transfer function that was used for the BPNN in this research was the sigmoid function as seen in Figure 11.4. The sigmoid transfer function is described by the following equation:

$$\phi(v) = \frac{1}{1+e^{-av}}$$

where

$\phi(\nu)$ = value of sigmoid function with input ν

a = slope parameter

Varying the slope parameter a yields sigmoid functions with different slopes. At the origin, the slope of the sigmoid transfer function is $a/4$. At the extremities of the function, as ν approaches infinity, the slope of the sigmoid function becomes infinitely small and training is very slow. For higher positive values of input ν, the transfer function scales the output value to 1, whereas for higher negative values of input ν, the transfer function scales the output value to 0. This scaling of values effectively squashes the input data within each neuron such that its output ranges from 0 to 1. This makes the larger numbers less significant and smaller numbers more significant such that the data fed into subsequent neurons can be more easily processed by the neural network. The bias neuron has a constant output of 1, which, when multiplied by the connection weight, acts as a translation term to shift the sigmoid activation function $\phi(\nu)$ such that it operates near its highest slope and therefore trains as quickly as possible.

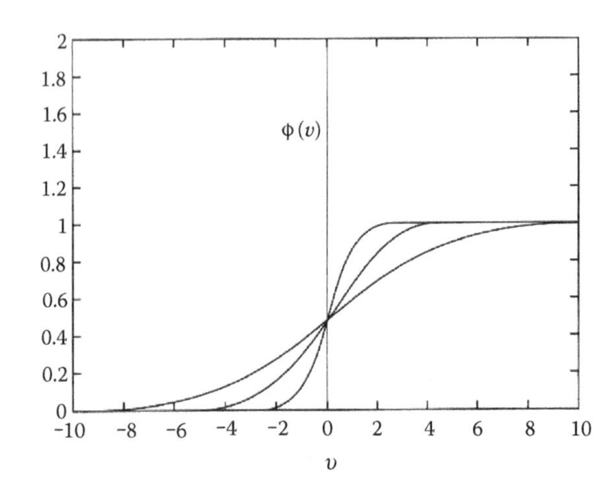

FIGURE 11.4
Sigmoid functions at different a values. (From Haykin, S.S., *Neural Networks: A Comprehensive Foundation*. Hamilton, ON, Canada: Macmillan, 1994.)

11.2.3 Fast Fourier Transform

A fast Fourier transform (FFT) maps a real image into the frequency domain, which enables the image to be filtered. This is accomplished by taking the FFT of each pixel in the image and transforming its value into a complex number in the frequency domain. The real image consists of the summation of several different sinusoids with different frequencies and amplitudes. When an image is transformed using an FFT, it represents each sinusoid as a pair of conjugate symmetric points, each with a frequency and a phase. Figure 11.5 shows the correlation between a sinusoidal image and its frequency transform. Measured with relation to the center of the FFT image, the distance to each point denotes the frequency of the sinusoidal wave. The angle that the position vector from the center of the FFT image to the point, measured from the horizontal, denotes the phase angle of the sinusoidal wave. To transform a real two-dimensional image into the frequency domain, the equation below must be calculated for each pixel in the real image:

$$F(x,y)=\sum_{m=0}^{M-1}\sum_{n=0}^{N-1} f(m,n)e^{-j2\pi\left(x\frac{m}{M}+y\frac{n}{N}\right)}$$

where

$F(x,y)$ = value of each pixel at position (x,y) in the frequency domain
$f(m,n)$ = value of each pixel at location (m,n) in the space domain
M = width of the image
N = height of the image

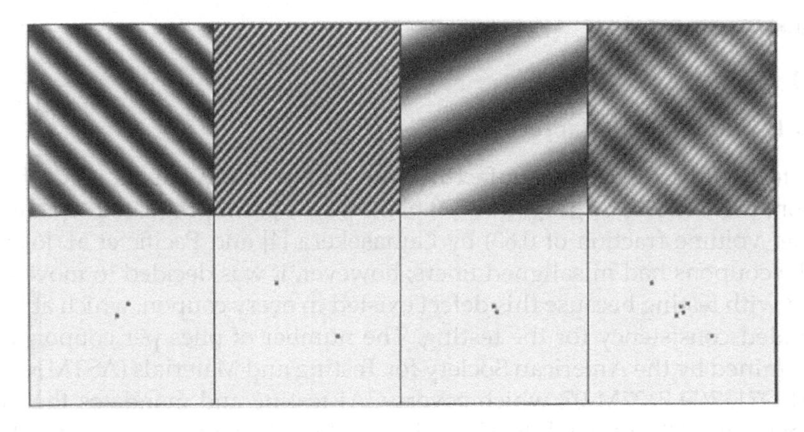

FIGURE 11.5
Sinusoidal pattern with corresponding frequency transform. (From Russ, J.C., *The Image Processing Handbook*. 4th Edition. Raleigh: CRC Press, Materials Science and Engineering Department, North Carolina State University, 2002.)

After the image has been filtered and the noise eliminated, the FFT must be inverted to return the transformed image to a real image in the spatial domain. By employing the equation below at each point and its corresponding conjugate in the frequency image, it is reverted back to a real image in the spatial domain:

$$f(m,n) = \frac{1}{MN} \sum_{m=0}^{M-1} \sum_{n=0}^{N-1} F(x,y) e^{j2\pi\left(x\frac{m}{M}+y\frac{n}{N}\right)}$$

It is important to notice in Figure 11.5 that as the distance of the conjugate symmetric points grows from the center point of the frequency image (bottom row), the frequency of the corresponding real image increases (top row). The importance of this is discussed later in the image filtering section, but an important generalization can be reached from this: the higher frequencies of the real image are focused on the edges of its corresponding frequency transform. Also, as a general rule, the lower frequencies determine the overall shape of the image, while the higher frequencies sharpen the edges and control the fine details. The two images on the far right column of Figure 11.5 are the result of the summation of the three signals in the preceding three columns. The image (top row) of column four is analogous to the images analyzed in this research; the C-scan image can be considered the summation of many different sinusoids of differing phases. The theory of noise cancellation of the C-scan images is predicated on the correct selection of the high-frequency noise and its cancellation.

11.3 Procedure

11.3.1 Coupon Manufacture

The test coupons were manufactured using six 24-ply panels made from Cycom 985 GF3070PW graphite 3070 plain weave preimpregnated tape (with a fiber volume fraction of 0.63) by Gunasekera [4] and Pacific et al. [6]. All of the coupons had misaligned fibers; however, it was decided to move forward with testing because this defect existed in every coupon, which at least provided consistency for the testing. The number of plies per coupon was determined by the American Society for Testing and Materials (ASTM) standard D7137/D 7137M-07, which covers CAI testing and mandates that test coupons be 0.20-in thick [7,8]. Considering the thickness of each individual ply and the thickness of the plain weave tape, it was determined that a 24-ply layup would yield a coupon thickness of the required 0.20 in. The composite panels were laid up in a wooden jig, and the resulting laminate panels were

FIGURE 11.6
C-scan image of misaligned fibers in test coupon.

cured at 355°F for 2 h while being clamped with four C-clamps between two aluminum caul plates to prevent warping. After curing, the oven was shut down, and the laminate panels were left to cool to room temperature in the oven. Once cooled, each panel was cut into four (4 in × 6 in) coupons using a diamond-tipped wet saw. Twenty-one of the 24 resulting coupons were useable for this research. Each coupon was labeled with a number and a letter designating from which plate the coupon originated, with different coupons coming from the same plate given the same number but different letters [4] (Figure 11.6).

11.3.2 Experimental Procedure

The 21 useable coupons were impacted at known energies of 10, 12, 14, 16, 18, and 20 J using an Instron Dynatup 9200 impacter with a blunt 0.5-in hemispherical tup (Figure 11.7) to create BVID. The impacter simulated a low-velocity impact similar to a tool dropping on the coupon. Pneumatic brakes on the impacter were used to avoid multiple impacts from the tup, since the impacter bounced after the initial impact. The coupon was marked with a silver metallic marker to more accurately determine the center of the coupon as the impact site for the tup. Pneumatic clamps held the sample in place to avoid movement during the test and ensure a precise impact location at the center of the coupon.

After impacting the samples, the BVID coupons were C-scanned using a Physical Acoustics Corporation (PAC) ULTRAPAC II water immersion C-scanner (Figure 11.7). The ULTRAPAC II system employed a 0.25-in-diameter piezoelectric crystal ultrasonic transducer with a characteristic frequency of 5 MHz for scanning. Figure 11.8 shows the hardware setup menu of the C-scan system, detailing all the settings used for this research. The

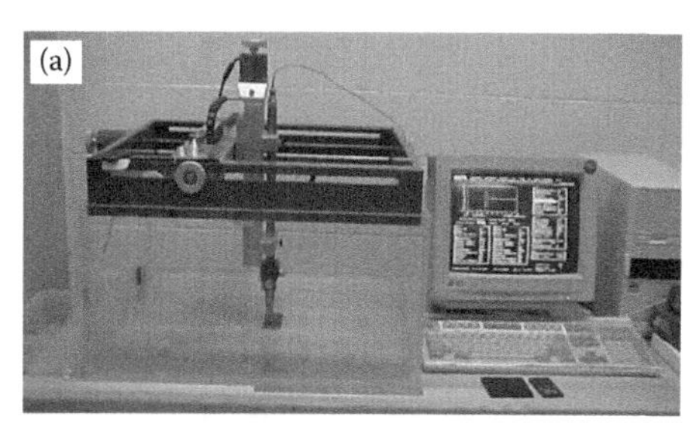
(a)

(b)

(c)

FIGURE 11.7
UltraPAC II C-scan imaging system (a); Boeing CAI test fixture (b); Instron Dynatup 9200 Impacter (c).

ULTRAPAC II system outputs several images, including amplitude and time-of-flight; however, only the amplitude image was used for this research. The horizontal lines on the A-scan output represent the gates of the scan. These gates control what information is recorded, which allows the computer to ignore the initial pulse as the wave enters the water couplant. The gate time selection also allows the user to select data specifically from the damaged layers of the composite, essentially looking inside the composite only at the depth of the damaged region.

The samples were then compressed to failure to determine their ultimate CAI loads using a Tinius-Olsen model 290 Lo Cap testing machine (Figure 11.7). A Boeing CAI test fixture was used to keep the coupons from buckling in accordance with ASTM standards D 7137/D 7137M-07 [9]. The coupons were secured in the Boeing CAI fixture and compressed to failure, with failure always occurring at the BVID impact location, indicating that the impact energies and associated levels of BVID selected had adequately compromised the strength of the composite.

Once the coupons were compressed to failure and testing had concluded, the C-scan images were cropped to a 100 × 100 pixel square around the damaged area (Figure 11.8). This damaged square image was a 256-color RGB

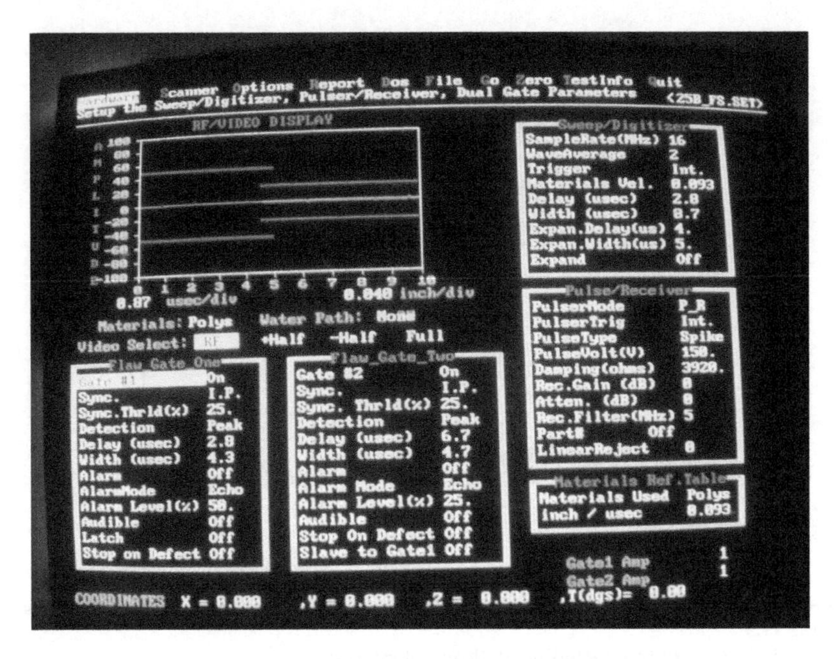

FIGURE 11.8
C-scan hardware setup screen.

image consisting of three layers—a red (R) layer, a green (G) layer, and a blue (B) layer—with the pixel of each layer assigned a number from 0 to 255. Finally, the image was input to a MATLAB processing program developed by Hess [2] and converted into a 300 row × 1 column matrix, which was used as the artificial neural network input for ultimate CAI load prediction.

11.3.3 Image Manipulation

The C-scan image files output from the UltraPAC II were in ".PCX" format and needed to be converted to ".BMP" format so they could be analyzed as 256-color RGB images. Once all the images were converted to 256-color BMP files, they were input to the MATLAB program. Figure 11.9 shows the MATLAB output for an impacted sample displaying the three color layers side by side. The graph below the image displays the color value of each pixel (AMP) and the x-position of the pixel. Positions 0 to 100 correspond to the red layer, positions 100 to 200 correspond to the green layer, and positions 200 to 300 correspond to the blue layer.

By observation, the red and the blue layers included mostly noise, whereas the green layer clearly displayed a dip corresponding to the impact location. Not all images were as free of noise as those shown in Figure 11.9. Several of the images that were C-scanned contained significant noise, such as sample 3B in Figure 11.10. When these noisy images were input to the BPNN,

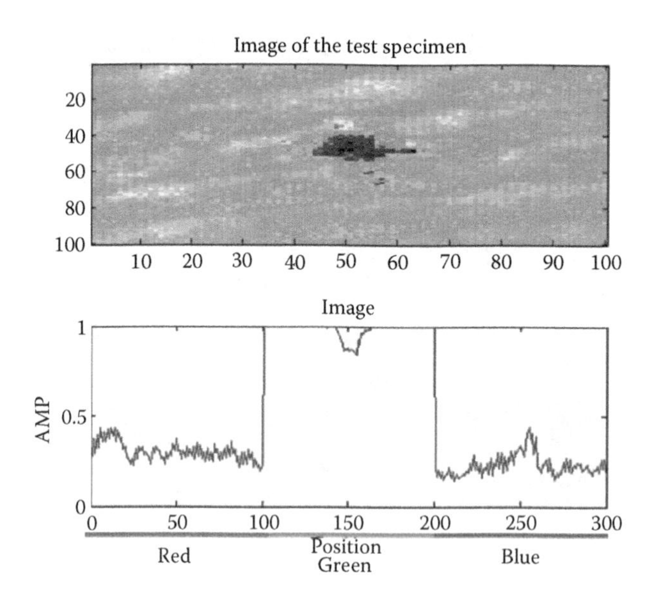

FIGURE 11.9
MATLAB program output including all three color layers.

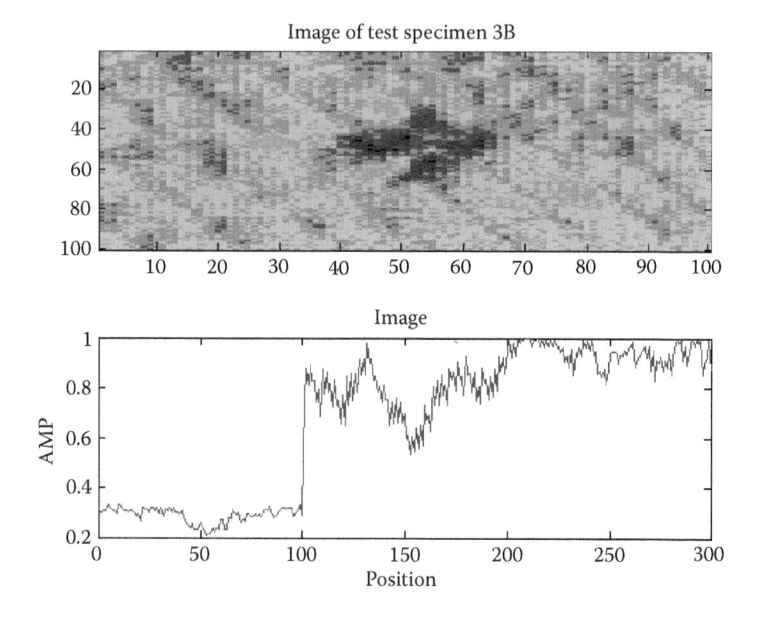

FIGURE 11.10
Noisy C-scan image.

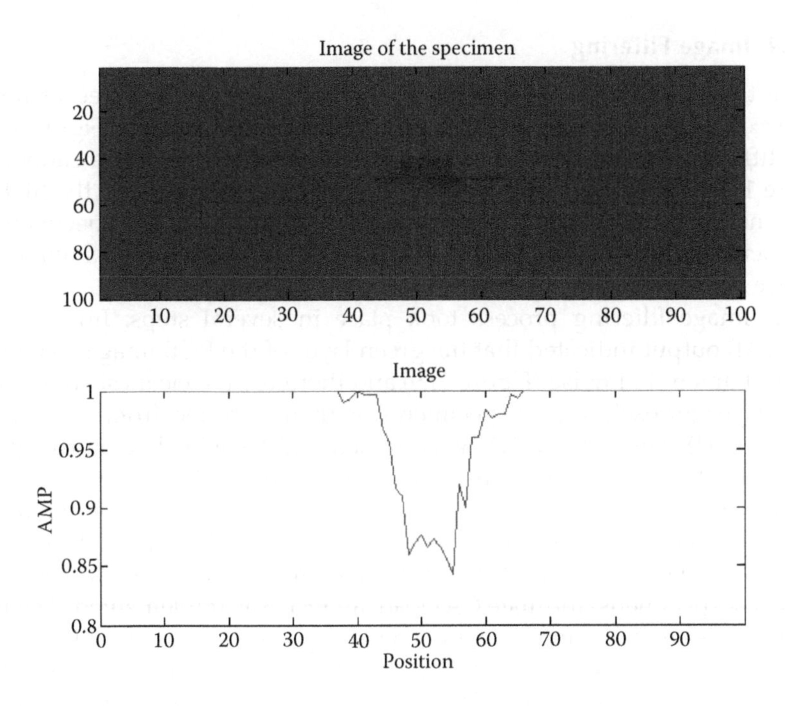

FIGURE 11.11
MATLAB program output containing green layer only.

the network had difficulty determining the location of the damaged section, thereby increasing the error significantly, as seen in Figure 11.10. In an effort to minimize noise, the red and blue layers were removed from each image (Figure 11.11), and only the green layer was used as input to the BPNN (Table 11.2).

TABLE 11.2
Worst-Case Error Using Red, Green, and Blue Layers

Coupon ID	Impact Energy (J)	Ultimate Compressive Load (lb_f)	Predicted Ultimate Compressive Load (lb_f)	Percent Error
27B	16	18,825	20,338.64	8.04%
25D	20	17,249	16,638.96	–3.54%
26C	18	20,729	20,079.15	–3.14%
3A	14	19,156	21,583.95	12.67%
24B	12	19,782	17,856.1	–9.74%
26A	10	22,190	23,274.83	4.89%

11.3.4 Image Filtering

While the majority of the green layer images were mostly free of noise, there existed a few images that were particularly noisy, such as Figure 11.10. It is difficult to identify the impact location and damage of the coupon in Figure 11.10; therefore, a filter had to be implemented to remove the higher-frequency noise. Figure 11.12 shows a side-by-side view of test specimen 3B, a particularly noisy image, after it was filtered and before it was filtered to remove the high-frequency noise.

This image filtering process took place in several steps. Initially, the MATLAB output indicated that the green layer of the RGB image contained the least amount of noise (Figure 11.9) and therefore the cleanest signal. The green layer for each impact specimen was then extracted from each image (Figure 11.11). Thus, the BPNN was trained and tested solely on the green layer of the C-scan images as discussed later in Section 11.4.2.

While the noise of most images was completely eliminated by removing the red and blue layers, significant noise still remained in some of the images, as can be seen in Figure 11.12b. In an effort to improve the BPNN's prediction of the test specimens' ultimate CAI load, an FFT was implemented. The FFT decomposes an image into real and complex parts that represent the image in the frequency domain. The FFT allows the higher-frequency noise to be eliminated fairly easily, removing the pixelation in the more noisy images and leaving the noiseless images relatively unchanged.

After the FFT of the image has been taken, the pixels that constitute the high-frequency noise in the image are located near the center of the FFT

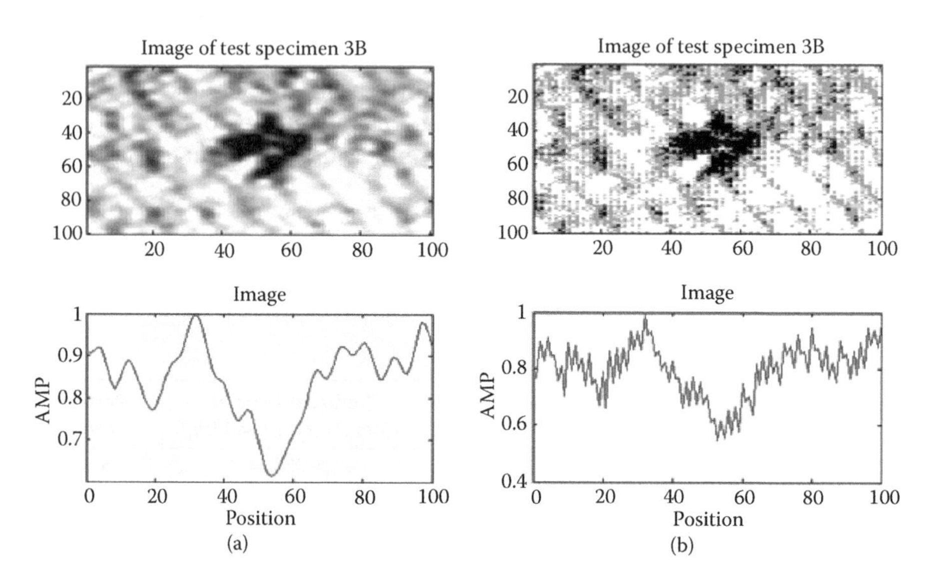

FIGURE 11.12
Green layer of noisy C-scan image: filtered (a); unfiltered (b).

(a)

(b)

FIGURE 11.13
Test specimen 3B after fast Fourier transform (a) and after shifting the fast Fourier transform (b).

image, while the low-frequency pixels are around the edges as seen in Figure 11.13a. The FFT was shifted to move the high-frequency noise to the edges of the FFT where they could then be filtered out (Figure 11.13b). While operating in the frequency domain, a simple MATLAB code was written to remove the high-frequency noise by setting their values equal to 0. The MATLAB code set the values of pixels to 0 that were outside of a square around the center of the image as seen in Figure 11.14.

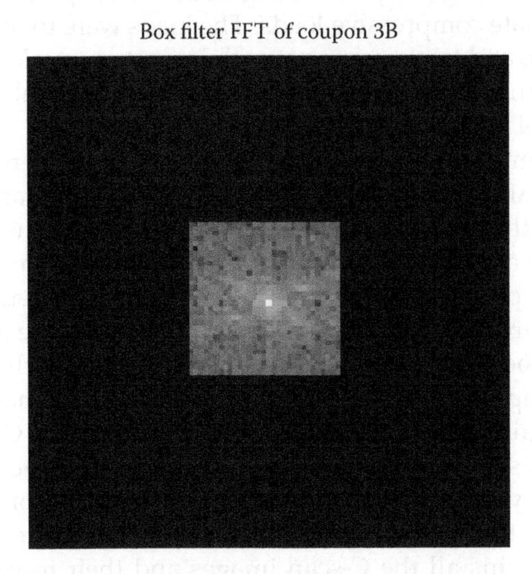

FIGURE 11.14
Fast Fourier transform of specimen 3B after filter was applied.

The size of the square around the center of the FFT controls how much of the image is filtered. A larger area square filters less of the image, while a smaller area square will filter more of the image. The square filter needs to be positioned at the center of the FFT image; its location is important because the FFT of a real image is "a conjugate symmetric." Conjugate symmetric refers to when a real signal is transformed to the frequency domain; each pixel is represented as a complex number with conjugate values located symmetrically at each corner of the image. To avoid getting a bad inverse, and therefore getting complex numbers as pixel values when the image is returned to the space domain, both of the pixels' conjugate values must be set equal to 0. This is accomplished by creating a square filter that is symmetric about the center of the image, as shown in Figure 11.14.

11.4 Results

11.4.1 Neural Network Training and Testing

The initial selection of coupons to be included in the training set and the testing set was random. This, however, led to testing a data set containing the highest ultimate CAI load, which confused the neural network because it encountered a higher load during testing than it had seen during training. This induced some error into the calculation and was therefore avoided in future training and testing sets. Subsequently, the selection of coupons for the training set began by including both the coupons with the highest and lowest ultimate compressive loads. The loads were then organized from highest to lowest, and two coupons at each impact energy level were selected for training, ensuring that the coupons selected were evenly spaced in order to give the BPNN an optimal selection of data points.

Because the sample size was small (only 21 coupons were C-scanned and compressed to failure), a method referred to as "bootstrapping data" was used to increase the apparent sample size for the BPNN. The method of bootstrapping data is a technique that uses the same composite coupon multiple times in random positions in the training file to fool the neural network into believing there are more coupons in the data set than are actually present. Bootstrapping does not skew the data; it only increases the size of a small data set. Bootstrapping data was employed in this experiment by using each coupon in the training set three times in a random order. Once the training and testing data sets and files were written, the neural network needed to be optimized to yield the lowest possible worst-case error. Appendix A of reference [9] identifies the coupons in the training and testing files, while Appendix B contains all the C-scan images and their respective MATLAB output files that became the BPNN input files that were used in this research.

11.4.2 Neural Network Optimization

The only way to find the optimal network architecture and parameter settings is through trial and error [10], which for this research was performed using the NeuralWorks Professional II Plus software package. The architecture chosen for the BPNN began with one hidden layer. Varying the number of hidden layer neurons, the hidden layer learning coefficient, the output layer learning coefficient, the momentum, the learning ratio, the F' offset, and the transition point through trial and error is required to arrive at the optimal neural network. This trial-and-error procedure can be very time consuming; thus, two methods were used to expedite the optimization process.

The two optimization techniques employed to find the optimal network architecture and parameters were series optimization and parallel optimization. Parallel optimization entailed varying all neural network parameters independently in different networks, then once the network was optimized, all the optimum parameters were included in the same neural network. Parallel optimization is the most time-efficient method of BPNN optimization; however, it can only be used when a team of people is available to optimize the neural network. For this project, it was found that both parallel and series optimization yielded identical BPNN parameters.

The first optimization technique employed herein was series optimization, which entailed varying all parameters individually, selecting the value that yielded the lowest worst-case error, and then implementing that parameter value in the BPNN. The first parameter value that was varied was the number of hidden layer neurons. Figure 11.15 shows the variation of the worst-case error as the number of hidden layer 1 neurons was increased. A closer view of the area of interest can be seen in Figure 11.16, which focused around 10 hidden layer 1 neurons; this was the optimal number of neurons that yielded an absolute value worst-case error of 11.61%.

With an absolute value worst-case error of 11.61%, the number of hidden layer 2 neurons was varied to explore the need for a two-hidden-layer

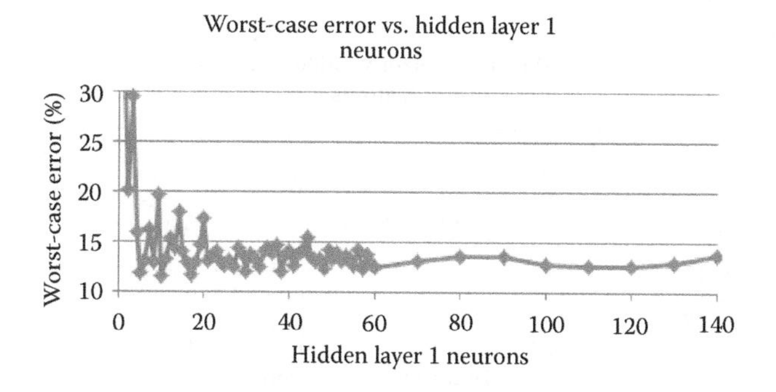

FIGURE 11.15
Hidden layer 1 worst-case error versus number of hidden layer 1 neurons.

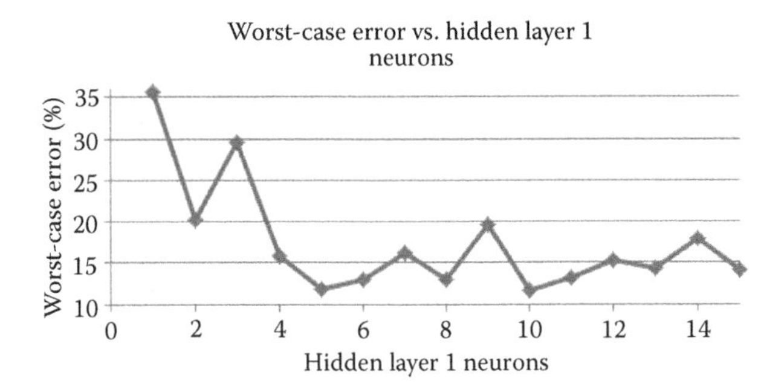

FIGURE 11.16
Zoomed in hidden layer 1 worst-case error versus number of hidden layer 1 neurons.

network. After varying the number of hidden layer 2 neurons between 10 and 100, Figure 11.17 shows that a second hidden layer was not necessary, as all the worst-case errors were greater than the 11.61% value obtained for one hidden layer.

As the network only had one hidden layer, only the hidden layer 1 learning coefficient needed to be varied. Hence, the hidden layer 1 learning coefficient was varied between 0.001 and 0.5, and the optimal value was found to be 0.001, as seen in Figure 11.18. This value yielded a slight reduction in the absolute value worst-case error of from 11.61% down to 11.5%.

The next logical step was to vary the output layer learning coefficient. This proved to be the most successful in decreasing the worst-case error. By decreasing the output layer learning coefficient to 0.017, the error reduced from 11.5% to 5.19%, a decrease of >50% (Figure 11.19).

Figures 11.20 through 23 show the variation of the remaining four parameters as they were modified and optimized using the same techniques.

Table 11.3 summarizes the optimal parameter settings found for the neural network. Using these values, the BPNN was able to predict the ultimate compressive load with a worst-case testing error of –5.16%, well within the ±10%

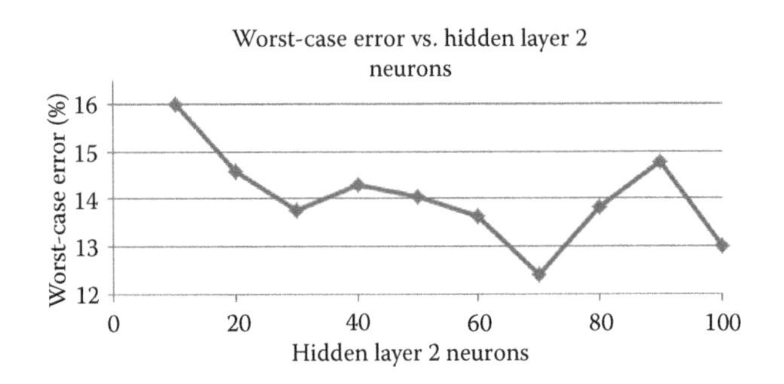

FIGURE 11.17
Worst-case error versus number of hidden layer 2 neurons.

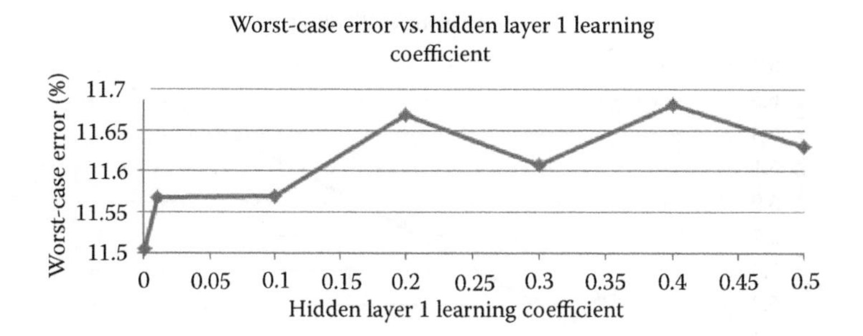

FIGURE 11.18
Worst-case error versus hidden layer 1 learning coefficient.

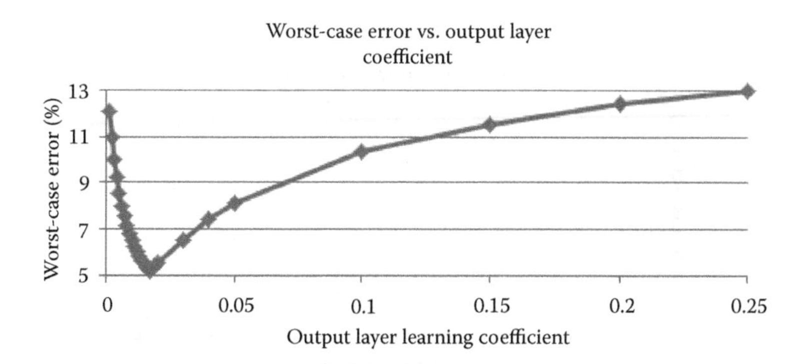

FIGURE 11.19
Worst-case error versus output layer learning coefficient.

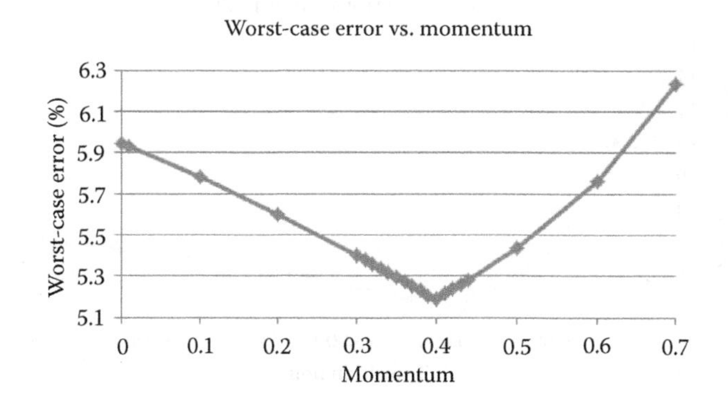

FIGURE 11.20
Worst-case error versus momentum.

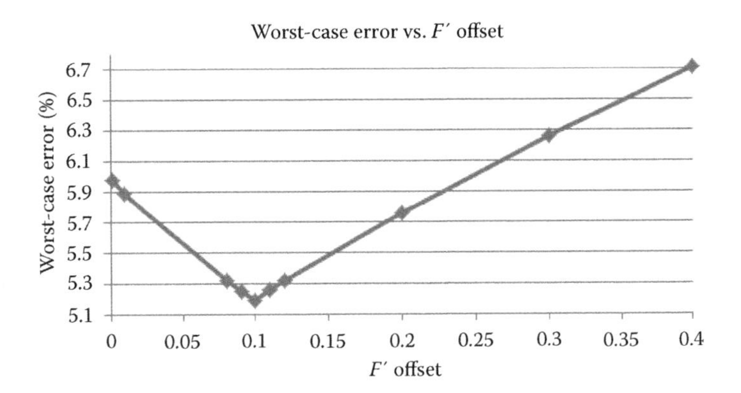

FIGURE 11.21
Worst-case error versus F' offset.

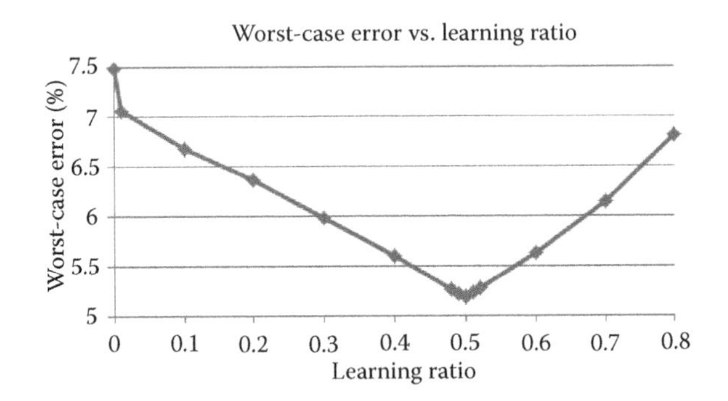

FIGURE 11.22
Worst-case error versus learning ratio.

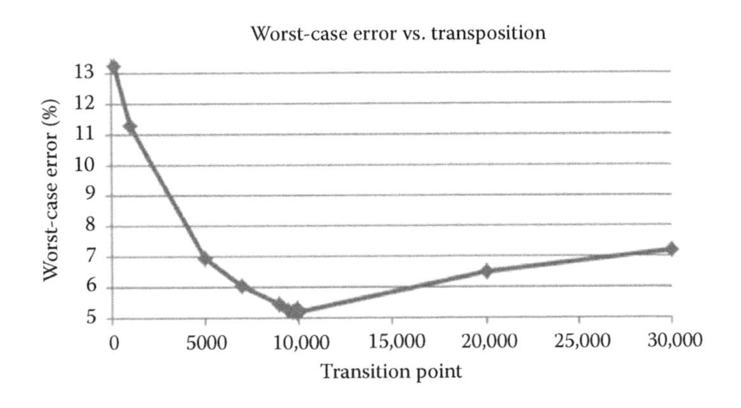

FIGURE 11.23
Worst-case error versus transition point.

TABLE 11.3

Optimized Neural Network Configuration

Number of input neurons	100
Hidden layer 1 neurons	10
Hidden layer learning coefficient	0.001
No. of output neurons	1
Output layer learning coefficient	0.017
Learning coefficient ratio	0.5
Momentum	0.4
Transition point	10,000
Transfer function	Sigmoid
Learning rule	Normalized cumulative delta
F' offset	0.1

target range. The NeuralWorks Professional II Plus software BPNN setup screen with optimal settings can be seen in Figure 11.24.

The compiled results, including both training and prediction errors, can be seen in Table 11.4. The rows in boldface are coupons that were included in the testing file; the other rows were coupons used in the training file. It should be noted that a worst-case testing error of within ±10% was the target of this research; unfortunately, by using optimal settings, the worst-case training error was found to be –12.52%, slightly outside of this range. Therefore, further optimization was sought.

The significant difference in testing and training errors can be attributed to the relatively small output layer learning coefficient, which could possibly

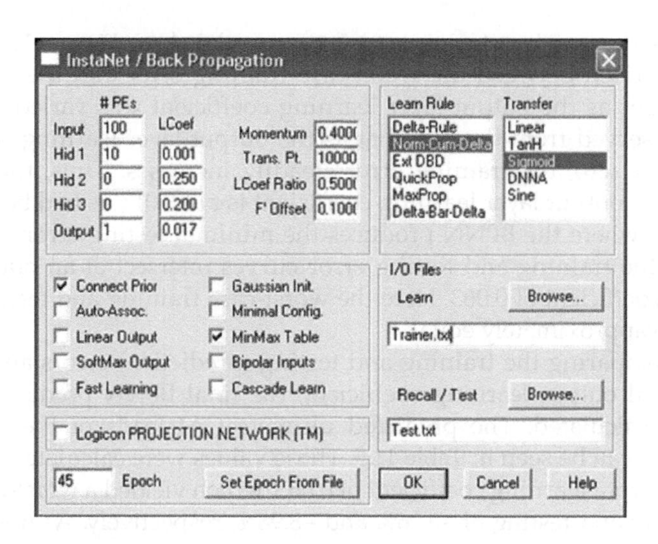

FIGURE 11.24
Optimal BPNN settings.

TABLE 11.4

Unfiltered Image BPNN Results

Coupon ID	Impact Energy (J)	Ultimate Compressive Load (lb_f)	Predicted Ultimate Compressive Load (lb_f)	Percent Error
26D	20	20,024	19,054.8145	–4.84%
24D	20	17,250	17,640.0879	2.26%
25D	**20**	**17,249**	**17,470.6836**	**1.29%**
26C	**18**	**20,729**	**19,663.1777**	**–5.14%**
25C	18	20,010	19,657.5039	–1.76%
27C	18	18,986	19,049.0488	0.33%
27B	**16**	**18,825**	**19,643.6719**	**4.35%**
27D	16	18,742	19,276.1270	2.85%
24C	16	17,944	18,925.3066	5.47%
3A	**14**	**19,156**	**19,802.0117**	**3.37%**
3D	14	19,152	21,107.6641	10.21%
2C	14	16,200	16,996.6348	4.92%
2A	12	21,750	21,078.8828	–3.09%
25B	12	21,749	20,173.1055	–7.25%
2B	12	18,900	20,036.0273	6.01%
3B	12	20,250	20,682.1895	2.13%
24B	**12**	**19,782**	**19,763.3281**	**–0.09%**
27A	12	17,249	19,385.4375	12.39%
24A	10	24,195	21,166.8926	–12.52%
26A	**10**	**22,190**	**21,044.6621**	**–5.16%**
25A	10	21,815	20,603.8965	–5.55%

have caused some overtraining to occur on the data. Figure 11.25 shows the comparison between the worst-case training error and the worst-case testing error as the output layer learning coefficient was varied. It can be clearly observed from the plot that as the output layer learning coefficient approaches zero, the training error steadily increases. Thus, the optimal value of the output layer learning coefficient is not 0.017, as had been earlier supposed, where the BPNN produces the minimal testing error, but rather is where the training and testing error curves intersect at an output layer learning coefficient of 0.063. Here the worst-case training and testing errors should be approximately equal.

After comparing the training and testing (prediction) errors and finding the optimal output learning coefficient, the final BPNN prediction errors could be calculated. The predicted ultimate CAI loads of the unfiltered image data can be seen in Table 11.5. These values were calculated using the optimal output learning coefficient of 0.063, which yielded a worst-case error for training and testing of –8.96% and –8.98%, respectively. As noted, these two values are approximately equal and both are within the ±10% worst-case prediction error goal.

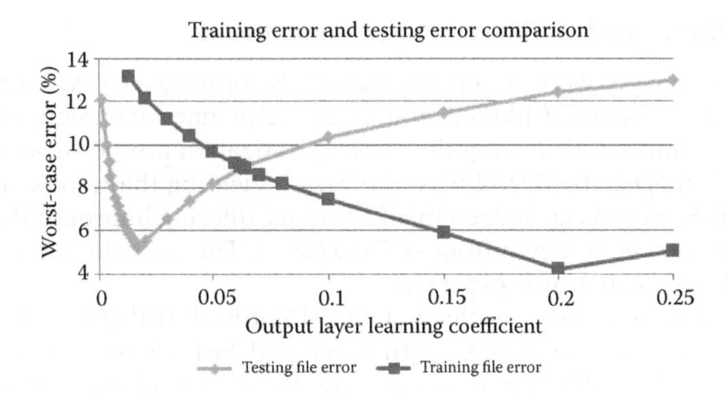

FIGURE 11.25
Comparison between training and testing (prediction) errors.

TABLE 11.5
BPNN Results Using 0.063 Output Layer Coefficient

Coupon ID	Impact Energy (J)	Ultimate Compressive Load (lb_f)	Predicted Ultimate Compressive Load (lb_f)	Percent Error
26D	20	20,024	19,764.19	–1.30%
24D	20	17,250	16,981.84	–1.55%
25D	**20**	**17,249**	**17,328.87**	**0.46%**
26C	**18**	**20,729**	**18,866.51**	***–8.98%***
25C	18	20,010	19,773.29	–1.18%
27C	18	18,986	19,326.75	1.79%
27B	**16**	**18,825**	**19,422.61**	**3.17%**
27D	16	18,742	18,726.62	–0.08%
24C	16	17,944	18,332.10	2.16%
3A	**14**	**19,156**	**20,328.60**	**6.12%**
3D	14	19,152	20,404.80	6.54%
2C	14	16,200	16,447.73	1.53%
2A	12	21,750	21,502.41	–1.14%
25B	12	21,749	21,371.35	–1.74%
2B	12	18,900	19,805.28	4.79%
3B	12	20,250	20,323.34	0.36%
24B	**12**	**19,782**	**19,218.11**	**–2.85%**
27A	12	17,249	18,452.20	6.98%
24A	10	24,195	22,026.11	*–8.96%*
26A	**10**	**22,190**	**21,571.86**	**–2.79%**
25A	10	21,815	21,028.18	–3.61%

11.4.3 Filtered and Unfiltered Image Comparison

The C-scan images were placed into the BPNN for ultimate CAI load prediction after they were FFT filtered, and a significant amount of high-frequency noise was removed. Following the same optimization procedure as with the unfiltered images, the BPNN was optimized, yielding the results tabulated in Table 11.6. Here it can be seen that FFT image filtering improved the worst-case prediction error down from –8.98 to 8.65%. This reduction in error was not nearly as much as was expected.

From the summarized results of Table 11.7, it is clear that the BPNN was able to predict accurately with both noisy and FFT filtered images. Using only its iterative optimization scheme, the BPNN was able to remove most of the high-frequency noise, as the weights of the respective neurons that contained most of the image noise approached zero. The FFT image filtering process provided some reduction in worst-case error; however, in this case, the rectangular filter used to remove the high-frequency noise may have

TABLE 11.6

Filtered Image BPNN Results

Coupon ID	Impact Energy (J)	Ultimate Compressive Load (lb_f)	Predicted Ultimate Compressive Load (lb_f)	Percent Error
26D	20	20,024	19,795.97	–1.14%
24D	20	17,250	17,045.25	–1.19%
25D	**20**	**17,249**	**18,306.68**	**6.13%**
26C	**18**	**20,729**	**18,942.68**	***–8.62%***
25C	18	20,010	19,858.97	–0.75%
27C	18	18,986	19,351.67	1.93%
27B	**16**	**18,825**	**18,478.77**	**–1.84%**
27D	16	18,742	18,471.83	–1.44%
24C	16	17,944	18,155.08	1.18%
3A	**14**	**19,156**	**20,735.68**	**8.25%**
3D	14	19,152	20,063.35	4.76%
2C	14	16,200	16,608.53	2.52%
2A	12	21,750	21,512.81	–1.09%
25B	12	21,749	21,400.50	–1.60%
2B	12	18,900	19,886.24	5.22%
3B	12	20,250	20,379.29	0.64%
24B	**12**	**19,782**	**18,632.73**	**–5.81%**
27A	12	17,249	18,741.61	*8.65%*
24A	10	24,195	22,124.94	–8.56%
26A	**10**	**22,190**	**21,316.96**	**–3.93%**
25A	10	21,815	20,981.32	–3.82%

TABLE 11.7

Filtered and Unfiltered Image Comparison

	Worst-Case Testing Error	Worst-Case Training Error
Unfiltered	–8.98%	–8.96%
Filtered	–8.62%	8.65%

removed too much frequency information in the image, thereby resulting in a higher-than-expected worst-case error.

11.4.4 Material Allowables

The tolerance interval within which the BPNN predictions should fall is referred to as the B-basis material allowables for the composite coupons. B-basis allowables are defined as the tolerance intervals within which there is a 95% confidence that 90% of all future ultimate CAI ultimate loads will fall [11]. The B-basis tolerance interval may be calculated from the following equation:

$$\text{Interval} = \pm K(n, P, c)s_x$$

where

K = factor dependent on the n, P, c parameters
n = number of samples in a group
P = fraction of population
c = confidence interval
s_x = sample standard deviation

It can be clearly observed in Figure 11.26 that all the ultimate compressive loads predicted by the optimized BPNN fall well within the B-basis allowables of the graphite-epoxy coupon sample group.

Increasing the number of coupons at each impact damage energy level would obviously decrease the B-basis allowables. If instead of the three or six samples available at each energy level, there were 30 coupons at each energy level, the K values would all decrease from 6.919 (three samples) or 3.723 (six samples), as seen in Table 11.8, to 2.140 (30 samples). Assuming the same mean and standard deviation values for the increased sample size, the B-basis allowables would decrease significantly to the values shown by the dashed lines in Figure 11.26. Note that all the BPNN predictions are well within these more conservative values as well. This is significant because the B-basis allowables are typically calculated for composites based on a sample size of 30 or more test specimens. Unfortunately, this research did not have the resources available to generate such a large sample size.

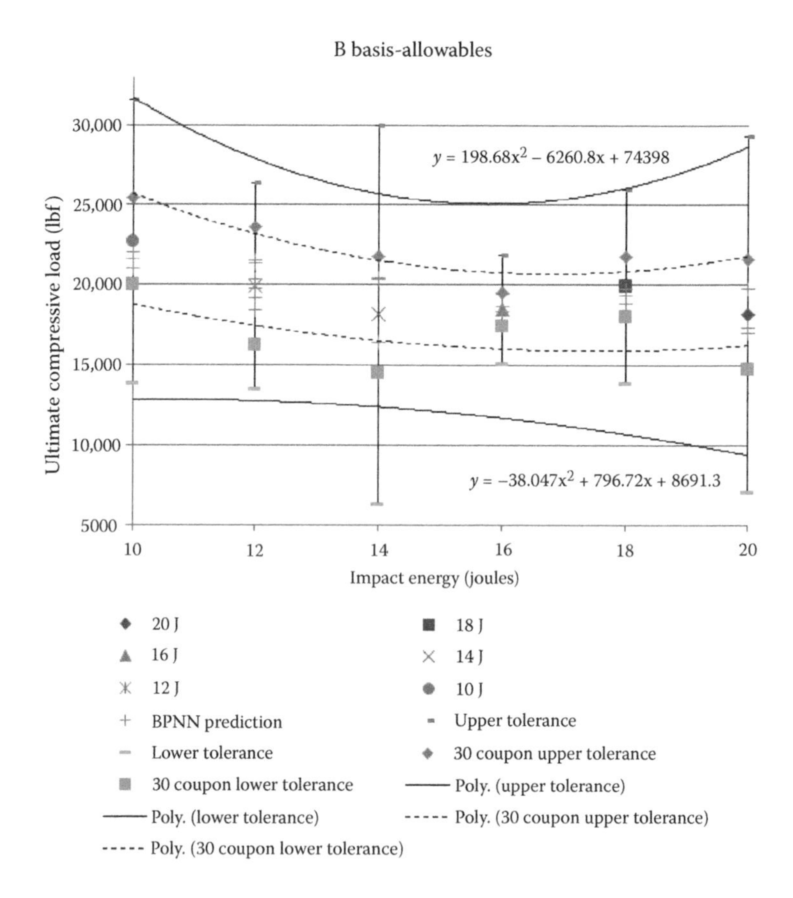

FIGURE 11.26
B-basis allowables.

TABLE 11.8
B-basis Allowables Tolerance Interval

Impact Energy (J)	No. of Coupons	Mean Ultimate Compressive Load (lb_f)	Standard Deviation (lb_f)	K Factor	B-basis Allowables Interval (lb_f)	Upper Limit (lb_f)	Lower Limit (lb_f)
10	3	22,733.33	1,279.65	6.919	±8,853.91	31,587.24	13,879.42
12	6	19,946.67	1,731.62	3.723	±6,446.80	26,393.47	13,499.86
14	3	18,169.33	1,705.49	6.919	±11,800.31	29,969.64	6,369.021
16	3	18,503.67	486.46	6.919	±3,365.81	21,869.48	15,137.86
18	3	19,908.33	875.94	6.919	±6,060.60	25,968.94	13,847.73
20	3	18,174.33	1,601.86	6.919	±11,083.26	29,257.59	7,091.075

11.5 Conclusions and Recommendations

11.5.1 Conclusions

After initial image preprocessing, a BPNN using the green layer only data from the ultrasonic C-scan image of BVID in graphite-epoxy composite laminates was able to accurately predict the ultimate CAI load with a worst-case error of –8.98%, which was within the ±10% goal for this research and comfortably within the B-basis allowable for composites. Because the FFT noise removal routine resulted in a slight improvement in the prediction capability of the BPNN, down from a worst-case error of –8.98% to 8.65%, it can be concluded that some high-frequency image noise was removed by the FFT, which aided the BPNN in making more accurate CAI load predictions. This research has demonstrated the viability of an ultrasonics based nondestructive evaluation technique that could save aircraft manufacturers and maintenance companies thousands of dollars on unnecessary repairs by giving a trained technician the ability to objectively evaluate the effect of BVID on any composite part and predict with confidence the effect of the damage on the ultimate CAI load.

11.5.2 Recommendations

Manually eliminating individual points in the FFT frequency image and setting their values to zero, as was done here, would be far too time consuming for an operational assessment of impact damage. Moreover, the square filter used in this research for FFT noise removal may have been too aggressive in eliminating pixels in the frequency domain. Future research might investigate the effect of different filters on image noise removal. Future research might also inquire into the possibility of using the raw reflectivity of the C-scan image, rather than using solely the green layer data, for image manipulation and ultimate CAI load prediction.

References

1. Quilter, Adam. Composites in Aerospace Applications. [Online] [Cited: April 12, 2011.] http://cis.ihs.com/NR/rdonlyres/AEF9A38E-56C3-4264-980C-D8D6980A4C84/0/444.pdf.
2. Hess, Christopher D. *Residual Compressive Strength Prediction of Carbon/Epoxy Laminates Subjected to Low Velocity Impact Damage*. Daytona Beach: Embry-Riddle Aeronautical University, 2003. MS Aerospace Engineering Thesis.
3. Nguyen, Tuan-Khoi Dang. *Damage Assessment and Strength Prediction in S2-Glass/Epoxy Laminates Subjected to Low Energy Impact*. Daytona Beach: Embry-Riddle Aeronautical University, 2005. MS Aerospace Engineering Thesis.

4. Gunasekera, Anthony M. *Compression After Impact Strength Prediction in Graphite/ Epoxy Laminates Using Acoustic Emission and Artificial Neural Networks*. Daytona Beach: Embry-Riddle Aeronautical University, 2009. MS Aerospace Engineering Thesis.
5. American Society for Nondestructive Testing. *Nondestructive Testing Handbook, Ultrasonic Testing*. [ed.] Gary L. Workman, Doron Kishoni and Patrick O. Moore. 3rd edition. Columbus: s.n., 2007. Vol. 7.
6. Pacific, Andrew B., Hill, Eric v.K., Geiselman, Nikolas L., Foti, Christopher J., Gonitzke, Matthew D., and Surber (now McCann), Hannah L. "Neural Network Prediction of Ultimate Compression After Impact Loads in Graphite-Epoxy Coupons from Ultrasonic C-Scan Images," Proceedings of the ASNT Fall Conference & Quality Testing Show 2010, American Society for Nondestructive Testing, Columbus, OH, 2010.
7. Standard Test Method for Compressive Residual Strength Properties of Damaged Polymer Matrix Composite Plates. D7137/D 7137 M. ASTM Standards. Conshohocken, PA: ASTM International, 2007.
8. Standard Test Method for Measuring the Damage Resistance of a Fiber-Reinforced Polymer Matrix Composite to a Drop-Weight Impact Event. D7136/D 7136 M. ASTM Standards. Conshohocken, PA: ASTM International, 2007.
9. Geiselman, Nikolas L. *Neural Network Prediction of Ultimate Compression After Impact Loads in Graphite-Epoxy Coupons From Ultrasonic C-Scan Images*. Daytona Beach: Embry-Riddle Aeronautical University, 2011. M.S. Aerospace Engineering Thesis.
10. Dorfman, Michele D. *Ultimate Strength Prediction in Fiberglass/Epoxy Beams Subjected to Three Point Bending Using Acoustic Emission and Neural Networks*. Daytona Beach: Embry-Riddle Aeronautical University, 2004. MS Aerospace Engineering Thesis.
11. Grégoire, Alexandre D. *Ultimate Compression After Impact Load Prediction in Graphite/Epoxy Coupons Using Neural Network and Multivariate Statistical Analyses*. Daytona Beach: Embry-Riddle Aeronautical University, 2011. MS Aerospace Engineering Thesis.

12

Distributed In Situ *Health Monitoring of Nanocomposite-Enhanced Fiber-Reinforced Polymer Composites*

Bryan R. Loyola, Valeria La Saponara, and Kenneth J. Loh

CONTENTS

12.1 Introduction

Fiber-reinforced polymer (FRP) composites have been used since the 1970s in a variety of engineering applications requiring materials with high stiffness/weight and strength/weight ratios. Unfortunately, they can experience complex damage modes, ranging from delamination and fiber cracks in monolithic composites made with prepreg composites, to shear core failure in sandwich composites or, in woven composites, macroscale damage with various features (transverse cracks in fill, shear failure in warp, cracks in pure-matrix regions, and inter-ply and intra-ply delamination). Damage may be triggered by stress

TABLE 12.1

Comparison of Nondestructive Testing Methods in Industry: Main Characteristics of NDT Methods Traditionally Used in Industry

Physics	Mechanical		Thermal	Magnetic	X-ray	Visual	
Techniques	**Ultrasound**	**Acoustic Emission**	**Thermography**	**Eddy Current**	**Tomography**	**Penetrant Testing**	**CCD Camera**
Suitable with metallic sample	+++	+++	+++	+++	−−	+++	+++
Suitable with composite sample	+++	+++	+++	−−−	+++	+++	+++
Direct test on structures	−−−	+++	+++	+++	−−−	+++	+++
Sample size	− Limited	+++ Unlimited	+++ Unlimited	+++ Unlimited	−−− Limited	+++ Unlimited	+++ Unlimited
Complex sample geometry	++	+	+	−	+++	−−−	−
Test by transmission	+++	+++	+++	−−−	+++	−−−	−−−
Test by reflexion	+++	+++	+++	+++	−−−	+++	+++
Field measurement	++	−−−	+++	++	++	−−	+++
Outer defect detection	+++	+	+++	+	+++	+++	+++
Inner defect detection	+++	+++	+++	+++	+++	−−−	−−−
Contact type	−− Fluid	− Fluid/solid	+++ Air	+ Solid	+++ Air	−−− Fluid	+++ Air
Respect of sample integrity	+	+++	+++	+++	+	−	+++
Environmental respect	+	+++	+++	+++	−−−	−−−	+++
Method implementation	−	+++	+	+++	−−−	+++	+++
Degree of possible automation	+++	−	+++	−	+	−−−	+++
Price of a standard equipment	−−	++	−	++	−−−	+++	+

Source: Kuhn, E., Valot, E., Herve, P., *Composite Structures*, 94, 1155, 2012. With permission.

Note: −−− : Very bad agreement with the considered characteristic, or even impossible to achieve. +++ : Very good agreement with the considered characteristic.

raisers around manufacturing defects (e.g., resin-rich areas, undesired inclusions, fiber clusters, excessive waviness, excessive porosity), impact (from tools, hail, debris, birds), overloads, hygro-thermo-mechanical static and fatigue loads (e.g., freeze–thaw cycles on a bridge sprayed by anti-icing), attack by chemical agents (e.g., common aerospace service fluids), and a combination of all of the above. All these factors cause a very complex service history for these structures, which is unpredictable during the design and certification phases.

Prevention of catastrophic failure and effective condition-based maintenance with reduced offline time require robust structural health monitoring (SHM), capable of discriminating also potentially serious damage onset from benign manufacturing defects. Desired traits of composite structures' SHM are easily deployable, low-cost, and easy-to-process methods that require noninvasive hardware and are able to capture the service-dependent health of critical load-carrying parts of a structure (which may be large).

This book offers an overview of different approaches and challenges to composites' SHM; therefore, an extensive literature review of SHM and nondestructive testing (NDT) methods will be redundant. We briefly cite infrared thermography; vibration-based methods; fiber optics, ultrasonics, acoustics, and wireless-based sensing; X-radiography; and direct current (DC) and electrical methods as procedures that have been studied for many years, some of which are discussed in this book, and/or in review articles and books (e.g., Chopra 2002, Montalvão et al. 2006, Giurgiutiu 2008, Wild and Hinckley 2008, Kuhn et al. 2012, Kupke et al. 2001 to name a few). Table 12.1 shows a comparison of common NDT methods in industry (Kuhn et al. 2012).

In this chapter, we will discuss the strain-sensing properties of layer-by-layer (LbL) and airbrushed nanocomposite films applied onto fiberglass-reinforced polymer composites (Section 12.2). These films were developed and studied with the purpose of characterizing the spatial distribution of health (or lack thereof) of a composite through changes of electrical resistance, electrical impedance spectroscopy (EIS) (Section 12.3), and electrical impedance tomography (EIT) (Section 12.4). Section 12.5 concludes the discussion, and some exercises are given at the end of the chapter.

12.2 Fabrication of Strain-Sensitive Films

The use of electrical resistance for characterizing changes in a structure requires that the structure must be conductive. Fiberglass/epoxy polymer (GFRP) composites are made of nonconductive fibers and a nonconductive matrix; hence, they need appropriate modifications, whether by making the epoxy conductive before infiltration (e.g., Thostenson and Chou 2008, Nofar et al. 2009, Yesil et al. 2010), or by adding a "patch" of (compatible) conductive material, which could be, for example, piezoelectric paint (Egusa and

Iwasawa 1998), or a film (e.g., Loyola et al. 2010, 2013a, Oliva-Avilés et al. 2011). In this chapter, we will present two different approaches for processing, manufacturing, and depositing strain-sensitive nanocomposites onto GFRP:

1. Layer-by-layer films: results are interpreted with electrical impedance spectroscopy. The strain sensitivity was characterized under selected axial static and dynamic loadings and also at different temperature and humidity conditions (Loyola et al. 2010, 2013a).
2. Airbrushed films: results are interpreted with electrical impedance tomography. The films' sensitivity to damage was investigated for damage due to impact, drilled holes, or cuts (Loyola et al. 2013b,c).

12.2.1 Layer-by-Layer Nanocomposite Films

Films built through an LbL self-assembly process have been developed for the past 25 years for a variety of applications, including structural damage detection (e.g., Loh et al. 2007), drug delivery (e.g., Thierry et al. 2005), self-healing anticorrosion coating (e.g., Andreeva et al. 2008), supercapacitors (e.g., Sarker and Hong 2012), etc.

In this chapter, we focus on the use of nanocomposite films as sensors for the SHM of GFRP composites. The process involved not only the careful selection and layering of polyelectrolytes (polymers whose monomer contains electrolyte groups, which would ionize when in a solvent), but also the preparation of the fiberglass weave. In our work (Loyola et al. 2010, 2013a), the preparation of the weave was carried out by immersing it in poly(vinyl alcohol), PVA, a polyelectrolyte. PVA adsorbs well to the as-received weave, which was treated by the manufacturer with a silane-based sizing agent to promote adhesion with epoxy. The glass weave with PVA deposited on it was then rinsed in deionized water and dried with nitrogen. This preparation not only removed contaminants, such as dust, but also changed the weave from an inherently negatively charged surface to a positively charged surface. Then, the weave (along with its adsorbed PVA monolayer) was immersed into a negatively charged solution of multiwalled carbon nanotubes (MWNTs), dispersed in poly(sodium 4-styrene sulfonate) or PSS. Rinsing and drying were performed again to remove loosely adsorbed nanotubes and polyelectrolytes. As shown in Figure 12.1, this procedure yielded a bilayer consisting of one monolayer of PVA and another one of MWNT–PSS. The LbL film is represented in this chapter using the notation $(\text{MWNT–PSS/PVA})_n$. The process was repeated to create a stronger film with n bilayers, where n was chosen depending on the goals of this study (e.g., the encoding of strain sensitivity in nanocomposites).

After deposition of a $(\text{MWNT–PSS/PVA})_n$ film onto a fiberglass substrate, scanning electron microscopy verified the quality of the deposition (Figure 12.2). One can see from Figure 12.2 that MWNTs remained adequately dispersed even after the LbL deposition and embedment within the PSS/PVA matrix. Then, electrodes were applied by using a conductive silver paint. The resulting

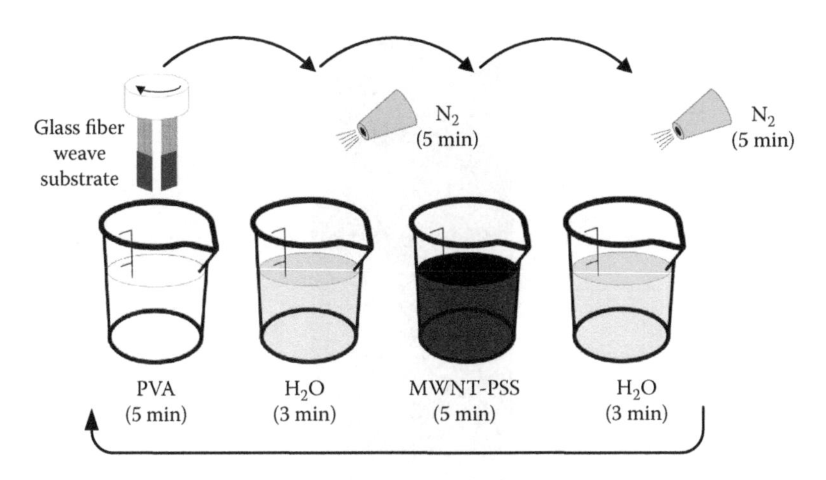

FIGURE 12.1
Sketch of layer-by-layer deposition of strain sensitive film onto fiberglass weaves. (From Loyola, B. R., La Saponara, V., Loh, K. J., *Journal of Materials Science*, 45, 6786, 2010. With permission.)

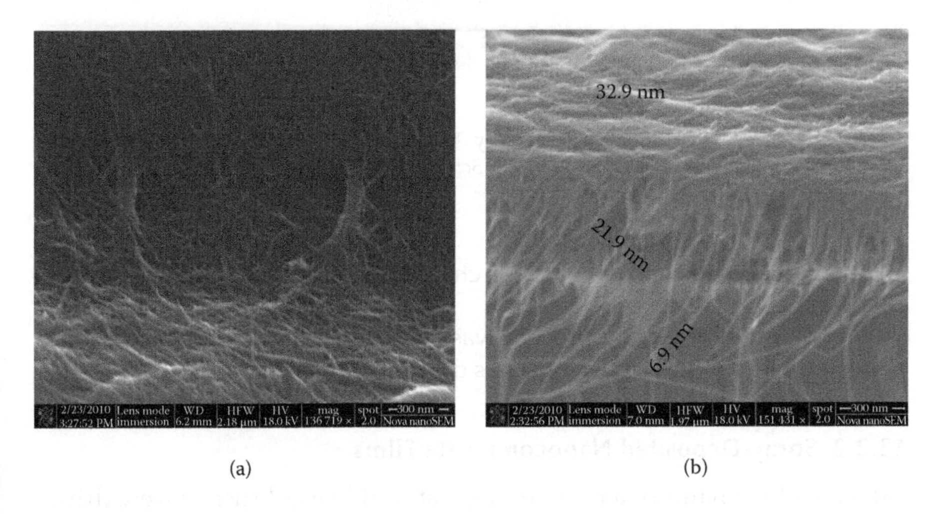

FIGURE 12.2
Scanning electron microscopy images of (a) surface and (b) cross-section of the carbon nanotube-based films show the random oriented nanotube network (Loyola et al. 2010).

specimen was infused with epoxy using conventional vacuum-assisted resin transfer molding. Specimens* had the required dimensions for tensile (Loyola et al. 2010) and bending tests (Loyola et al. 2013a), as shown in Figure 12.3.

* The epoxy was Proset 125/237, the cure cycle was 15 hours at room temperature followed by 8 hours at 82 deg. C. The electrodes were prepared with 28 American Wire Gauge (AWG) single-stranded wire. Tabs for the tensile specimens were prepared with G10 fiberglass sheets bonded with a Hysol 907 adhesive that was cured at 60°C for 2 h. The tabs covered the electrodes (see Figure 12.3).

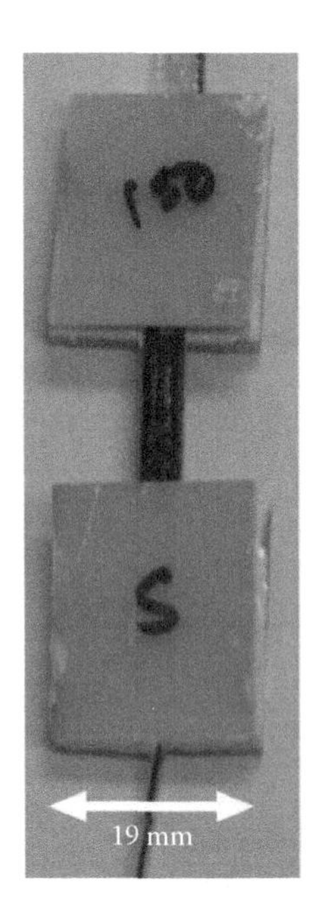

FIGURE 12.3
Tabbed fiberglass/epoxy specimen covered by $(MWNT–PSS/PVA)_{100}$ film. (From Loyola, B. R., La Saponara, V., Loh, K. J., *Journal of Materials Science*, 45, 6786, 2010. With permission.)

The methods and results of these characterization tests are discussed in Section 12.3.1.

Since the LbL assembly process was time consuming, an alternate film deposition method was developed, as discussed in Section 12.2.2.

12.2.2 Spray-Deposited Nanocomposite Films

Advances for a nanocomposite sensor that could be airbrushed were driven by the desire to achieve high-throughput fabrication, low-cost, portability, and rapid field deployment (ideally on large surfaces). More details of the work presented in this section can be found in Loyola et al. (2013b,c).

A solution of PSS in deionized water was prepared, to which the appropriate quantities of MWNTs and a polar solvent (*N*-methyl-2-pyrrolidone, NMP) were added. The MWNTs in the PSS/NMP solution were dispersed through tip sonication in an ice bath, with the PSS and the NMP both aiding the dispersion process. Afterward, a second solution was prepared, containing a latex of 150-nm spherical particles of polyvinylidene fluoride (PVDF) in a solution of surfactant and deionized water (the NMP assisted with the coalescence of PVDF). The two solutions were mixed vigorously together, creating a network of MWNTs that did not penetrate the PVDF particles. The paint that was sprayed with an airbrush onto woven fiberglass specimens had 13% of solids weight content and was then dried in an oven for 10 min.

12.3 Characterization and Processing

This section presents different techniques used to characterize and interpret the data acquired from the nanocomposite films.

12.3.1 Characterization of Layer-by-Layer Films

The strain-sensitive behavior of the $(MWNT–PSS/PVA)_n$ films was assessed in Loyola et al. (2010), by studying their electrical response due to an increase of applied quasi-static tensile strain, using two main techniques: time-domain DC surface resistance measurements, and alternating current (AC) frequency-domain electrical impedance spectroscopy.

Quasi-static tensile loading was applied with a loading frame at a displacement rate of 1 mm·min^{-1} and between 0 and ~100,000 microstrain (με) or until sample failure, to capture their linear elastic behavior in the lower strain regime (0–10,000 με) and transverse cracking in the higher strain regime. The load frame was paused and held its load/displacement during electrical measurements of the samples.

The DC resistance measurements were obtained with a two-point probe method, by directly measuring the resistance at the electrodes with a multimeter. One potential problem may be contact resistance, which should be minimized by a four-point probe (or Kelvin) measurement method (e.g., Schroder 2006, Kuphaldt 2006), where current is injected into two electrodes and voltage is measured at two other electrodes adjacent to the first pair. In the case of the $(MWNT–PSS/PVA)_n$ films discussed in this chapter, the space for electrodes' application was limited, instead silver paint was used to reduce contact resistance in the two-point probe measurements.

EIS is a common technique widely used for electrochemistry-driven characterization studies and applications, from fuel cells (Wang et al. 1999, He et al. 2007, Cañas et al. 2013) to biosensors (Ruan et al. 2002, Jahnke et al. 2013), to nanocomposite sensors for SHM (e.g., Pohl et al. 2001, Kang et al. 2006). The system under study was interrogated with an impedance analyzer at a range of frequencies ω, with output $Z(\omega) = Z'(\omega) + iZ''(\omega)$, where $i = \sqrt{-1}$. The data produced a so-called Nyquist plot, where the real part of the impedance Z was plotted against its negative imaginary part. Figure 12.4 plots a complex impedance response for an $(MWNT–PSS/PVA)_n$ film (Loyola et al. 2010).

A Nyquist plot may not show explicitly the frequencies at which the data has been calculated (note the higher frequencies appearing on the left part

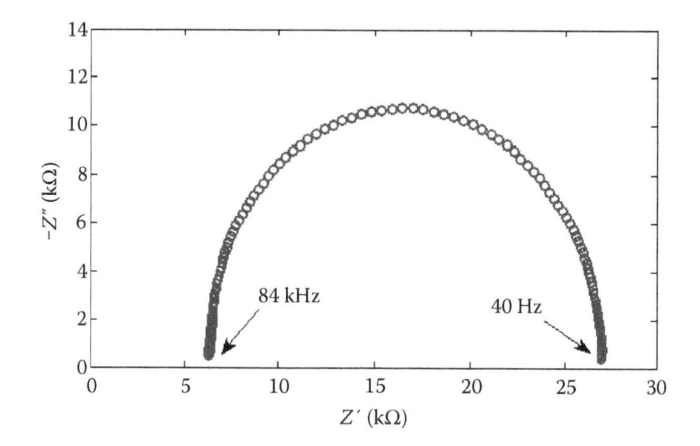

FIGURE 12.4
Example of Nyquist plot from the $(MWNT\text{-}PSS/PVA)_n$ films study, with $n = 50$ (Loyola et al. 2010).

of the plot), in contrast with a Bode plot. One could model the system under study by building an equivalent circuit model whose impedance matches the measured Nyquist and Bode plots. The fact that Figure 12.4 shows a semicircular Nyquist response suggested that a single circuit element (e.g., resistor, capacitor, or inductor) cannot account for this type of impedance response, and thus a more complicated circuit model was needed. For the $(MWNT\text{–}PSS/PVA)_n$ films, Loyola et al. (2010) derived an equivalent circuit model shown in Figure 12.5, which was also used in the literature for other types of electrolyte, such as zirconia–yttria solid electrolytes (Bauerle 1969). This model was adopted for the AC electrical properties of $(MWNT\text{–}PSS/PVA)_n$ films, and its impedance response is mathematically shown in Equation 12.1:

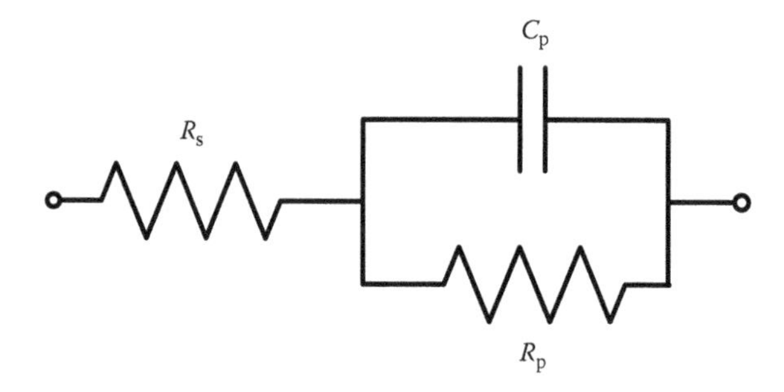

FIGURE 12.5
Equivalent electrical circuit for the $(MWNT\text{–}PSS/PVA)_n$ films. (From Loyola, B. R., La Saponara, V., Loh, K. J., *Journal of Materials Science*, 45, 6786, 2010. With permission.)

$$Z(\omega) = Z'(\omega) + jZ''(\omega) = \left(R_s + \frac{1/R_p}{(1/R_p)^2 + \omega^2 C_p^2} \right) - j\left(\frac{\omega C_p}{(1/R_p)^2 + \omega^2 C_p^2} \right) \quad (12.1)$$

The physical interpretation of the circuit elements was proposed to be the following, extending Bauerle (1969)'s interpretation of these parameters: the parallel resistor (R_p) and parallel capacitor (C_p) represented, respectively, the resistance and capacitance of the internanotube junctions, while R_s modeled the bulk resistance of the film. These hypotheses were tested by assessing how the behavior of these equivalent circuit elements varied with applied load, and/or by microscopy of the damaged samples, if applicable.

Figure 12.6 shows the effect of axial strain on the real and imaginary parts of the impedance. Axial strain reduced the imaginary parts and increased the

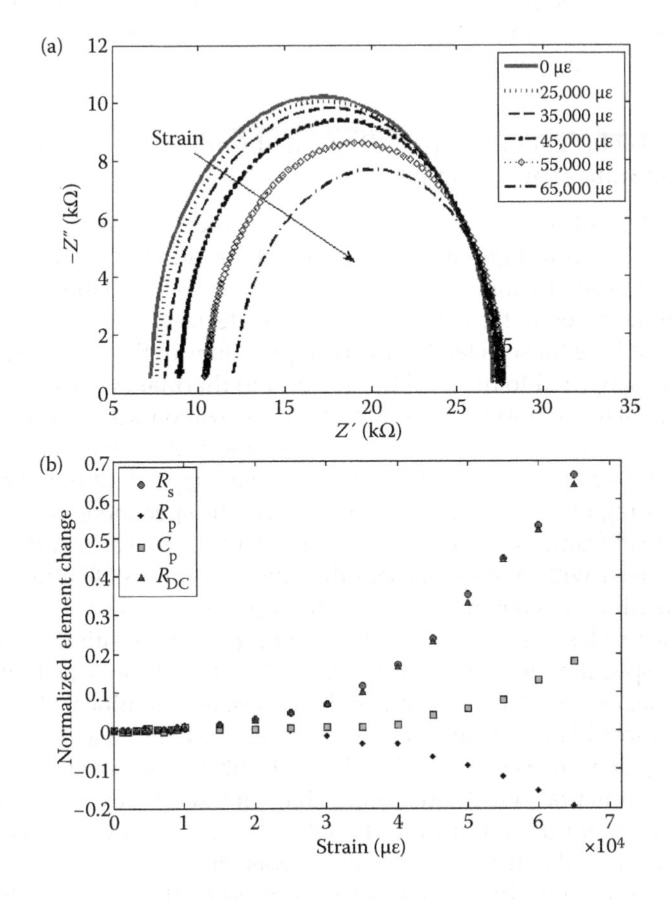

FIGURE 12.6
(a) impedance variation and (b) variation of the equivalent circuit elements, as strain increases. (From Loyola, B. R., La Saponara, V., Loh, K. J., *Journal of Materials Science*, 45, 6786, 2010. With permission.)

real components (for all frequencies) part, which are each a nonlinear function of R_s, R_p, and C_p. Hence, these variations could not be ascribed to a single circuit element. The impedance changes were in turn reflected into the behavior of R_s, R_p, and C_p as strain increased (Figure 12.6). These trends could be fitted with (in this case) smoothing curves (cubic splines, locally weighted scatter plot smoothing, etc.), to allow for the calculation of average strain responses and of the gauge factor. The gauge factor or strain sensitivity (*S*) ties the variation of resistance ΔR to the variation in strain Δε through $S = \frac{\Delta R / R_0}{\Delta \varepsilon}$. Here, the changes of resistance were assessed in terms of the resistive elements of the equivalent circuit, and R_0 is the initial resistance. The investigation on the strain sensitivity of the films showed distinct behaviors: linearity at low strains (0–10,000 με) and nonlinearity at higher strain levels, which could be explained with cracking of the polymeric matrix and of the films. This unique bifunctional strain sensitivity was advantageous for monitoring strain and crack formation in FRP composites.

12.3.2 Characterization of Spray-Deposited Carbon Nanotube–Latex Thin Films

When film deposition does not require high precision thicknesses but instead is to be applied to a large area, it is desirable to adopt a spray deposition method as opposed to LbL. However, the sensitivity of these films to mechanical and environmental factors should be characterized. Methods and results are discussed here for selected testing configurations in this study, in particular for films subjected to uniaxial tension and to thermal cycling.

The uniaxial tension strain sensitivity characterization was conducted using an acrylic substrate cut to form a dog-bone shape, with dimensions outlined in ASTM D638-2010. The MWNT–PVDF film was deposited* onto both sides of the center region of the dog-bone specimens. The specimens were tested in a uniaxial load frame at a rate of 3 mm·min^{-1}, and strain was measured using an extensometer within the gauge length of the MWNT–PVDF film. The resistance of the films was obtained via two-point probe resistance measurements between electrodes of silver epoxy. The resulting strain sensitivity from a representative specimen is presented in Figure 12.7. The strain sensitivity of the film was linear through 0.7% strain, and then became nonlinear. This nonlinearity was most likely due to cracking from the underlying acrylic substrate, thus tearing the film, as witnessed with GFRP substrates (Loyola et al. 2013a).

The environmental sensitivities were characterized by exposing the films to cycling temperatures. With the thought of applying these films to aerospace structures, the temperature range was determined to be –70°C to +80°C, for simulating extreme temperatures present at flying altitude and on

* The film dimensions were 3.1 × 12.7 mm^2, with the long dimension in the longitudinal direction of the specimen.

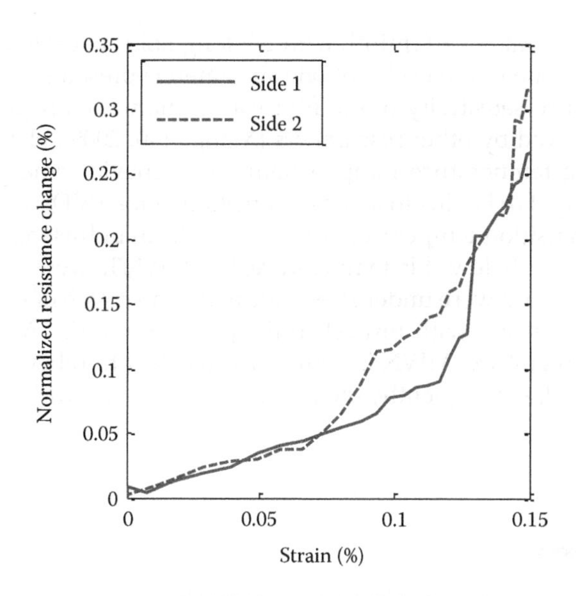

FIGURE 12.7
Resistance changes versus strain for MWNT-PVDF film under bending (one side is in compression, the other in tension).

the tarmac. The test was performed with MWNT–PVDF films sprayed onto glass microscope slides, to minimize effects from thermal expansion of the substrate. Four-point probe resistance measurements were obtained to eliminate the issue of contact resistance. Finally, the specimens were placed in an environmental chamber, and the specimens were cycled between –70°C to

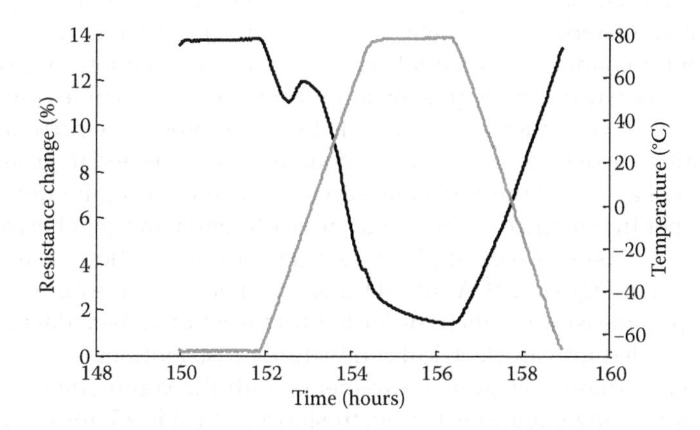

FIGURE 12.8
Temperature sensitivity of MWNT–PVDF films. Temperature *T* (gray) and resistance (black) are plotted versus time.

+80°C at a ramp rate of 1°C·min^{-1}, with 2-h-long holds at each of the extreme temperatures. One of the cycles of one specimen is presented in Figure 12.8. The temperature sensitivity of the film was a line with a negative slope, as was also observed by other researchers (Xiang et al. 2009). However, during the increasing temperature ramp, a nonlinear behavior emerged at –35°C. This was thought to be due to a phase transition of the PVDF polymer, which has a glass transition temperature of –42°C (Laiarinandrasana et al. 2009). It was believed that below this temperature, the MWNTs were locked into the polymer matrix and were under stress due to the mismatch of their coefficient of thermal expansion with respect to the polymer matrix. Above the glass transition temperature, MWNTs could then reorient to relieve this accumulated stress and, consequently, the resistance of the network would behave nonlinearly.

12.4 Structural Monitoring via Spatially Distributed Conductivity Measurements

To achieve spatially distributed sensing with these films, it is necessary to implement a measurement method that can capture their spatially distributed conductivity.

12.4.1 Electrical Impedance Tomography

EIT is a soft-tomographic imaging method that can map the electrical conductivity across a two- or three-dimensional body. Unlike hard-tomographic methods that use penetrating rays that propagate in a straight line through interrogated materials, EIT measures how an electrical current spreads throughout a conductive material from its point of injection to ground. A typical EIT measurement begins by depositing a set of electrodes around the boundary of the conductive area that is to be monitored. A current is then injected into one of these electrodes and another is connected to ground. The resulting voltage distribution is measured at the remaining boundary electrodes, either in reference to ground or in a differential manner between two adjoining electrodes. An example of a single current injection measurement is illustrated in Figure 12.9. A full EIT measurement requires numerous current injection measurements in order to build a set of voltage data that will be reconstructed into an electrical conductivity distribution.

To correlate these voltage measurements with the conductivity distribution, the governing equations for electrostatics, which is a Laplace equation, is applied (Equation 12.2):

$$\nabla \cdot \sigma \nabla u = 0 \tag{12.2}$$

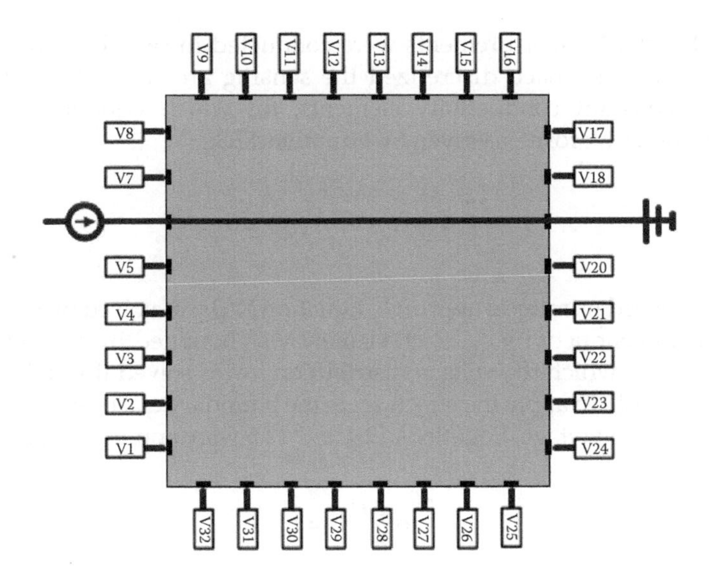

FIGURE 12.9
Schematic of one current injection measurement as part of the larger EIT measurement, where current is injected into electrode 6 and electrode 19 is grounded.

The terms σ and u of Equation 12.2 represent, respectively, the spatial conductivity distribution across the medium and the corresponding voltage distribution that results from the selected current source and ground configuration. Typically, the method for solving this type of Laplace equation is in the forward manner, where the conductivity distribution is known and the voltage distribution is solved for across the medium. However, for EIT measurements, the boundary voltage measurements are known and the conductivity distribution is sought. This requires the use of an inverse approach with only boundary value inputs, which is an ill-posed problem, with more variables than equations.

An approach to solve the EIT inverse problem was not solved until 1980 by Calderon (1980, paper reprinted in 2006). This initial solution led the way for more optimized EIT solutions that can solve a variety of cases: (a) linear (Breckon 1990, Cheney et al. 1990, Graham and Adler 2006, Polydorides et al. 2002) and nonlinear (Polydorides 2002, Breckon 1990, Horesh et al. 2006, Li and Oldenburg 1999, Soleimani and Lionheart 2005) conductivities; (b) actual conductivity distribution (Polydorides 2002); (c) changes of conductivity distributions (Polydorides 2002, Vauhkonen 1997, Adler and Guardo 1996); (d) isotropic (Polydorides 2002, Vauhkonen 1997) or (e) anisotropic conductivity distribution (Abascal et al. 2008, El Badia 2005); (f) with iterative solution methods (Polydorides et al. 2002, Hua et al. 1991, Yorkey et al. 1987) or (g) one-step (Cheney et al. 1990, Adler and Guardo 1996) solution methods.

As few analytical solutions are known for Equation 12.2, the typical solution method for this equation is to use finite elements. In the work of Loyola et al. (2013b,c), EIT measurements were conducted in two dimensions. The finite elements approach discretized the sensing area and electrodes into piecewise constant conductivity elements, for which Equation 12.2 was solved in the weak form, as given by Equation 12.3:

$$\iint_{\Omega} \nabla\phi \cdot \sigma_{\Omega} \cdot \nabla u \, dx \, dy = 0 \tag{12.3}$$

With the selected numerical approach, Equation 12.3 was solved over each element Ω with conductivity σ_{Ω}. Each element was bounded by a set of points called nodes, at which the voltage distribution u was solved through a linear shape function ϕ. To apply the effects from the boundary electrodes associated with current and voltage, Equations 12.4 and 12.5 were applied, respectively:

$$\int_{E_l} \sigma \frac{\partial u}{\partial \upsilon} \, ds = I_l \tag{12.4}$$

$$u + z_l \sigma \frac{\partial u}{\partial \upsilon} = V_l \tag{12.5}$$

Equation 12.4 was used to apply a current-related boundary condition or measure current through one of the boundary electrodes. E_l corresponds to the distance along the boundary that correlates to the l-th electrode through which I_l current is flowing. For example, if a current injection measurement is performed with a current of 1 mA in the configuration presented in Figure 12.9, the boundary conditions would be given by I_6 equal to 1 mA, and I_{1-5}, I_{7-18}, and I_{20-32} are equal to zero. $\frac{\partial u}{\partial \upsilon}$ is the derivative of the voltage normal (υ) to the electrode, and ds is an infinitesimal length along the boundary of the electrode in contact with the sensing area, over which the integral is performed. In addition, Equation 12.5 was used to govern or measure the voltage on each boundary electrode. To account for the grounded electrode, Equation 12.5 was applied to the l-th electrode, where the grounded electrode V_l would equal zero. In the example given above, V_{19} would equal zero to mandate that this electrode is grounded. For the remaining electrodes, Equation 12.5 extracted the boundary voltage distribution from the model (Paulson et al. 1992, Polydorides 2002, Vauhkonen 1997).

For SHM applications, EIT could monitor changes in the conductivity pattern that are used as an indication of onset of damage. For instance, EIT can map the spatial conductivity distribution of a strain-sensitive film embedded in GFRP. Since film conductivity is calibrated to structural strain, the resulting conductivity map from EIT will be equivalent to a strain distribution map of the GFRP structure. For this purpose, the work presented in this chapter uses a one-step differential solver created by Adler and Guardo (1996) called maximum *a posteriori* (MAP). MAP is a one-step solution method, making it an optimal solution scheme for real-time monitoring. The MAP scheme implemented two important characteristics of the reconstruction. The first characteristic consists in differential voltage measurements being taken rather than the absolute measurements. During each current injection measurement, the voltage difference between adjacent electrodes was taken rather than the voltage of each electrode in reference to ground. This stabilized the solution to the forward problem, as described previously. The second characteristic is that the MAP method reconstructs the normalized change in the spatially distributed conductivity over time using corresponding normalized changes in voltage. This correlation is governed by the MAP reconstruction algorithm, given in Equation 12.6:

$$\frac{\Delta\sigma}{\sigma_0} = (H^T WH + \lambda R)^{-1} H^T W \left(\frac{\Delta V}{V_0}\right) \tag{12.6}$$

The main component of the MAP algorithm is the sensitivity matrix (H), which contains the Jacobian of the forward solution, with considerations for the normalized differential inputs: in other words, the sensitivity matrix correlates how a small change in conductivity in each element affects the changes in the differential boundary voltage measurements. The benefit of performing normalized differential reconstruction is that the initial conductivity distribution of the sensing area is not necessary for the calculation of the sensitivity matrix. To take into account the (postulated) Gaussian white noise of the voltage measurements, the variance matrix (Equation 12.7) was incorporated into the reconstruction algorithm:

$$W_{i,i} = \frac{1}{\alpha_i}, \; W_{i,j} = 0 \text{ for } i \neq j \tag{12.7}$$

where α_i represents the variance of the i-th measurement.

As stated previously, the inverse boundary value problem is ill posed, which necessitates the inclusion of additional parameters to stabilize a solution. For this reason, the reconstruction of the conductivity matrix is regularized (smoothed). In the MAP approach, this is accomplished by incorporating a Gaussian high-pass filter matrix (R) into the algorithm. Normally, a high-pass filter would have a sharpening effect on the algorithm; however, because it is located within the portion of the equation that undergoes an inverse, it acts like a low-pass filter that has a smoothing effect. In addition, the amount of smoothing must be specified using a parameter λ, called the hyperparameter. The hyperparameter was determined using a metric called the noise figure (NF), which is given in Equation 12.8:

$$\mathrm{NF} = \frac{\mathrm{SNR_V}}{\mathrm{SNR}_\sigma(\lambda)} \tag{12.8}$$

The noise figure is the ratio of the signal-to-noise ratio (SNR) of the voltage measurements from a specific conductivity distribution versus the SNR of the reconstructed conductivity distribution using a given hyperparameter. The noise figure was computed with a calibration process, using a specimen with a conductivity pattern that has a uniform conductivity across the sensing region with a small area (roughly 5% of the total area) having a 50% decrease in conductivity. Using the collected voltage measurements, the reconstructions were conducted for an array of hyperparameters until a noise figure of 1 was found. This ensured that the SNR of the conductivity reconstruction was the same as the SNR of the voltage measurements.

12.4.2 Examples of EIT for Spatially Distributed Health Monitoring

By combining the electrically conductive thin films that are sensitive to mechanical loading and damage effects with EIT, a spatially distributed SHM system can be achieved. The simplest case under which to perform EIT reconstructions is where a conductive film is applied to an existing structure and the conductivity of the film is isotropic (direction independent).

As a first example of the application of EIT for spatial detection using the methods discussed earlier, the change in conductivity patterns within geometrically identical specimens was detected. Specifically, two specimens were manufactured with square sensing areas of 78×78 mm^2 and eight 3-mm-wide electrodes equally spaced on each side. This was the geometry of the sensing areas used for the rest of this chapter unless otherwise specified. For this example, MWNT–PVDF films were spray-deposited on GFRP substrates. The first specimen had a film of uniform thickness across the sensing region. The second specimen had a uniform thickness across the sensing region except for a small region (6×6 mm^2) to the left of center that was less thick by 50% than the rest of the specimen. A photograph of the patterned specimen can be seen in Figure 12.10a. A differential EIT image was

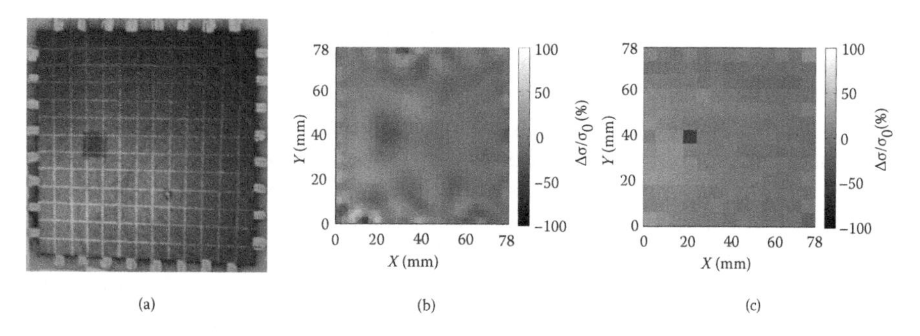

FIGURE 12.10
(a) A photograph of a specimen with a film that has an area of reduced conductivity, and (b) the corresponding EIT reconstructed conductivity distribution in reference to a pristine film. The validation is with (c), the differential conductivity distribution obtained by distributed four-point probe conductivity measurements.

taken of the two specimens with the pristine film as the initial image and the patterned specimen as the contrasting image. The EIT-reconstructed conductivity distribution is provided in Figure 12.10b. To validate this response, four-point probe conductivity measurements were taken at the center of a grid (3 × 3 mm^2 in this case), as can be seen in Figure 12.10a. The spatial distribution of these four point-probe measurements is in Figure 12.10c.

There are three aspects of the EIT reconstruction to comment on (Figure 12.10b). First, the EIT reconstruction captured reasonably well the conductivity distribution, as illustrated by the similarity of the reconstruction to the four point-probe distribution. However, the smoothing that was implemented as part of the regularization gave a low resolution. Second, there was a ring of increased conductivity, located around the contrast in the EIT reconstruction, which does not show up in the four-point probe distribution. This is a known phenomenon in reconstructions using smoothing for regularization (Graham and Adler 2006; Adler and Guardo 1996): in areas of sharp contrast; the reconstruction would return a ringed area around the contrast of the opposite magnitude, to provide a smooth transition to the appropriate conductivity value rather than a sharp transition. The third observation is that the spot of large decrease in the bottom left corner was an artifact from the contact resistance changing at the electrode that was second from the left on the lower boundary. Although this example is being used as an instructive illustration of the potential artifacts and interpretation problems, typically the contact resistance of that electrode would be adjusted to remove this artifact during data collection.

Typically, differential EIT measurements were performed on the same specimen at numerous points in time, so as to determine a change of state in reference to a past measurement. One example of this would be to map the strain to which a composite material was subjected. To perform distributed strain measurements, GFRP specimens were coated with an MWNT–PVDF film at the center of the specimens, in a 18 × 18 mm^2 area with four 2-mm electrodes equally spaced on each side. The specimens were subjected to four-point bending. The film was placed into tension or compression by conducting the test with the film face down or face up, respectively. An example of a film in compression is presented in Figure 12.11a. The load frame was paused so as to hold strain values at increments of 1000 $\mu\varepsilon$ up to $\pm$5000 $\mu\varepsilon$. EIT measurements were taken at each pause of the load frame, and differential reconstructions were performed in reference to the unstrained state. Owing to the uniform strain state across the films, the conductivity response was averaged across the surface and plotted against strain in Figure 12.11b. The strain response from one EIT measurement was compared with the response of two-point probe conductivity measurements of a MWNT–PVDF film in uniaxial tension. As illustrated, the response measured by EIT closely matched that of the two-point probe measurements. At high levels of tensile strain, the MWNT–PVDF began to tear due to cracking in the underlying substrate, which caused the nonlinearity captured in Figure 12.11b, and was also previously reported by Loyola et al. (2013a).

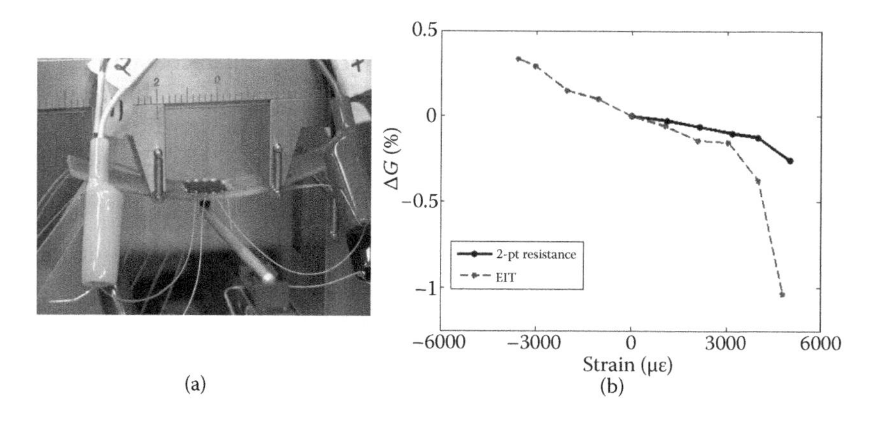

FIGURE 12.11
(a) GFRP specimen undergoing four-point bending with the sensing film under compression. (b) Measured tensile and compression strain sensitivity of the films as captured by EIT, and the corresponding tensile strain sensitivity as measurement by two-point probe conductivity measurements.

12.4.3 Spatially Distributed Health Monitoring of Smart GFRP Composites

When considering the construction of a smart composite structure, it is advantageous to embed the sensing capability inside the structure, to enable SHM where damage tends to manifest and is typically inaccessible to visual inspection. To demonstrate this capability, MWNT–PVDF films were embedded within the structure of a GFRP laminate. One such specimen is presented in Figure 12.12a. Not only did this enable embedded sensing but it also protected the film from adverse environmental effects. When the MWNT–PVDF film was spray deposited on a substrate, such as a glass fiber mat, the underlying geometry of the fiber mat might cause the conductivity of the film to be anisotropic in nature (or directionally dependent). It was found that the conductivity in the direction of the fibers was twice that of the transverse direction. To accommodate for this, the conductivity was modeled as an array with the directional conductivities along the diagonal. Equations 12.2 through 12.6 became matrix equations rather than scalar, although the results of the matrix algebra were scalar values (Abascal et al. 2008, Loyola et al. 2013b). One strategy to account for this within the algorithm was to reconstruct for a nominal scalar value and use a matrix multiplier to account for the anisotropic nature of the conductivity. Equation 12.9 presents how anisotropic conductivity was addressed by the reconstructions algorithm in this work:

$$\tilde{\sigma} = \begin{bmatrix} \sigma_x & 0 \\ 0 & \sigma_y \end{bmatrix} = \begin{bmatrix} 1 & 0 \\ 0 & \dfrac{\sigma_y}{\sigma_x} \end{bmatrix} \sigma_x \qquad (12.9)$$

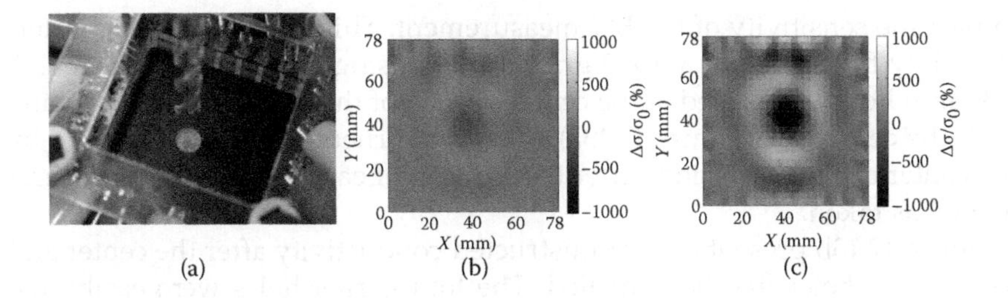

FIGURE 12.12
(a) Photograph of GFRP specimen with a 12.7-mm hole drilled in the center of the sensing region. The corresponding EIT conductivity reconstructions for other specimens with (b) 6.35-mm and (c) 12.7-mm holes.

The ratio between the conductivity in the x-direction and y-direction was assumed to be constant across the entire sensing region. During the baseline EIT measurement, two-point probe conductivity measurements were obtained in both directions, to solve for the ratio in the right-hand side of Equation 12.9. In the reconstruction, the changes in the magnitude of the x-direction conductivities of each element were determined, a process that halved the number of reconstruction unknowns. The reconstruction could easily have been performed with respect to the y-direction conductivities.

To demonstrate the applicability of EIT measurements for embedded sensing, and to validate the anisotropic reconstruction approach, size and spatial distribution sensitivity characterizations were performed with GFRP specimens with embedded films. For size characterization, progressively larger holes were drilled at the center of the sensing area, with diameters ranging from 1.59 to 12.7 mm. Figure 12.12a shows this approach, where a fixture with spring-loaded pins made electrical connections to every electrode, to enable the EIT measurements. While the specimen was in the fixture, progressively larger holes were drilled, and EIT measurements were taken after the drilling of each hole. Reconstructions were performed for each measurement with respect to the measurement of the pristine sample. The reconstructions for the 6.36 and 12.7 mm response are pictured in Figures 12.12b,c, respectively.

One aspect of these reconstructions should be noted. Figure 12.12b,c indicated that a larger drilled hole corresponded to an increased response in the EIT reconstruction. However, the magnitude of the indicated damage was >100%, which was physically impossible. It is speculated that this was due to how the model accounted for the anisotropic conductivity of the embedded MWNT–PVDF film (Loyola et al. 2013b). Despite this, the EIT method was still able to detect a 3.17-mm hole drilled at the center, which was theoretically the most insensitive section of the sensing area, due to the distance from the electrodes (Polydorides 2002).

In addition to understanding the effect of damage size on the EIT response, it was important to realize how the location of sustained damage would impact the sensitivity of the EIT measurement. This dependence was determined by first drawing a 3 × 3 grid in the sensing area of the specimen. A 6.35-mm hole was drilled at the center of each of these grids spaces, starting with the center cell (Figure 12.13a). The four corners were next, followed by the center cell on each side. After the drilling of each hole, an EIT measurement was taken.

Figure 12.13b presents the reconstructed conductivity after the center and four corner holes had been drilled. The four corner holes were easily discernible, as they were located near the electrodes. The conductivity response from the center hole was much lower due to the higher effects of the corner holes, but a negative change in conductivity was still present at the location of the center hole. Figure 12.13c contains the EIT response from a hole in every cell. The response from the top and bottom center cells was present, but the responses from the cells in the left and right cells of the center row were less distinguishable. In fact, the responses from those holes appeared to be shifted toward the center of the response area. This was likely due to how the model accounts for the anisotropic conductivity of this specimen, which was described earlier.

As a last characterization example, the sensitivity of EIT to actual distributed damage was assessed. Several EIT specimens were subjected to impact damage, where the sensing area was on the surface opposite to the impact event and was also where the damage was most significant and possibly visible. An example of this damage is presented in Figure 12.14a from a specimen subjected to 100 J of impact energy.

EIT measurements were taken before and after each impact event, and the reconstructions were performed with the resulting data. Whether or not the opposite face of the composite had visible signs of damage, the embedded film had still undergone high levels of strain that were previously shown to damage the electrical network. This damage led to a decrease in conductivity, thus indicating damage. This was the case for the response shown in Figure 12.14b,

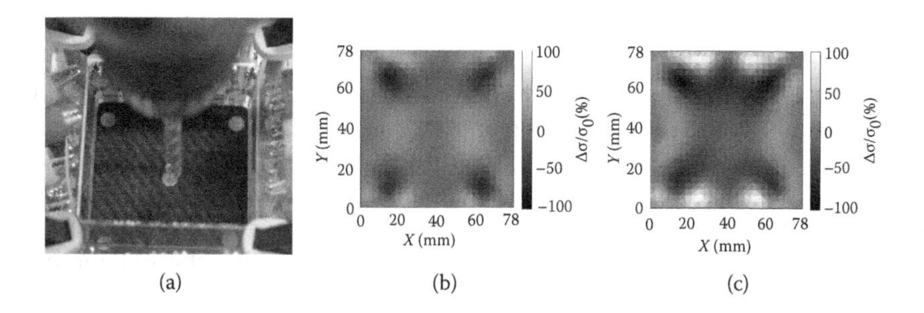

FIGURE 12.13
(a) Image of holes being drilled in a "X" pattern, with (b) the corresponding EIT reconstruction of the conductivity distribution; (c) the EIT reconstruction corresponding to the sensing area with nine holes drilled within it.

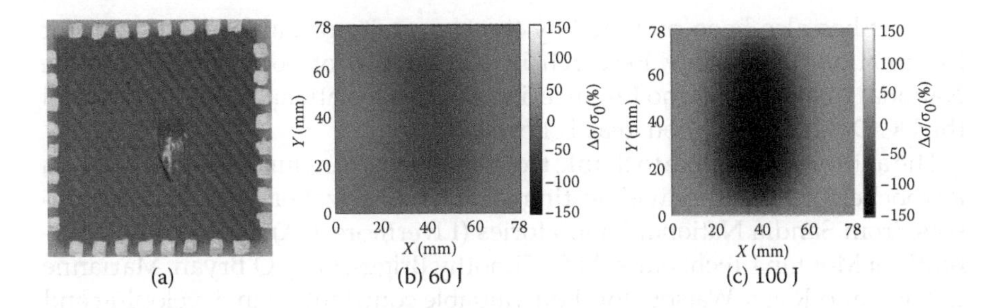

FIGURE 12.14
(a) The back surface of a 78 × 78 mm² sensing area in a GFRP specimen after a 100 J impact; the reconstructed conductivity patterns from specimens that have been subjected to (b) 60 J and (c) 100 J impacts.

where the specimen was subjected to a 60 J impact without visible damage. The response to higher energy impacts led to more pronounced responses from the conductivity response (see Figure 12.14c, which is for a specimen subjected to a 100 J impact event). This specimen had visible signs of damage on the impacted surface. The damage to the electrical network in the impact location was captured by a 100% reduction in conductivity in that region.

In summary, EIT proved to be effective in detecting, locating, and determining the severity of the damage that was sustained within the sensing regions.

12.5 Conclusions

This chapter has presented an overview of recent research results for nanocomposite films that can be embedded in polymer matrix composites for *in situ* SHM and for detecting structural damage. While there are several remaining challenges (e.g., seamless embedment of wires inside a composite structure, potential artifacts resulting from signal processing techniques, and anisotropy of the host structure, among others), this work offers a potentially transformative technique for effective health monitoring of FRP composite structures. The implementation of this technology in realistic and large-scale structures is currently under way.

Acknowledgments

This research is supported by the National Science Foundation (NSF) under grant nos. CAREER CMMI-0642814 and CMMI-1200521. Additional support has also been provided by the University of California, Center for Information Technology Research in the Interest of Society (CITRIS), the National Institute of Nano Engineering at Sandia National Laboratories, and the UC Davis Dissertation Year Fellowship.

The authors would like to thank, from UC Davis, Yingjun Zhao and Luciana Arronche, for assistance with testing and characterization work of these sensors; from Sandia National Laboratories (Livermore, CA), Jack Skinner (currently at Montana Tech, Butte, MT), Timothy Briggs, Greg O'Bryan, Marianne LaFord, and Roger Watson, for their valuable contribution in developing and characterizing the spray-deposited nanocomposite sensor.

Sandia National Laboratories is a multiprogram laboratory managed and operated by Sandia Corporation, a wholly owned subsidiary of Lockheed Martin Corporation, for the US Department of Energy's National Nuclear Security Administration under contract DE-AC04-94AL85000.

References

Abascal, J.-F. P. J., Arridge, S. R., Atkinson, D., Horesh, R., Fabrizi, L., De Lucia, M., Horesh, L., Bayford, R. H., Holder, D. S., "Use of anisotropic modelling in electrical impedance tomography; description of method and preliminary assessment of utility in imaging brain function in the adult human head," 2008, *Neuroimage*, 43(2), pp. 258–268.

Adler, A., Guardo, R., "Electrical impedance tomography: Regularized imaging and contrast detection," 1996, *IEEE Transactions on Medical Imaging*, 15(2), pp. 170–179.

Andreeva, D. V., Fix, D., Möhwald, H., Shchukin, D. G., "Self-healing anticorrosion coatings based on pH-sensitive polyelectrolyte/inhibitor sandwichlike nanostructures," 2008, *Advanced Materials*, 20, pp. 2789–2794.

Bauerle, J. E.,"Study of solid electrolyte polarization by a complex admittance method," 1969, *Journal of Physics and Chemistry of Solids*, 30(12), pp. 2657–2670.

Breckon, W. R., "Image reconstruction in electrical impedance tomography," 1990, PhD Thesis, Manchester Institute for Mathematical Sciences/School of Mathematics, The University of Manchester, Manchester, UK.

Calderón, A. P., "On an inverse boundary value problem," 2006, *Computational and Applied Mathematics*, 25(2–3), pp. 133–138 [reprinted from 1980 article, published by the Brazilian Mathematical Society (SBM) in the Proceedings of the *Seminar on Numerical Analysis and its Applications in Continuum Physics of SBM*, Rio de Janeiro, pp. 65–73].

Cañas, N. A., Hirose, K., Pascucci, B., Wagner, N., Friedrich, K. A., Hiesgen, R., "Investigations of lithium-sulfur batteries using electrochemical impedance spectroscopy," 2013, *Electrochimica Acta*, 97, pp. 42–51.

Cheney, M., Isaacson, D., Newell, J. C., Simske, S., Goble, J., "NOSER: An algorithm for solving the inverse conductivity problem," 1990, *International Journal of Imaging Systems and Technology*, 2(2), pp. 66–75.

Chopra, I., "Review of state of art of smart structures and integrated systems," 2002, *AIAA Journal*, 40(11), pp. 2145–2187.

Egusa, S., Iwasawa, N., "Piezoelectric paints as one approach to smart structural materials with health monitoring capabilities," 1998, *Smart Materials and Structures*, 7, pp. 438–445.

El Badia, A., "Inverse source problem in an anisotropic medium by boundary measurements," 2005, *Inverse Problems*, 21(5), pp. 1487–1506.

Giurgiutiu, V., *Structural Health Monitoring with Piezoelectric Wafer Active Sensors*, 2008, Academic Press, Boston.

Graham, B. M., Adler, A., "Objective selection of hyperparameter for EIT," 2006, *Physiological Measurement*, 27(5), pp. S65–S79.

He, B.-L., Dong, B., Li, H.-L., "Preparation and electrochemical properties of Ag-modified TiO_2 nanotube anode material for lithium-ion battery," 2007, *Electrochemistry Communications*, 9, pp. 425–430.

Horesh, L., Schweiger, M., Arridge, S. R., Holder, D. S., "Large-scale non-linear 3D reconstruction algorithms for electrical impedance tomography of the human head," 2006, in *World Congress on Medical Physics and Biomedical Engineering*, Seoul, Korea, pp. 3862–3865.

Hua, P., Woo, E. J., Webster, J. G., Tompkins, W. J., "Iterative reconstruction methods using regularization and optimal current patterns in electrical impedance tomography," 1991, *IEEE Transactions on Medical Imaging*, 10(4), pp. 621–628.

Jahnke, H.-G., Heimann, A., Azendorf, R., Mpoukouvalas, K., Kempski, O., Robitzki, A. A., Charalampaki, P., "Impedance spectroscopy—An outstanding method for label-free and real-time discrimination between brain and tumor tissue *in vivo*," 2013, *Biosensors and Bioelectronics*, 46, pp. 8–14.

Kang, I., Schulz, M. J., Kim, J. H., Shanov, V., Shi, D., "A carbon nanotube strain sensor for structural health monitoring," 2006, *Smart Materials and Structures*, 15, pp. 737–748.

Kuhn, E., Valot, E., Herve, P., "A comparison between thermosonics and thermography for delamination detection in polymer matrix composites," 2012, *Composite Structures*, 94, pp. 1155–1164.

Kuphaldt, T. R., "Lessons in electric circuits, vol. 1-DC," *Chapter 8: DC Metering Circuits*, 2006, available under openbookproject.net/electricCircuits.

Kupke, M., Schulte, K., Schüler, R., "Non-destructive testing of FRP by d.c. and a.c. electrical methods," 2001, *Composites Science and Technology*, 61, pp. 837–847.

Laiarinandrasana, L., Besson, J., Lafarge, M., Hochstetter, G., "Temperature dependent mechanical behaviour of PVDF: Experiments and numerical modelling," 2009, *International Journal of Plasticity*, 25(7), pp. 1301–1324.

Li, Y., Oldenburg, D. W., "3-D inversion of DC resistivity data using an L-curve criterion," 1999, in *Society of Exploration Geophysicists Annual Meeting*, Houston, TX.

Loh, K. J., Kim, J., Lynch, J. P., Kam, N. W. S., Kotov, N. A., "Multifunctional layer-by-layer carbon nanotube polyelectrolyte thin films for strain and corrosion sensing," 2007, *Smart Materials and Structures*, 16(2), pp. 429–438.

Loyola, B. R., La Saponara, V., Loh, K. J., "*In situ* strain monitoring of fiber-reinforced polymers using embedded piezoresistive nanocomposites," 2010, *Journal of Materials Science*, 45, pp. 6786–6798.

Loyola, B. R., Zhao, Y., Loh K. J., La Saponara, V., "The electrical response of carbon nanotube-based thin film sensors subjected to mechanical and environmental effects," 2013a, *Smart Materials and Structures*, 22, 025010 (11pp), doi:10.1088/0964-1726/22/2/025010.

Loyola, B. R., Briggs, T. M., Arronche, L., Loh, K. J., La Saponara, V., O' Bryan, G., Skinner, J. L., "Detection of spatially distributed damage in fiber-reinforced polymer composites," 2013b, *Structural Health Monitoring*, 12(3), pp. 225–239, doi: 10.1177/1475921713479642.

Loyola, B. R., La Saponara, V., Loh, K. J., Briggs, T. M., O'Bryan, G., Skinner, J. L., "Spatial sensing using electrical impedance tomography," 2013c, *IEEE Sensors Journal*, 12(6), pp. 2357–2367, doi: 10.1109/JSEN.2013.2253456.

Montalvão, D., Maia, N. M. M., Ribeiro, A. M. R., "A review of vibration-based structural health monitoring with special emphasis on composite materials," 2006, *Shock and Vibration Digest*, 38(4), pp. 295–324.

Nofar, M., Hoa, S. V., Pugh, M. D., "Failure detection and monitoring in polymer matrix composites subjected to static and dynamic loads using carbon nanotube networks," 2009, *Composites Science and Technology*, 69(10), pp. 1599–1606.

Oliva-Avilés, A. I., Avilés, F., Sosa, V., "Electrical and piezoresistive properties of multi-walled carbon nanotube/polymer composite films aligned by an electric field," 2011, *Carbon*, 49(9), pp. 2989–2997.

Paulson, K., Breckon, W., Pidcock, M., "Electrode modelling in electrical impedance tomography," 1992, *SIAM Journal on Applied Mathematics*, 52(4), pp. 1012–1022.

Pohl, J., Herold, S., Mook, G., Michel, F., "Damage detection in smart CFRP composites using impedance spectroscopy," 2001, *Smart Materials and Structures*, 10, pp. 834–842.

Polydorides, N., "Image reconstruction algorithms for soft-field tomography," 2002, PhD Thesis, Department of Electrical Engineering and Electronics, University of Manchester Institute of Science and Technology, Manchester, UK.

Polydorides, N., Lionheart, W. R. B., McCann, H., "Krylov subspace iterative techniques: On the detection of brain activity with electrical impedance tomography," 2002, *IEEE Transactions on Medical Imaging*, 21(6), pp. 596–603.

Ruan, C., Yang, L., Li, Y., "Immunobiosensor chips for detection of *Escherichia coli* O157:H7 using electrochemical impedance spectroscopy," 2002, *Analytical Chemistry*, 74(18), pp. 4814–4820.

Sarker, A. K., Hong, J.-D., "Layer-by-layer self-assembled multilayer films composed of graphene/polyaniline bilayers: High-energy electrode materials for supercapacitors," 2012, *Langmuir*, 28(34), pp. 12637–12646.

Schroder, D. K., "Semiconductor material and device characterization," *Chapter 3: Contact Resistance and Schottky Barriers*, 2006, Third edition, John Wiley & Sons, Inc., Hoboken, New Jersey.

Soleimani, M., Lionheart, W. R. B., "Nonlinear image reconstruction for electrical capacitance tomography using experimental data," 2005, *Measurement Science and Technology*, 16, pp. 1987–1996.

Thierry, B., Kujawa, P., Tkaczyk, C., Winnik, F. M., Bilodeau, L., Tabrizian, M., "Delivery platform for hydrophobic drugs: Prodrug approach combined with self-assembled multilayers," 2005, *Journal of the American Chemical Society*, 127(6), pp. 1626–1627.

Thostenson, E. T., Chou, T.-W., "Carbon nanotube-based health monitoring of mechanically fastened composite joints," 2008, *Composites Science and Technology*, 68(12), pp. 2557–2561.

Vauhkonen, M., "Electrical impedance tomography and prior information," 1997, PhD Thesis, Department of Physics, University of Kuopio, University of Kuopio, Finland.

Wang, G. X., Bradhurst, D. H., Liu, H. K., Dou, S. X., "Improvement of electrochemical properties of the spinel $LiMn_2O_4$ using a Cr dopant effect," 1999, *Solid State Ionics*, 120, pp. 95–101.

Wild, G., Hinckley, S., "Acousto-ultrasonic optical fiber sensors: Overview and state-of-the-art," 2008, *IEEE Sensors*, 8(7), pp. 1184–1193.

Xiang, Z.-D., Chen, T., Li, Z.-M., Bian, X.-C., "Negative temperature coefficient of resistivity in lightweight conductive carbon nanotube/polymer composites," 2009, *Macromolecular Materials and Engineering*, 294(2), pp. 91–95.

Yesil, S., Winkelmann, C., Bayram, G., La Saponara, V., "Surfactant-modified multiscale composites for improved tensile fatigue and impact damage sensing," 2010, *Materials Science and Engineering A*, 527, pp. 7340–7352.

Yorkey, T. J., Webster, J. G., Tompkins, W. J., "Comparing reconstruction algorithms for electrical impedance tomography," 1987, *IEEE Transactions on Biomedical Engineering*, BME-34(11), pp. 843–852.

Section I Exercises

Chapter 1: Field Coupling Analysis in Electrically Conductive Composites

1. Investigate several applications of multifunctional composites and find the possible couplings between different fields in each application.
2. Find all the coupling terms in the equations of the Lorentz force (Equation 1.5). What are the electromagnetic parameters affecting the terms in a coupled problem?
3. Derive Equations 1.33 and 1.34 from the Maxwell equations and write all the assumptions considered.
4. Derive the components of the Lorentz force (Equations 1.40 through 1.42) from the equation of the Lorentz force for a transversely isotropic plate (Equation 1.5) for a single-layer composite plate shown in Figure 1.4.
5. Why are N_{xz} and N_{yz} in Equation 1.26 not expanded in terms of the derivatives of the middle plane displacements similar to other resultants? Note how they are computed for a laminated plate.
6. Apply the numerical procedure introduced in this chapter (MOL, Newmark's scheme, and quasilinearization) to the third expression of Equation 1.86.
7. To investigate how different the governing equations of a single-layer and a laminated plate are, derive the equation for dN_{xy}/dy for a general laminated composite plate and compare it with the third expression of Equation 1.86.

Chapter 2: Design and Characterization of Magnetostrictive Composites

8. What are the main benefits of incorporating giant magnetostrictive materials in composite form?
9. A magnetostrictive material with a rectangular shape is tested under uniform magnetic field. Assume that the relative permeability of the material remains constant for the interested magnetic field range as $\mu_r = 6$ and magnetostriction remains linear for the range with a differential strain of 0.01 ppm/(A/m). If the magnetic flux density $B_1 = 0.12$ (T) is in a direction perpendicular to the ends of the material and is very close to the ends: (a) What is the magnitude of magnetic flux density B_2 at the ends inside the material? (b) What is the magnetic field intensity H_2 corresponding to B_2 in (a)? (c) What should be the value of magnetostrictive strain in parallel with H_2?
10. Given the magnetic field intensity $H_1 = 50$ kA/m in a direction parallel to the side of the material and very close to the side: (a) What is the magnetic field intensity H_2 near the side inside the material? (b) What should be the value of magnetostrictive strain in parallel with H_2?
11. Assuming a Poisson's ratio of 0.5, what are the magnetostrictive strains perpendicular to H's in Question 10?
12. For a 4-cm-long magnetostrictive medium under a nonuniform magnetic field, we attach a 1-cm-long strain gauge in parallel with the magnetic field and obtain a reading of 200 microstrain. Could we find the magnetostrictive expansion along the field direction?

Chapter 3: Graphitic Carbon Nanomaterials for Multifunctional Nanocomposites

13. Name three allotropes of carbon, and specify all the types of bonds that exist between their atoms.
14. Rank the following in terms of thermal stability (choose from the following: poor, med, high)

    a. _______Pure graphene with very high in-plane dimensions

b. _______Highly defective single-wall CNT with a diameter of ~0.5 nm

c. _______A multiwalled CNT with an outer diameter of 10 nm

15. A CNT-reinforced composite is stretched *in situ* under Raman excitation. Does the Raman peak down shift or up shift as a result of applied strain? Plot the relative location of the Raman peak as a function of applied strain.
16. Rank the following in terms of electrical conductivity:

    d. _______Graphite, through the thickness

    e. _______Graphite, in plane

    f. _______Graphene

    g. _______SWNT

    Assume all the structures have similar defect density.
17. In a CNT-reinforced polymer composite with electron tunneling–based piezoresistivity, by increase in the CNT content from near the percolation threshold to 10 times higher, will the material show more or less piezoresistivity?

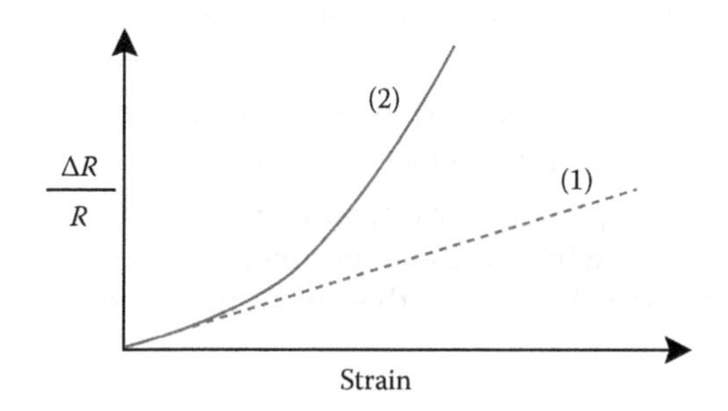

18. In a CNT nanocomposite with a percolation threshold of V_p, how would you expect the resistivity to change with strain for the volume fraction $V >> V_p$? Select the proper curve.
19. Assuming that we can perfectly control the diameter of CNTs and their number of shells, which one of the following would have been a more clever choice as CNT nanoreservoirs for structural self-healing?

    h. SWNTs with the diameter of 0.3 nm

    i. SWNTs with the diameter of 3 nm

    j. SWNTs with the diameter of 6 nm
20. Explain the actuation mechanism in CNT bucky paper/Nafion actuators.

Chapter 4: Active Fiber Composites: Modeling, Fabrication, and Characterization

21. Using the one-dimensional micromechanics model, calculate and plot the effective piezoelectric coupling of the active structural fiber $\left(d_{31}^{\text{multi}}\right)$ as a function of various aspect ratios ($\alpha = 0.2, 0.4, 0.6, 0.8$, and 0.9). The materials' properties are listed below:

    $d_{31} = -320$ pm/V

    $Y^p = 50$ GPa

    $Y^f = 250$ GPa

    $Y^m = 4$ GPa

22. Using the same material properties in question 21, calculate and plot the effective piezoelectric coupling of the active lamina $\left(d_{31}^{\text{Lam}}\right)$ as a function of ASF volume fraction ($v^p + v^f$) for various aspect ratios ($\alpha = 0.2, 0.4$, and 0.6).
23. In the EPD process, what are the experimental parameters controlling $BaTiO_3$ coating thickness on SiC fiber? Briefly explain the mechanism of EPD (why $BaTiO_3$ nanoparticles will be deposited onto the cathode SiC fiber).
24. What is the purpose of sintering after the EPD process? Why are the heating and cooling rates set to 6°C/min?
25. For the effective piezoelectric strain coupling testing of active single fiber lamina, why are the experimental results for high-aspect-ratio samples ($\alpha = 0.60$) lower than what the micromechanics model predicted?

Section II Exercises

Chapter 5: Modeling and Characterization of Piezoelectrically Actuated Bistable Composites

1. Consider the three piezoelectric actuators in the table below. In all cases, the piezoactuator dimensions are 100, 100, and 1 mm in the *x*-, *y*-, and *z*-directions, respectively. For actuator 1, the electrodes are placed at the top and bottom of the piezoelectric. For actuator 2, the electrodes are placed at two edges. For actuator 3, an interdigitated electrode (IDE) structure is used with an electrode separation of 1 mm. Since the polarization of each actuator is undertaken using its own electrode structure, the electric field and polarization direction is as shown in column 2. The piezoelectric material has a d_{33} and d_{31} (strain per unit electric field) of 600 and –300 pm/V, respectively, and the maximum operational electric field is 1.67 kV/mm.

   Calculate the maximum free displacement of each actuator in the *x*-direction. In addition, calculate the voltage required to achieve such a displacement and then calculate and compare the "displacement per unit voltage." The electric field in each of the actuators can be assumed to be voltage/electrode separation.

   Table of electrode and polarization geometries of actuators 1 to 3:

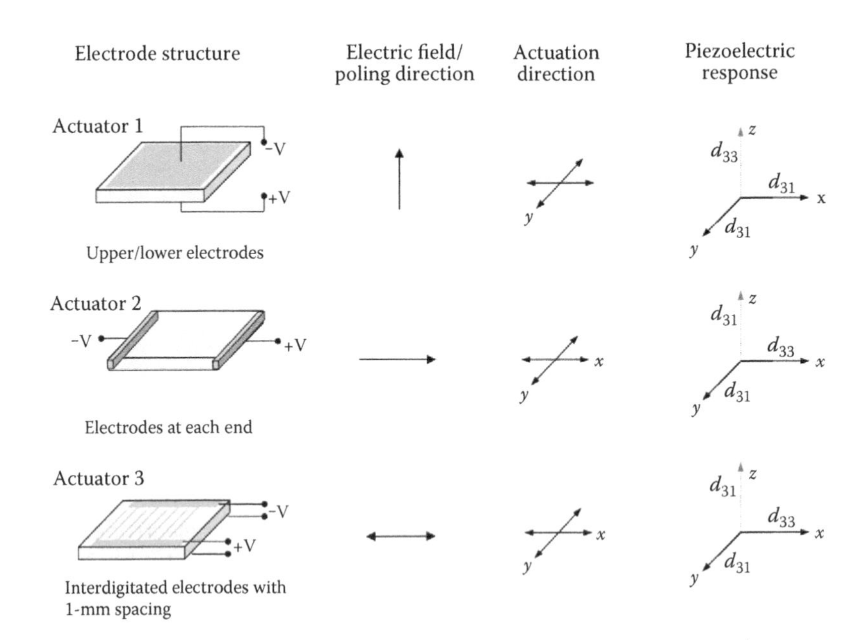

Comments

Actuator 3 achieves highest displacement per unit voltage by using the higher d_{33} coefficient and a small electrode separation to generate a high electrode field at low voltages. A smaller IDE separation (e.g., 0.5 mm) could improve the displacement per unit voltage even further. In reality, an IDE produces a more complex electric field distribution compared with the simple planar electrodes of actuators 1 and 2, and using voltage/electrode separation to determine the electric field is an oversimplification. Nevertheless, this example demonstrates why IDEs are used in devices such as the MFCs discussed in this chapter.

Chapter 6: Wing Morphing Design Using Macrofiber Composites

2. List two out of three aerodynamic effects of variation of camber in an aircraft wing.
3. List the three aerodynamic benefits of the variable-camber airfoil compared with a conventional airfoil that was shown both theoretically

and experimentally in this chapter. Briefly discuss how these observed benefits are related to your answers to question 1. (Hint: Remember the lift and drag coefficient plots and lift-to-drag ratio plots.)

4. List two issues that arise specifically owing to the use of a piezoelectric actuator in a wing morphing application. Discuss each issue and provide methods for addressing them. (Hint: Discuss about characteristics, models, technological advances, etc.)

Chapter 7: Analyses of Multifunctional Layered Composite Beams

5. A cantilever beam of length 200 mm is made of PZT and aluminum layers (see figure below, $w = 10$ mm, $t = 2$ mm). The beam is subjected to a uniform temperature change $\Delta T = 100°C$. The PZT and aluminum have the following properties:

$$\text{Aluminum: } E = 70 \text{ GPa and } \alpha = 25 \times 10^{-6}/°C$$

$$\text{PZT: } E = 80 \text{ GPa}, \alpha = 5 \times 10^{-6}/°C, d_{311} = -200 \times 10^{-12} \text{ m/V}, \text{ and } d_{333} = 450 \times 10^{-12} \text{ m/V}.$$

(a) Determine the lateral (transverse) deformation and axial thermal stresses.

(b) The PZT is polarized through its thickness as shown below. Determine the magnitude and direction of electric field that should be applied to minimize the lateral deflection at the tip (free end) due to the temperature change.

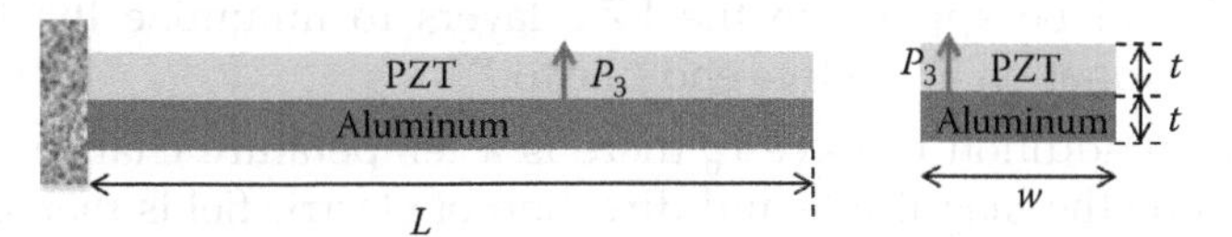

6. A composite beam is composed of six layers where each layer has thickness h (shown below). The following materials will be used for the composite beam: aluminum with elastic modulus E of 70,000 MPa and coefficient of thermal expansion (CTE) α of $20 \times 10^{-6}/°C$; fiber reinforced polymer (FRP) with E of 72,000 MPa and α of $8 \times 10^{-6}/°C$; polymeric foam with E of 22 MPa and α of $50 \times 10^{-6}/°C$.

(a) To minimize the transverse deformation due to bending, how would you arrange the materials? Explain your answer; it is not necessary to numerically/symbolically solve the problem.

(b) If the beam is subjected to a uniform temperature change ΔT, in order to minimize the thermal deformation, how would you arrange the materials? Explain your answer; it is not necessary to numerically/symbolically solve the problem.

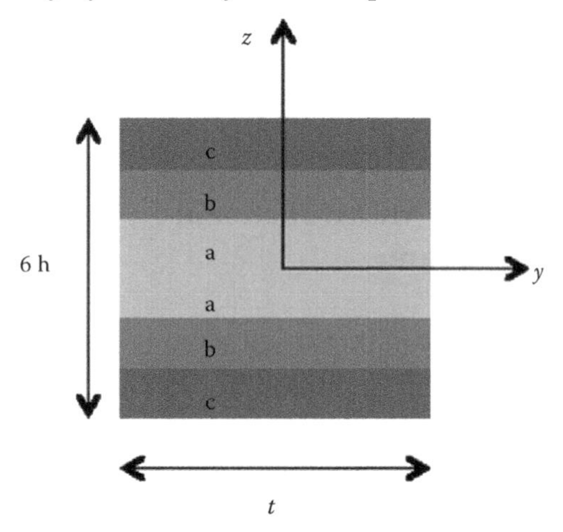

7. A piezoelectric composite cantilever beam, comprising an elastic layer sandwiched between PZT layers at the top and bottom parts, is subjected to a concentrated force F_o. The elastic layer has thickness $2t_s$ and each PZT layer has thickness t_p. The beam has uniform width of w and the PZT layers are polarized through the thickness. The elastic layer has the following elastic modulus and coefficient of thermal expansion (CTE), E_{el} and α_{el}, respectively, and the PZT layers have the following elastic modulus, CTE, and piezoelectric constant, E_{PZT}, α_{PZT}, and d_{311}, respectively.

(a) Determine the magnitude and direction of electric fields that should be applied to the PZT layers to minimize the lateral deformation at the free end due to F_o.

(b) If in addition to force F_o there is a temperature change, determine the magnitude and direction of electric fields that should be applied to the PZT layers to minimize the lateral deformation at the free end.

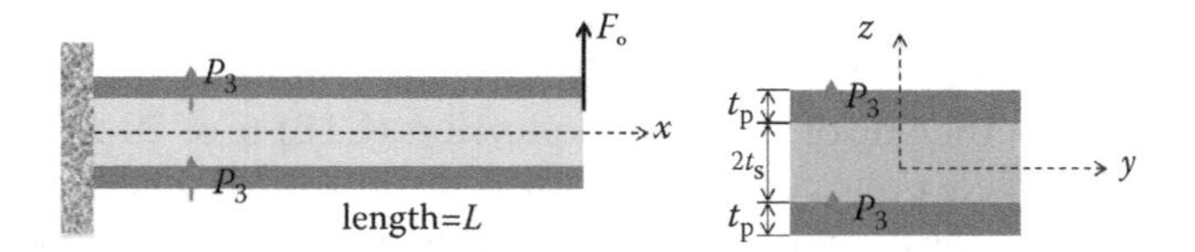

Section III Exercises

Chapter 8: Wireless Health Monitoring and Sensing of Smart Structures

1. Data buffering. Single-precision floating point numbers can be represented using 4 bytes of data. Assume that each data packet broadcast over the wireless communication medium requires 12 bytes of overhead data and can hold a payload from 0 to 128 bytes. How much channel use efficiency is gained by buffering data locally until a full 128-byte payload is ready versus transmitting a single floating point number at a time?
2. Channel allocation—lossless network. A network of 40 wireless sensors is distributed on a structure. Assuming that each wireless sensor can communicate data over the air at a rate of 250 kbps (kilobits per second), what sampling rate can a single communication channel support and still collect data in real time? Each data sample can be represented by a 2-byte integer. Assume that no collisions or retransmissions happen and that approximately 40% of the bandwidth used is lost on overhead and delays. If you need to collect data at a rate of 1000 samples per second, how many communication channels should be used to support all 40 sensor channels?
3. Channel allocation—lossy network. Repeat problem 8.2 but assume that there is a 2% chance that any given data packet will be dropped and must be retransmitted (for this problem, you may assume that the probability of success of any transmission is independent of all others). What failure conditions will you consider for your sensor network? Is it enough to have an expectation that the data will be transmitted in real time, or should the probability of network failure be smaller still?
4. Embedded processing—channel quality. Consider the 40 wireless sensor network of problem 8.2. Assume that the sensors can interrogate data internally instead of broadcasting it all to a central server. By doing so, a compression ratio of 500 to 1 is achieved. How many channels are required to support 40 sensors collecting data at 1000 Hz?
5. Embedded processing—power usage. One wireless sensor can transmit a single data point to a central server expending 1.5 μW of

energy. Assume that the sensor collects 2^{16} data points for an SHM application that can be transmitted to a central server for processing or processed locally. If local computing consumes 12 nW of power per operation, how many operations can you perform in an engineering algorithm and still save sensor power versus full transmission of raw sensor data? Assume that the transmission power requirements for the interrogated results are negligible compared with the raw data stream.

Chapter 9: Acoustic Emission of Composites: A Compilation of Different Techniques and Analysis

6. Describe the principles, advantages, and limitations of AE testing.
7. What is the main problem that causes an approximate localization by means of acoustic emission of a fiber crack on a composite laminate $(0°/90°)_{sym}$ assuming that the plate is monitored by using acoustic emission sensors?
8. Is a narrowband transducer resonant at 50–150 kHz feasible to monitor a composite structure?

Chapter 10: Neural Network Nondestructive Evaluation of Composite Structures from Acoustic Emission Data

9. (T/F) An artificial neural network can be trained *only* if a relationship exists between the inputs and the proposed output.
10. (T/F) The neural network can be thought of as an automated *optimization* problem solver.
11. (T/F) Backpropagation neural networks (BPNNs) are used primarily for performing data classification.
12. (T/F) Kohonen self-organizing map (SOM) neural networks are used primarily to classify data into clusters having similar properties.
13. What is the purpose of the BPNN employed in the tutorial? What data are used as the input? What is the output?
14. Referring to question 13, how many inputs are there in the amplitude histogram and what do they represent?

15. Again referring to question 13, how many hidden layers are required to solve this problem? How many hidden layer neurons are optimal?
16. What is bootstrapping? What was the purpose of bootstrapping the BPNN input data from the seven beams used to train the network? How many times was the data set repeated? What was the epoch size for the BPNN?
17. Describe the backpropagation process using 21 sample inputs.
18. What can be said concerning the magnitude of the output errors of the trained BPNN versus those produced by the trained BPNN on previously unseen data?

Chapter 11: Prediction of Ultimate Compression after Impact Loads in Graphite–Epoxy Coupons from Ultrasonic C-Scan Images Using Neural Networks

19. What is BVID, and why is it an issue for polymer matrix composites?
20. What is CAI, and why is it an issue for polymer matrix composites?
21. Does an ultrasonic testing (UT) C-scan image provide a qualitative or a quantitative measure of the effect of BVID on composite CAI ultimate load?
22. Why is a quantitative measure of the effect of BVID on structural integrity desirable?
23. What is the purpose of the backpropagation neural network (BPNN) employed herein? What data are used as the input? What is the output?
24. Why was the green (G) layer data chosen for BPNN processing?
25. Why is a fast Fourier transform (FFT) applied to the UT C-scan image data? What is the effect of FFT on the prediction accuracy of the BPNN? Why?
26. What is bootstrapping? What was the purpose of bootstrapping the BPNN input data from the 15 CAI coupons used to train the network? How many times was the data set repeated? What was the epoch size for the BPNN?
27. Describe the backpropagation process using 45 sample inputs.
28. What can be said concerning the magnitude of the output errors of the trained BPNN versus those produced by the trained BPNN on previously unseen data?

Chapter 12: Distributed *In Situ* Health Monitoring of Nanocomposite-Enhanced Fiber-Reinforced Polymer Composites

29. A two-point probe technique was used to measure the electrical properties of a rectangular carbon nanotube–based thin film of uniform thickness (3.5 μm). DC current of 15 μA was passed through the two electrodes. A voltmeter measured the potential drop to be 15.75 mV. The distance between the two electrodes was 2 cm, and the width was 0.3 cm. Calculate the film's electrical resistivity (ρ) and conductivity (σ).
30. The same type of film in problem 29 was tested, but the width was 0.5 cm (instead of 0.3 cm, and everything else remained the same). What was the film's resistance and resistivity? Assume that contact resistance was negligible.
31. A carbon nanotube–based thin film strain sensor was used as part of a quarter bridge Wheatstone bridge circuit. The initial or unstrained resistance of the thin film was 354 Ω. The other three resistors in the bridge circuit were 350 Ω and insensitive to strain. The strain sensitivity or gauge factor of the film was 10. The Wheatstone bridge circuit was powered by a 10-V DC source. What was the bridge circuit's output voltage when the film was strained to 1% strain?
32. A prototype thin film strain sensor had a nominal resistance of 105 Ω. You mounted the thin film in a load frame and conducted monotonic uniaxial tensile testing. The film's resistance was measured at different levels of applied strains and is shown here:

Strain (mm·mm^{-1})	Resistance (Ω)
0	105
0.001	105.1886
0.002	106.1309
0.003	106.0612
0.004	106.3323
0.005	106.7696
0.006	107.4553
0.007	107.7349
0.008	108.0313
0.009	108.8084
0.010	109.0057
0.011	109.3710

Estimate the sensitivity or gauge factor of the thin film strain sensor.

33. The equivalent circuit model of a nanocomposite was shown to be a resistor (R) connected to a capacitor (C) in series. Derive and write the equation that characterizes the nanocomposite's complex electrical impedance response (Z) as a function of cyclic frequency (ω).
34. The equivalent circuit of the nanocomposite in problem 33 comprised a resistor of 2000 Ω and a capacitor of 25 pF. What is the nanocomposite's AC impedance at 12.1 MHz?

Index

Page numbers followed by f and t indicate figures and tables, respectively.

B

C

F

M

O

P

Q

R

X

Y

Z

For Product Safety Concerns and Information please contact our EU representative GPSR@taylorandfrancis.com Taylor & Francis Verlag GmbH, Kaufingerstraße 24, 80331 München, Germany